Learning and Analytics in Intelligent Systems

Volume 46

The main aim of the series is to make available a publication of books in hard copy form and soft copy form on all aspects of learning, analytics and advanced intelligent systems, and related technologies. The mentioned disciplines are strongly related and complement one another significantly. Thus, the series encourages cross-fertilization highlighting research and knowledge of common interest. The series allows a unified/integrated approach to themes and topics in these scientific disciplines which will result in significant cross-fertilization and research dissemination. To maximize dissemination of research results and knowledge in these disciplines, the series publishes edited books, monographs, handbooks, textbooks, and conference proceedings.

Indexed by EI Compendex.

Elissaios Sarmas · Vangelis Marinakis ·
Haris Doukas

Artificial Intelligence for Energy Systems

Driving Intelligent, Flexible and Optimal Energy Management

Springer

Elissaios Sarmas
Athens, Greece

Haris Doukas
Department of Electrical and Computer Engineering
National Technical University of Athens
Athens, Greece

Vangelis Marinakis
Department of Electrical and Computer Engineering
National Technical University of Athens
Athens, Greece

ISSN 2662-3447 ISSN 2662-3455 (electronic)
Learning and Analytics in Intelligent Systems
ISBN 978-3-031-85208-4 ISBN 978-3-031-85209-1 (eBook)
https://doi.org/10.1007/978-3-031-85209-1

This Springer imprint is published by the registered company Springer Nature Switzerland AG
The registered company address is: Gewerbestrasse 11, 6330 Cham, Switzerland

Series Editor's Foreword

Over the last two centuries, the Industrial Revolution has driven a dramatic surge in global energy consumption, a trend that continues unabated in the 21st century. Estimates suggest global energy use has risen from less than 6000 TWh in 1800 to over 120,000 TWh in 2000, and over 180,000 TWh in 2023 [1]. Projections indicate that energy consumption will likely exceed 250,000 TWh by 2050 [2].

This increase reflects significant advancements in human civilization and improvements in global living standards. However, it also brings substantial challenges, including serious environmental consequences. Carbon dioxide (CO_2) emissions and other greenhouse gases, primarily resulting from fossil fuel use, are major contributors to climate change and global warming [3]. Emissions have risen from approximately 28 million tons in 1800 to nearly 35 billion tons in 2022 and over 35 billion tons in 2023 [4, 5].

Governments worldwide, including the European Union, are prioritizing efforts to combat climate change [6]. The transition from fossil fuel-based energy production to renewable and environmentally sustainable sources, such as solar, wind, hydrogen, and lithium-ion technologies, has become a shared goal. However, fossil fuels still account for about 83% of global energy consumption, while the adoption of greener energy sources progresses at a slower-than-necessary pace [2], raising concerns even from industry leaders like BP [7, 8].

Fortunately, transformative technologies—including artificial intelligence (AI), big data, the Internet of Things (IoT), and blockchain—offer powerful tools to enhance efforts, accelerate the energy transition, and foster greater engagement from energy stakeholders in decision-making processes related to energy production, distribution, and management.

As the Chief Editor of Springer's *Learning and Analytics in Intelligent Systems (LAIS)* series, I am delighted to introduce this groundbreaking book, *Artificial Intelligence for Energy Systems: Driving Intelligent, Flexible and Optimal Energy Management*. Authored by internationally distinguished researchers Elissaios Sarmas, Vangelis Marinakis, and Haris Doukas—renowned for their expertise in energy sector decision-making and policy design—this book addresses and fills in a significant gap in the literature. It explores the integration of advanced technologies, including AI and big data, within the energy sector, offering insights into recent advancements.

More specifically, this comprehensive volume presents a comprehensive library of artificial intelligence models and optimization algorithms that can significantly enhance the efficiency and effectiveness of energy management, renewable energy integration, and building energy efficiency. It is divided into four parts, namely (i) *Fundamentals* (three chapters), (ii) *Forecasting Energy Production and Consumption* (two chapters), (iii) *Energy Efficiency* (two chapters), and (iv) *Energy Management* (two chapters). Key topics encompass the roles and analytical needs of energy sector stakeholders, advanced machine learning models, optimization algorithms tailored to complex energy problems, and the development of solutions within defined constraints.

The book is expertly written and serves as a valuable resource for both field experts and newcomers. Experts will gain updated insights into challenges and solutions for energy transition and the transformation of the energy sector into a more sustainable industry. Newcomers will acquire in-depth knowledge and practical skills, while readers from other disciplines will find it an excellent introduction to this critical scientific field.

As Series Editor, I am proud to welcome this path-breaking monograph into Springer's *Learning and Analytics in Intelligent Systems* series. I extend my congratulations to the authors for their exceptional work, confident that this book will empower readers to both understand and implement its methodologies in addressing energy transition challenges. I encourage the editors to continue their research in this vital domain and keep the scientific community informed of their progress, perhaps via additional monographs that the LAIS series certainly welcomes.

January 2025

George A. Tsihrintzis
Department of Informatics
University of Piraeus
Piraeus, Greece

References

1. https://ourworldindata.org/energy-production-consumption
2. Global Energy Consumption: The World Counts
3. Climate Change: Greenhouse Gases—YouTube
4. https://ourworldindata.org/co2-emissions
5. https://www.energyinst.org/statistical-review
6. https://unfccc.int/process-and-meetings/the-paris-agreement?gad_source=1&gclid=CjwKCAiA9vS6BhA9EiwAJpnXw2w1to4U1cJxfRdKnNKbfPUJvbYIDdlEYi05gmA3sIw9BU5GnguvlRoCFNsQAvD_BwE

7. Spencer Dale, BP Chief Economist, 2019: *"There is a growing mismatch between societal demands for action on climate change and the actual pace of progress, with energy demand and carbon emissions growing at their fastest rate for years. The world is on an unsustainable path."*
8. https://www.bp.com/en/global/corporate/energy-economics/energy-outlook/foreword.html

Preface

The energy sector is currently undergoing significant and unprecedented changes due to various factors, including growing concerns about the planet's sustainability. The Paris Agreement of 2015, which mandates the design and implementation of sustainable, robust, and socially acceptable policies to combat climate change globally, serves as a clear indication of these concerns. Addressing this challenge requires a coordinated effort to transform and shift the energy sector from fossil fuel-based systems of production and consumption to alternative energy sources such as solar, wind, hydrogen, and lithium-ion batteries, among others. This energy transition process is complex and carries multiple social, technological, and environmental implications beyond the goal of achieving carbon neutrality.

The objective of this monograph is to contribute to this global effort by developing advanced learning methods and mathematical optimization algorithms for the energy and building sectors, with a particular focus on three major categories: (1) renewable energy sources, (2) distributed energy resources at the microgrid level, and (3) energy efficiency in buildings. Through the development of artificial intelligence models and mathematical optimization, this monograph presents solutions to problems in each of these areas, offering a comprehensive analysis of the key parameters and the interactions between them.

More specifically, this monograph presents a comprehensive library of artificial intelligence models and optimization algorithms that can significantly enhance the efficiency and effectiveness of energy management, renewable energy integration, and building energy efficiency. An integrated methodology is proposed for selecting models, algorithms, and input features for energy-related problems across different categories and for integrating them into a unified framework. In addition to contributing to the optimization of the energy and building sectors, new techniques are introduced in the application of artificial intelligence, as transfer learning, reinforcement learning, and meta-learning are presented as novel approaches to addressing challenges related to renewable energy sources, distributed energy resources, and energy efficiency in buildings. These techniques enable the development of adaptable models, which can optimize the energy sector more efficiently and effectively, paving the way for future research in this area. Additionally, an in-depth analysis is conducted of the interdependencies between the three main application areas, which is essential for developing effective models and algorithms. Finally, the proposed models and algorithms are validated on real-world data and test cases.

The monograph makes a meaningful contribution to the global efforts against climate change by supporting the energy transition. The proposed solutions aim to ease the transition to sustainable energy systems, tackling key challenges related to decarbonization, digitalization, decentralization, and energy democratization. While the energy sector is complex, we have strived to cover it comprehensively, both broadly and deeply. We hope we have succeeded, at least in part. However, some aspects and applications may not have been fully addressed in here, but we aim to explore them in future editions, particularly as the climate crisis intensifies and society demands a faster and more efficient energy transition. We anticipate that the research in this monograph will have a significant and lasting impact on humanity, advancing the understanding of energy systems and buildings, and offering practical solutions to today's critical challenges.

Athens, Greece

Elissaios Sarmas
Vangelis Marinakis
Haris Doukas

Competing Interests The authors have no competing interests to declare that are relevant to the content of this manuscript.

Contents

Part I
Fundamentals

Chapter 1
The Climate Crisis and the Four Pillars of Energy Transition: Decarbonization, Digitization, Decentralization, and Democratization

Abstract The global energy sector is undergoing a transformative shift driven by the imperative to combat climate change and the proliferation of advanced technologies, including Artificial Intelligence (AI), the Internet of Things (IoT), and blockchain. Four foundational pillars-Decarbonization, Digitization, Decentralization, and Democratization-collectively referred to as the 4 D's, define and propel the energy transition. These pillars address the multifaceted challenges and opportunities associated with reshaping energy production, consumption, and management. In addition to decarbonization, the energy transition encompasses social equity, technological innovation, and environmental sustainability, integrating concepts such as the 5 Vs of Big Data (Volume, Variety, Velocity, Veracity, and Value), decentralized energy systems, and democratized energy communities. The interplay among these pillars and emerging technologies fosters renewable energy integration, optimizes energy efficiency in buildings, and enables the development of smart microgrids. AI plays a pivotal role in managing distributed energy systems, addressing data scarcity, and enhancing decision-making. By focusing on the creation of sustainable, resilient, and inclusive energy systems, this chapter establishes a foundation for exploring AI-driven solutions and optimization algorithms aimed at addressing critical energy challenges and accelerating the transition toward a sustainable future.

Keywords Artificial intelligence · Machine learning · Optimization · Renewable energy sources · Digitization · Decarbonization · Decentralization · Democratization

1.1 Climate Crisis and Energy Transition

The energy sector is undeniably undergoing fundamental and unprecedented changes. These changes are attributed to multiple causes,the most significant of which is the growing concern regarding the sustainability of the planet in its current state, as articulated in the Paris Agreement formulated in 2015 [1]. The Agreement focuses on the design and implementation of sustainable, robust, socially acceptable, and fair policies to combat climate change within a global framework. At the same

E. Sarmas et al., *Artificial Intelligence for Energy Systems*, Learning and Analytics in Intelligent Systems 46, https://doi.org/10.1007/978-3-031-85209-1_1

time, the ever-increasing momentum of modern technologies such as the Internet of Things (IoT), AI, and Blockchain technologies [6, 27] creates a unique opportunity to accelerate the energy transition. The energy transition can be defined as the process of transforming and shifting the energy sector from systems of energy production and consumption reliant on fossil fuels-such as oil, natural gas, and coal-to alternative energy sources, including solar energy, wind energy, hydrogen, and lithium-ion batteries.

It is important to note that the issue of the energy transition encompasses, but is not limited to, the goal of decarbonization; it entails multiple social, technological, and environmental implications. The success, speed, and social acceptance of the energy transition are challenging objectives, as they largely depend on four components known as the 4 Ds of the energy sector: Decarbonization, Digitization, Decentralization, and Democratization of Energy, which are illustrated in Fig. 1.1. Consequently, the 4 Ds can serve as the driving force behind the anticipated energy transition; however, they also present ongoing challenges and risks that could jeopardize the efforts toward energy transition within a reasonable time frame. Therefore, the 4 Ds can undoubtedly be the driving force behind the anticipated energy transition, but they simultaneously present ongoing challenges and risks that could jeopardize the energy transition effort within a reasonable time frame.

In recent decades, significant advancements in digital technologies have played a pivotal role in shaping habits and behavioral patterns, regulating performance and strategies, and influencing the functioning of organizations, communities, and society at large. It is not an exaggeration to assert that we are currently experiencing the era of digitization. Indeed, digitization acts as a principal catalyst for continuous growth within the field of energy innovation, which has also been substantially impacted by these transformative changes.

More specifically, the term digital technologies refers to any electronic tools, systems, devices, and resources that generate, store, or process data and information. Although digital technologies emerged decades ago, their rapid development

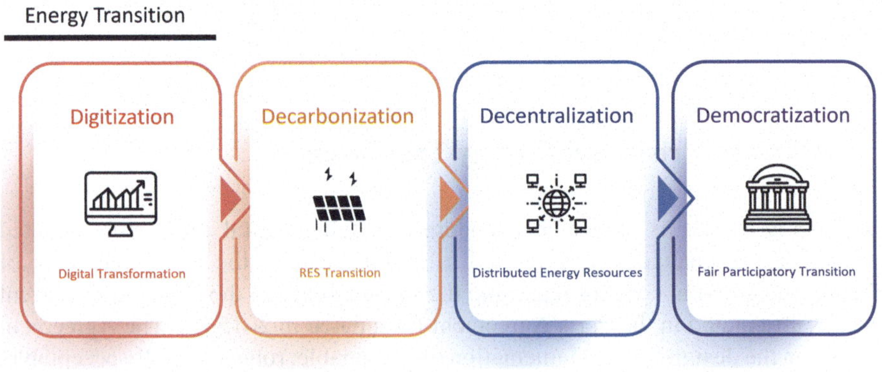

Fig. 1.1 The *4 'D's* of the energy transition

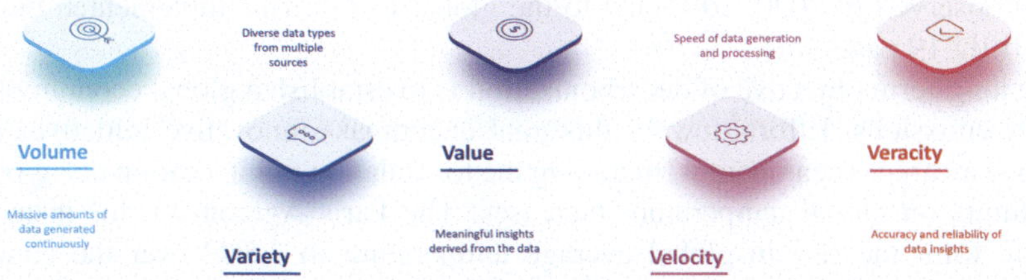

Fig. 1.2 The 5 main attributes of Big Data

in recent years has gradually transformed the energy sector into the era of Energy 4.0. According to the International Energy Agency (IEA), digitization in the energy sector is viewed as the increasing application of information and communication technologies (ICTs), contributing to the convergence of the digital and physical worlds [7].

Furthermore, the availability and quality of data have improved significantly, facilitating the optimization of operations and daily processes at the building level, whether in residential environments, workplaces, communities, or even cities. The available data are characterized by a high volume, with thousands of terabytes generated cumulatively by smart sensors every second; great variety, including both structured and unstructured, real-time, and asynchronous data; rapid generation of new data and updating of existing data (velocity); and increased reliability (veracity). These four properties of big data create opportunities for leveraging cutting-edge digital technologies that aim to add value to the data. Consequently, the 5 Vs (volume, variety, velocity, veracity, and value) of Big Data are formed, as shown in Fig. 1.2. The value of data translates into the influence that its analysis and use have on an organization to optimize operations and support the decision-making process.

1.1.1 Decarbonization: The Need to Convert to Renewable Energy Sources

Recent advances highlighting the intensifying effects of climate change, coupled with increasing carbon dioxide (CO_2) emissions, underscore the urgent need for transformative action at both national and global levels to accelerate decarbonization in the energy sector. Decarbonization refers to the gradual transition from carbon-based energy systems to alternatives that significantly reduce CO_2 emissions. By 2019, global CO_2 emissions had surged to levels over 60% higher than those recorded in 1990, the year when international climate negotiations commenced. This stark increase underscores CO_2 emissions as the primary driver of global warming. Notably, 2019 marked a record high for emissions, with a temporary decline observed in 2020, attributed to the global lockdowns implemented during the COVID-19 pandemic.

The long-term objective of decarbonization is to establish a global economy free of CO_2 emissions. Efforts toward this goal encompass innovative initiatives and extensive awareness campaigns, including the formulation of internationally agreed-upon limits on global temperature increases. The Paris Agreement, for instance, seeks to limit the rise in global average temperature to 1.5°C over the coming decades. Nonetheless, current progress remains significantly insufficient to achieve these critical targets.

A widely recognized pathway to decarbonization is the large-scale adoption of Renewable Energy Sources (RES). In this context, the International Renewable Energy Agency (IRENA) has advocated for an ambitious plan to achieve net-zero carbon emissions by 2050. Numerous nations have already integrated substantial proportions of renewable energy, including wind and solar power, into their energy systems. However, the transition to a renewable energy mix remains uneven globally due to disparities in economic development, legal frameworks, and social conditions among countries.

1.1.2 Decentralization: Distributed Energy Resources

The extensive technological advancements across various sectors over the past decade have been accompanied by the mobilization of consumers, contributing to the decentralization of the energy system. A decentralized energy system is characterized by the placement of energy production facilities near points of energy consumption, contrasting with a centralized energy system, which is defined by large-scale electricity generation at central power plants. The creation and development of decentralized energy systems rely on the adoption and widespread use of RES to reduce dependency on fossil fuels, as analyzed in the previous subsection.

However, the continuous development of systems based on RES, along with the existence of active consumers capable of demand-side management, presents significant challenges for the operation of the energy system and the electricity market. In general, there is a trend toward transforming the structure of the electricity market from a centralized model focused on suppliers to a more decentralized model centered on consumers, with households playing a crucial role by producing and storing the energy they require [13].

1.1.3 Democratization: Knowledge Spread in Energy Communities

It is therefore evident that the goal of decentralization is today more relevant than ever, leading to the gradual transformation of energy systems while leveraging technical and technological advancements. As a result, distributed, peer-to-peer energy transactions are being promoted, based on community engagement, aiming to create a consumer-centric model [4]. Additionally, predictive models and optimization algorithms are required to ensure the successful operation of such innovative microgrids and decentralized systems [12].

The decentralized model described in the previous paragraphs relies heavily on individuals' ability to produce, store, and manage the energy they consume [24]. This model can be implemented through the extensive installation of energy production equipment, such as rooftop photovoltaic (PV) systems, energy storage systems-despite their persistently high costs-and intelligent systems for managing and monitoring the resulting microgrid. The term commonly used to describe individuals who produce the energy they consume is prosumer (a combination of the words producer and consumer). A prosumer is generally defined as any electricity consumer capable of generating a portion of their electricity needs through their own means of production [28]. Prosumers have the capacity to feed surplus production into the grid, when available, and conversely, draw from the grid when their self-production is inadequate.

The democratization of energy requires universal access to affordable and clean energy, enabling everyone to utilize RES according to their needs [3]. This condition fosters the creation of local economic value, as development opportunities increase for prosumers and small- and medium-sized enterprises due to reliable access to electricity. Therefore, individuals using energy should actively participate in decision-making processes regarding how energy is promoted and distributed to maximize the economic benefits of the energy they produce [14].

In this context, the full democratization of energy data and services aspires to provide access to data, information, smart systems, and services for the average end user. The aim of democratizing data and applications is to empower members of the energy community to collect and analyze data from their energy production systems, consumption patterns, and flexible loads without requiring external assistance from specialists.

1.1.4 The Role of Artificial Intelligence

The preceding paragraphs have delineated the challenges associated with the energy transition and its interconnection with technological advancements, which have generated a substantial volume of available data. This situation presents significant opportunities for the design, implementation, and development of intelligent systems that can facilitate optimal energy management and support decision-making across various issues. A key role in the development of these systems is played by AI, which can be used to optimize energy networks by managing energy flows between homes, businesses, storage batteries, RES, microgrids, and the electrical grid. This reduces energy waste while increasing consumer participation in energy management. Within this framework, AI is already being applied in numerous energy-related applications, ranging from RES to building energy efficiency and smart energy management at the microgrid level, which includes distributed energy resources.

Regarding RES, AI is used across almost all types, such as wind, solar, geothermal, hydroelectric energy, and hybrid forms of these, for tasks including design, optimization, forecasting, management, distribution, and policymaking [10]. A review of the existing literature indicates that most scientific research focuses on AI applications in wind and solar energy systems, though there are studies exploring hybrid energy sources. Key research areas include developing predictive models for electricity production from RES, fault detection in RES systems, and optimizing the sizing and geographic placement of RES installations.

In the field of building energy efficiency, AI is employed at all stages of a building's lifecycle. Specifically, AI methods are used in building design, transforming how architects envision and create structures [26]. During construction, AI has been successfully applied in cost analysis, construction management, defect detection, and building information modeling (BIM) [9]. Furthermore, significant applications have been developed for building operation and maintenance, with the objective of maximizing occupant comfort while enhancing the building's energy efficiency [8].

As the energy sector grows increasingly complex, there is a need for intelligent applications and mechanisms to effectively manage systems and make timely decisions [2]. Emerging fields, such as Machine Learning (ML) and Deep Learning (DL), alongside neural networks, genetic algorithms, and reinforcement learning, are among the most prominent artificial intelligence techniques employed in the development of intelligent energy management applications aimed at optimizing the utilization of distributed energy resources (DERs) [15]. Nevertheless, the current developmental stage of such applications remains immature, with many functionalities operating at a rudimentary level of automation within microgrids, often failing to effectively leverage electricity production.

Although AI and ML have been extensively used in these three domains (RES, building energy efficiency, and distributed energy management in microgrids), several limitations and gaps have been identified, leaving room for further research.

1. Lack of Standardized Methodologies: Most studies are still in the research and development phase, with few adopted by the industry. A major reason for this is the absence of standardized methodological frameworks for each problem, including clear requirements for data input formats and the features that each developed model should satisfy.
2. Inconsistent Data and Comparisons: Different AI methods and models are trained and tested on varying datasets, making fair and reliable comparisons or evaluations

among models impossible [9]. Thus, clear evaluation metrics are needed for each problem to enable comparisons across different datasets.
3. Interpretability of AI Results: A significant limitation of AI lies in the interpretability of results. Most AI fields, including ML and DL, adopt data-driven approaches that function as black boxes, making it often unclear-even to experts-how a particular result was derived [5].

The design and development of structured learning models and optimization algorithms for energy management problems, RES, or building energy efficiency remain a scientific field requiring continuous research and innovative ideas. The ultimate goal is to support the energy transition in a fair, sustainable, and democratized manner within an inclusive society that ensures equal rights to energy access for all.

1.2 Objectives and Contributions of the Monograph

The subject of the monograph is the development of an integrated series of learning models and optimization algorithms to support decision-making in the energy and building sectors. Specifically, the monograph aims to provide modern solutions to critical energy challenges faced by the scientific community by developing innovative models, learning procedures, optimization algorithms, and comprehensive methodologies. These address problems across three key axes: (a) distributed energy production, (b) flexibility at the microgrid level, and (c) building energy efficiency. Therefore, the monograph's scope is two-fold: it aspires to provide a tailored and comprehensive solution to each problem, on the one hand, and a generalized methodology for tackling a wide range of issues, adaptable to new energy challenges through appropriate modifications, on the other hand.

The monograph aims to bridge the scientific "gap" identified regarding the development of modern learning methods and mathematical optimization algorithms for solving various problems in the energy and building sectors. This scientific field is in a state of continual evolution, as new methodologies, algorithms, and models emerge rapidly, thereby heightening the necessity for ongoing monitoring of developments and contributions within the domain. Specifically, the monograph focuses on the development of models and algorithms for problems categorized into three broader areas: (a) Distributed Production and RES, (b) Flexibility at the Microgrid Level, and (c) Building Energy Efficiency. Consequently, AI models and mathematical optimization algorithms are developed and presented for applications related to each of these categories through a thorough analysis of the fundamental parameters of each application, as well as the interactions between them.

The contribution of this monograph lies in the development of a series of AI models and optimization algorithms, focusing on specific problems in the fields of energy management, renewable energy sources (RES), and building energy efficiency. More specifically, the contribution of the monograph is distinguished at four (4) levels, as depicted in Fig. 1.3 and analyzed in the following paragraphs.

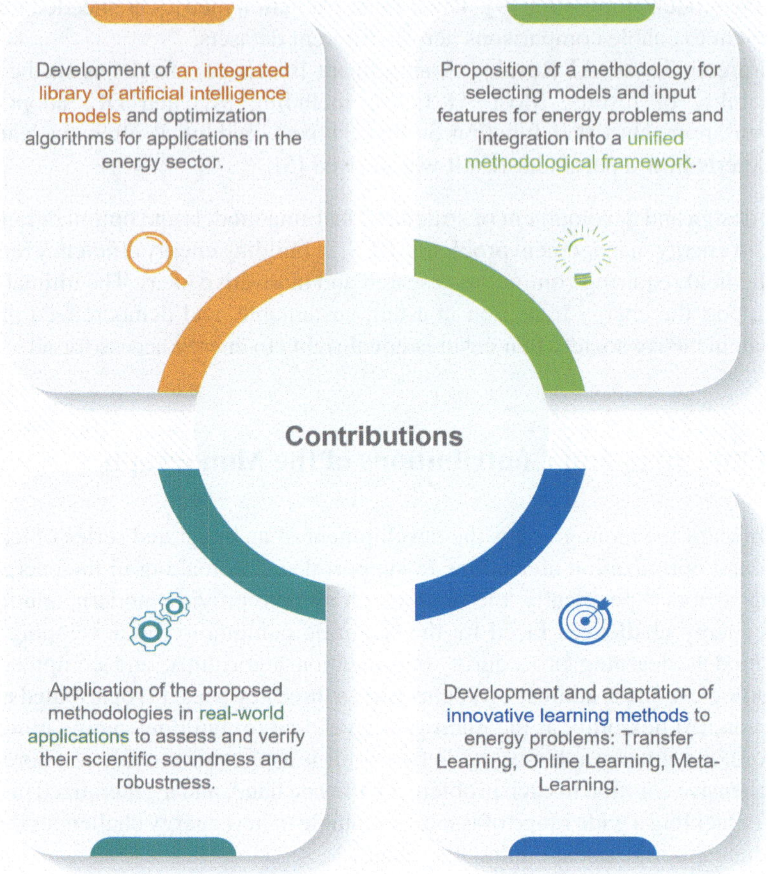

Fig. 1.3 Monograph Contributions

1.2.1 Development of Innovative AI Models and Optimization Algorithms for a Series of Applications in the Energy Sector

A series of AI models is presented, specifically ML and Bayesian Models, along with optimization algorithms developed based on a comprehensive methodological framework (Chap. 3). The starting point for selecting specific research areas is the identification of particular problems for which there are either limited available solutions or where existing solutions require further investigation.

The first category includes the following problems: (a) the problem of evaluating investments related to building renovations aimed at improving energy efficiency, (b) the problem of estimating energy savings from renovation actions before their implementation (a priori), and (c) the problem of peak load smoothing in decentralized microgrids through the optimization of flexible load scheduling, such as irrigation pumps. For these specific problems, it has been identified that there are limited algorithmic solutions that take advantage of the abundance of available data.

The second category includes the following problems: (a) the problem of forecasting energy production from PV systems, (b) the problem of forecasting energy consumption at the building level, and (c) the problem of optimizing microgrid operation through the scheduling of electric vehicle charging stations and energy storage systems. These problems already have several available solutions in the literature; however, the proposed solutions either achieve better prediction accuracy or approach the problems from a different perspective by exploiting available data.

The solutions proposed in this monograph address problems that can be categorized into three subcategories, as shown in detail in Fig. 1.4. Specifically, these categories are (a) distributed production, (b) building energy efficiency, and (c) microgrid flexibility, which are presented in further detail below:

➢ **Distributed Production and RES (AI-4-RES)**: This category focuses on the development of predictive models for electricity production. Specifically, the monograph focuses on the problem of forecasting energy production from PV

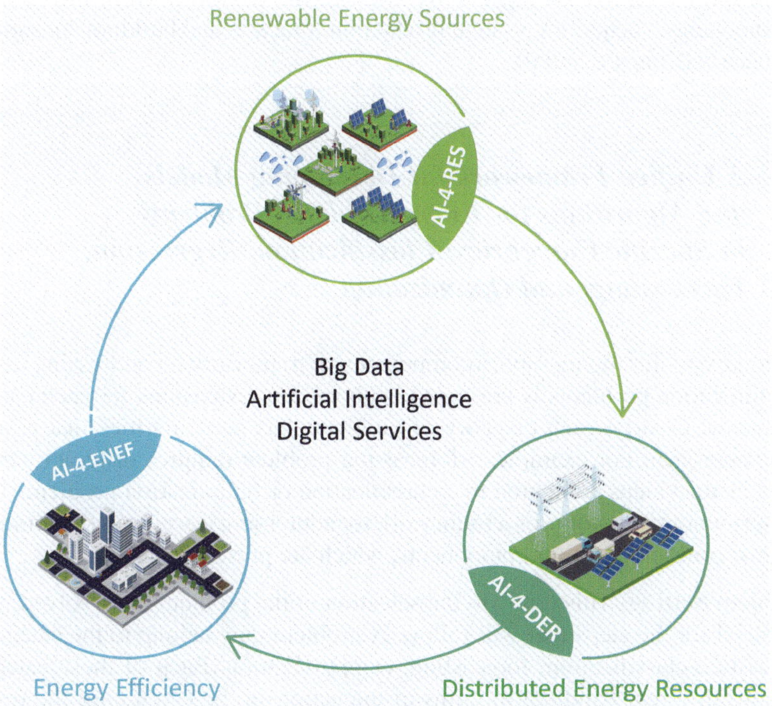

Fig. 1.4 The three distinct categories of AI models in the energy sector

systems (Chaps. 4, 8 and 9). Different ML and DL models, various prediction time horizons (short-term and medium-term forecast models), and the most suitable features to be incorporated into the models are examined. Additionally, the specific case of forecasting PV system production under data scarcity is addressed using transfer learning.

➢ **Building Energy Efficiency (AI-4-ENEF)**: This category deals with the development of models aimed at evaluating energy efficiency measures in buildings. These models branch into those that predict energy savings before the action is implemented (a-priori) based on historical data, and those that estimate the actual savings of a renovation action once it has been carried out (ex-ante). In the context of ex-ante analysis, predictive models for building energy consumption are developed. Furthermore, an investment evaluation model for building renovations is presented, which takes into account both the energy savings achieved and the economic aspects of the problem (Chaps. 5, 6 and 7).

➢ **Microgrid Flexibility (AI-4-FLEX)**: This category focuses on the development of optimization algorithms for the optimal use and utilization of distributed energy resources, such as electric vehicle charging stations, energy storage systems, irrigation pumps, and heating and ventilation systems in buildings, among others. The goal of these algorithms is to propose the optimal scheduling of the operation of these resources to achieve smart energy management at the microgrid level. These algorithms leverage the outcomes from the other two components, such as PV system production forecasts and building consumption forecasts (Chaps. 8 and 9).

1.2.2 A Unified Framework for Developing Models and Algorithms for Energy-Related Problems in Specific Categories: Classification, Regression, Forecasting, and Optimization

In the context of this monograph, a comprehensive framework for addressing learning and optimization problems is presented, with specific extensions for each category of problems. Clearly, each category of problems has particularities that require a specific approach. For example, a forecasting problem requires different handling and has distinct steps compared to a classification or optimization problem. Therefore, this monograph proposes a framework for addressing learning and optimization problems, consisting of four components, which are presented below:

- **Problem Setting**: This refers to the selection of the problem to be solved, which is crucial for the steps that will follow. A problem may belong to the category of regression, classification, forecasting, or optimization. Each of these categories requires different handling in terms of the actions to be taken until the result is produced. This component also includes the introduction of the data on which the problem to be solved is based.

- **Feature Engineering**: This involves selecting the most suitable input features for the problem to be solved. According to the proposed methodology, the selection of features or input variables is based on individual steps, including data visualization, statistical analysis of the data, and the study of correlations between different variables. All these steps ultimately lead to the selection of the most appropriate features to improve the performance of learning models and optimization algorithms.
- **Pre-processing and Fine-tuning**: This involves the steps that must be performed before training the models. These steps include pre-processing the data to ensure it is in the appropriate format for input into the learning model or optimization algorithm, as well as the process of finding the suitable hyperparameters for the models. This is a very important step to ensure the optimal functioning of the models and, consequently, to achieve good accuracy in the results. Additionally, this component includes finding the appropriate metrics for evaluating the models, which also differ depending on the category of the problem.
- **Model Training and Predictions**: This involves the process of training the learning model or executing the optimization algorithm. After selecting the models, input variables, model hyperparameters, and evaluation metrics, according to the proposed framework, the models are trained, compared, and used for predictions on unknown data.

1.2.3 Application of the Proposed Methodologies to Real-World Problems with Real Data for Verification of Their Scientific Correctness

The process of developing machine learning models and optimization algorithms is closely linked to the validation and verification of the results produced by the models in applications involving real-world data. In fact, as mentioned in the previous sections, the availability of this data has inspired the design of innovative applications. Therefore, all the models developed within the framework of this monograph are trained and evaluated on real-world data.

In each of the six chapters that present the applications, a comprehensive analysis of the data utilized in the experimental validation process is incorporated. In this context, various visualizations of the data are provided, along with information regarding their statistical distribution, details concerning their pre-processing, and noteworthy findings that may have arisen during the analysis. Moreover, although the models were initially developed with a particular dataset in consideration, the final form of the developed applications is structured to facilitate generalization to other datasets of a similar nature.

The data utilized encompasses a broad spectrum of available information within the energy sector. The datasets are heterogeneous in nature and encompass both static and dynamic data. Consequently, they fulfill all components of the five Vs previously mentioned. Time-series data, static data, datasets of varying sizes, and cases with incomplete data, among others, have been utilized. Moreover, it is important to highlight that the data referenced in this monograph originates from five different countries (Greece, Italy, Portugal, Latvia, and Spain), in addition to European Union databases.

The available data belong to the following categories: time-series production data from large PV systems, time-series production data from rooftop PV systems, time-series data on electricity and thermal consumption in buildings, catalogs of energy efficiency investments, irrigation system data, time-series data on network consumption, electric vehicle data, and energy storage system data. Each category possesses distinct characteristics and necessitates varied processing methods, which will be elaborated upon in the forthcoming chapters of the monograph. Nonetheless, the integration of these diverse data sets enables the development of intelligent solutions and applications for optimization in the realm of energy management and efficiency.

1.2.4 Development and Adaptation of Innovative Learning Methods to Energy and Building Sector Problems: Meta-Learning, Online Learning, Transfer Learning

AI addresses the problem of creating computational models that improve automatically through experience, making it one of the fastest-growing technical fields of our time. AI lies at the intersection of computer science, statistics, and data science [11]. Over the last decade, significant progress has been made in all areas of AI due to the development of new algorithms and learning theories. The recent progress in ML is attributed to both the development of new algorithms and learning theories, along with the ongoing explosion in data availability and the reduction in computing costs and time [25]. In this context, the fourth (and final) level of contribution of this monograph involves exploring, developing, and adapting innovative learning methods to address challenging energy problems. Specifically, the techniques of *(meta-learning)*, *online learning*, and *transfer learning* are applied and thoroughly evaluated.

The technique of *meta-learning* involves the use of ML algorithms that learn how to optimally combine predictions from several other ML algorithms. This means that meta-learning models require the development of other learning models that have already been trained on data. Meta-learning is used to increase the predictive accuracy of models and has been successfully applied in many fields. In this monograph, the meta-learning technique is used for forecasting production from PV systems (Chap. 4), forecasting electricity consumption (Chap. 5), and evaluating building renovation investments (Chap. 6) [20].

Online/incremental learning is a branch of ML that involves processing datasets that enter as real-time data streams, with little to no knowledge of the data distribution. In online learning, input data is continuously used to expand the knowledge of the existing model, helping to further train the model incrementally. In this monograph, the online learning technique is evaluated for the prediction of PV system production and building consumption (Chap. 9), and the performance of the incrementally trained models is compared with models that have been trained once and used offline [23].

The technique of *transfer learning* is applied in cases where there is insufficient training data for a particular problem. As a result, data from similar problems are examined, models are developed for those problems, and these models are transferred to the problem with insufficient data. Transfer learning has generally been applied to problems in many fields. In the context of this monograph, the transfer learning technique is examined for PV system forecasting problems in data-scarce conditions (Chap. 4). The results of this research have been very encouraging, highlighting transfer learning as a highly reliable technique for this specific problem [16].

Finally, all the models were developed using the *Python* programming language and with the help of ML and DL libraries, such as *Scikit-Learn*, *Tensorflow Keras* and *PyTorch*. Similarly, the optimization models were developed with the help of corresponding libraries, such as the *MIP* package in *Python*.

1.3 Monograph Structure

The monograph consists of the current introductory chapter and an additional nine (9) chapters, as shown in Fig. 1.5. This chapter to Chap. 3 present the theoretical background, methods, and proposed methodology. In contrast, Chaps. 4–9 include the experimental applications across six (6) different problems. Finally, Chap. 10 presents the conclusions of this work and highlights some of its potential future extensions. More specifically:

This chapter is entitled "The Climate Crisis and the Four Pillars of Energy Transition: Decarbonization, Digitization, Decentralization, and Democratization". It briefly presents the problems addressed in this monograph and establishes the need for the development of AI methods and optimization algorithms to support decision-making in energy management and energy efficiency. Furthermore, Chap. 1 outlines the subject, objective, and contribution of this monograph.

Chapter 2 is entitled "The Role of Artificial Intelligence in Transforming the Energy Sector: A Comprehensive Review" and provides a detailed analysis of the background of energy management problems that can be solved using AI methods. Specifically, this chapter presents the major energy problems addressed with the help of AI and classifies them into the three pillars outlined in a previous section. Additionally, for each problem, the related literature is reviewed, along with the practices followed and the ways in which AI has contributed to the more effective handling of the respective problem.

Fig. 1.5 Monograph contents and structure

Chapter 3 is entitled "Scalable Framework for Intelligent System Architecture to Address Challenges in the Energy Sector" and presents the methodological framework for the development of AI algorithms and models for problems related to distributed energy production, energy management, microgrid flexibility, and building energy efficiency. The proposed framework is designed to support different stakeholders in the energy sector during the development of effective algorithms and models. The framework covers the stages of model and algorithm development, as well as the architecture of AI models, including all necessary levels such as data management, model management, and result handling.

Chapter 4 is entitled "Deep Learning Models for Short-Term Forecasting of Photovoltaic Energy Production" and presents a meta-learning methodology that dynamically combines the core predictions of multiple deep learning models to identify the optimal performance conditions for each model [22]. This learning process leads to improved overall prediction accuracy. Specifically, four baseline Long Short-Term Memory (LSTM) models with different architectures are developed, and energy production forecasts for PV systems are generated over a short-term horizon without using numerical weather forecasts. The aim of this approach is to accelerate the training process and enhance the reliability of the proposed solution. Furthermore, the accuracy of the meta-learning model is evaluated using three PV systems located in Lisbon, Portugal. Our findings confirm that different baseline models demonstrate varying performance across diverse PV systems, suggesting that meta-learning can significantly improve accuracy, particularly during peak PV production hours. This chapter also explores the technique of transfer learning. In particular, transfer learning is employed for both initializing the weights of the LSTM model and for feature extraction, utilizing distinct transfer learning approaches in each instance. These transfer learning strategies are compared with a conventional learning model that does not employ transfer, as well as with the smart persistence model, which is widely regarded as the most accepted benchmark [16]. The application is assessed in the context of forecasting the hourly production of six PV systems, and the results demonstrate that transfer learning models significantly outperform conventional models.

Chapter 5 is entitled "Machine Learning-Driven Energy Consumption Forecasting for Building Profiling" and examines the use of AI models to forecast energy consumption in buildings. The primary aim is to estimate energy savings resulting from renovation actions. Accurate estimation of energy savings is essential for the effective implementation of energy-saving measures. Therefore, we propose an ensemble model for the precise estimation of baseline energy consumption, based on tree algorithms, namely Random Forest, XGBoost, and LightGBM. It is noteworthy that our model emphasizes explainability, providing transparency and insights into the key factors influencing baseline energy consumption [17]. To validate its accuracy, we conduct experimental evaluations on a building cluster in Latvia. The results demonstrate the superiority of the proposed ensemble model over individual models and even a BM network tailored for energy consumption estimation, offering a robust and interpretable solution for estimating energy savings in the building sector.

Chapter 6 is entitled "Meta-Learning Approaches for Assessing Energy Efficiency Investments in Buildings" and presents a methodology aimed at bridging the financing gap for energy efficiency projects. The energy efficiency of buildings is crucial for achieving global energy and climate objectives; however, it necessitates substantial investments. The absence of mature decision-support systems, coupled with traditional investment mechanisms that prioritize the economic aspects of energy efficiency projects while neglecting their environmental impacts, may impede the ability of such projects to secure funding. In this chapter, various classification methods are evaluated and subsequently integrated through a meta-learning model designed to enhance overall classification performance and determine the appropriate funding for each investment based on its specific characteristics [20]. The proposed methodology is evaluated using a dataset of more than 300 completed renovation projects in buildings in Latvia. The results indicate that the meta-learning model outperforms all baseline ML models, successfully identifying projects with high and medium potential while differentiating those with low potential.

Chapter 7 is entitled "Ensemble Machine Learning Models for Estimating Energy Savings from Efficiency Measures in Buildings" and introduces a ML-based framework for predicting energy savings resulting from various energy efficiency renovations. The framework utilizes three widely used tree-based ensemble algorithms, namely Random Forest, XGBoost, and LightGBM [18]. Predictions from these models are combined through an ensembling layer with equal weights to mitigate uncertainty and enhance forecasting accuracy. The performance of both the individual models and the ensemble is evaluated using a comprehensive database of energy efficiency renovation investments. The findings demonstrate that the ensemble model consistently outperforms the individual machine learning models with respect to four accuracy metrics. Furthermore, the analysis indicates that certain types of energy efficiency measures exhibit greater predictability than others, providing valuable insights for stakeholders evaluating investment risks. Significantly, none of the energy efficiency categories displayed markedly poor predictive performance.

Chapter 8 is titled "Optimization Model for Irrigation Pump Scheduling to Smooth Load Peaks" and introduces an energy management method that optimizes electricity consumption by smoothing consumption peaks through the scheduling of optimal operating hours for an irrigation pumping system on an isolated island. A set of ML models is initially used to accurately predict electricity consumption at the island level, as well as the energy produced by RES on an hourly basis. These forecasts are subsequently utilized in a heuristic optimization algorithm that allocates the operating hours of the pumps to minimize consumption peaks while considering the constraints associated with pump operation. The performance of the proposed method is evaluated using the case of the remote Greek island Tilos [21]. This island is equipped with an energy management system that facilitates the monitoring and control of local water pumping stations, thereby supporting water supply and irrigation for households. The results indicate that the intelligent scheduling of water pumps in a small-scale island environment can reduce the daily and weekly deviation of electricity consumption by more than 15% at no monetary cost. It is also concluded that the potential benefits of the proposed approach are directly related to the amount of load that can be shifted each day, the accuracy of the forecasts used, and the amount of electricity produced by PV systems.

Chapter 9 is entitled "Optimization Model for Electric Vehicle and Energy Storage System Operation for Microgrid Energy Autonomy" and presents an integrated methodology to enhance energy autonomy at the microgrid level while addressing the issue of reverse power flow. Reverse power flow is defined as the continuous flow of electricity in the opposite direction of the normal power flow within a grid. This phenomenon typically occurs in microgrids when the energy produced by distributed electricity generation facilities exceeds local load demand. Such occurrences pose various risks, including inefficient operation, potential damage to equipment, grid instability, and energy losses. To reduce reverse power flow in microgrids and support energy autonomy, we introduce an algorithm based on predictive models. The framework relies on deep learning models that forecast the electricity produced by PV systems and consumed by the microgrid, alongside an optimization algorithm that schedules its flexible loads, specifically electric vehicles, based on these forecasts [19]. We conduct a payback study to assess the impact of optimized scheduling and energy storage on the microgrid's autonomy, while also exploring the effects of varying battery capacities and electric vehicle fleet sizes. Our results indicate that load shifting based on forecasts can significantly reduce reverse power flow, particularly with larger amounts of flexible loads. Additionally, we find that energy storage can complement load shifting. Nonetheless, these improvements are dependent on the accuracy of the forecasts and the capabilities of the storage systems.

Finally, Chap. 10 is entitled "Future Directions of Intelligent Energy Management and the Role of Generative AI" and presents a summary of the achievements of the proposed approach, derived from both the analysis of the previous chapters and the empirical results of practical application. This chapter also includes a series of suggestions and ideas for further research activities related to the application of AI in energy management and building energy efficiency problems.

References

1. P. Agreement, Paris agreement, in *Report of the Conference of the Parties to the United Nations Framework Convention on Climate Change (21st Session, 2015: Paris). Retrived December*, vol. 4 (2015), p. 2017. HeinOnline
2. S.S. Ali, B.J. Choi, State-of-the-art artificial intelligence techniques for distributed smart grids: a review. Electronics **9**(6), 1030 (2020)
3. A. Arsenopoulos, E. Sarmas, A. Stavrakaki, I. Giannouli, J. Psarras, A data-driven decision support tool at the service of energy suppliers and utilities for tackling energy poverty: a case study in Greece, in *2021 12th International Conference on Information, Intelligence, Systems & Applications (IISA)* (IEEE, 2021), pp. 1–6
4. S. Carley, D.M. Konisky, The justice and equity implications of the clean energy transition. Nat. Energy **5**(8), 569–577 (2020)
5. D.V. Carvalho, E.M. Pereira, J.S. Cardoso, Machine learning interpretability: a survey on methods and metrics. Electronics **8**(8), 832 (2019)
6. I.C. Dimitriou, G.P. Trachanas, A. Arsenopoulos, E. Sarmas, Future heat risk index: The case of Attica regions, Greece, in *2024 15th International Conference on Information, Intelligence, Systems & Applications (IISA)* (IEEE, 2024), pp. 1–7

7. H. Doukas, A. Nikas, M. González-Eguino, I. Arto, A. Anger-Kraavi, From integrated to integrative: delivering on the Paris agreement. Sustainability **10**(7), 2299 (2018)
8. Z. Du, B. Fan, X. Jin, J. Chi, Fault detection and diagnosis for buildings and HVAC systems using combined neural networks and subtractive clustering analysis. Build. Environ. **73**, 1–11 (2014)
9. T. Hong, Z. Wang, X. Luo, W. Zhang, State-of-the-art on research and applications of machine learning in the building life cycle. Energy Build. **212**, 109831 (2020)
10. S.K. Jha, J. Bilalovic, A. Jha, N. Patel, H. Zhang, Renewable energy: present research and future scope of artificial intelligence. Renew. Sustain. Energy Rev. **77**, 297–317 (2017)
11. M.I. Jordan, T.M. Mitchell, Machine learning: trends, perspectives, and prospects. Science **349**(6245), 255–260 (2015)
12. W. Lyu, J. Liu, Artificial intelligence and emerging digital technologies in the energy sector. Appl. Energy **303**, 117615 (2021)
13. J. Markard, The next phase of the energy transition and its implications for research and policy. Nat. Energy **3**(8), 628–633 (2018)
14. D. Mourtzis, J. Angelopoulos, N. Panopoulos, A collaborative approach on energy-based offered services: energy 4.0 ecosystems. Proc. CIRP **104**, 1638–1643 (2021)
15. C. Ramos, C.-C. Liu, Ai in power systems and energy markets. IEEE Intell. Syst. **26**(2), 5–8 (2011)
16. E. Sarmas, N. Dimitropoulos, V. Marinakis, Z. Mylona, H. Doukas, Transfer learning strategies for solar power forecasting under data scarcity. Sci. Rep. **12**(1), 1–13 (2022)
17. E. Sarmas, A. Forouli, V. Marinakis, H. Doukas, Baseline energy modeling for improved measurement and verification through the use of ensemble artificial intelligence models. Inf. Sci. **654**, 119879 (2024)
18. E. Sarmas, E. Spiliotis, N. Dimitropoulos, V. Marinakis, H. Doukas, Estimating the energy savings of energy efficiency actions with ensemble machine learning models. Appl. Sci. **13**(4), 2749 (2023)
19. E. Sarmas, E. Spiliotis, V. Marinakis, M.A. Bucarelli, F. Santori, H. Doukas, Revving up energy autonomy: a forecast-driven framework for reducing reverse power flow in microgrids. Sustain. Energy, Grids Netw. **38**, 101376 (2024)
20. E. Sarmas, E. Spiliotis, V. Marinakis, T. Koutselis, H. Doukas, A meta-learning classification model for supporting decisions on energy efficiency investments. Energy Build. **258**, 111836 (2022)
21. E. Sarmas, E. Spiliotis, V. Marinakis, G. Tzanes, J.K. Kaldellis, H. Doukas, Ml-based energy management of water pumping systems for the application of peak shaving in small-scale islands. Sustain. Urban Areas **82**, 103873 (2022)
22. E. Sarmas, E. Spiliotis, E. Stamatopoulos, V. Marinakis, H. Doukas, Short-term photovoltaic power forecasting using meta-learning and numerical weather prediction independent long short-term memory models. Renew. Energy **216**, 118997 (2023)
23. E. Sarmas, S. Strompolas, V. Marinakis, F. Santori, M.A. Bucarelli, H. Doukas, An incremental learning framework for photovoltaic production and load forecasting in energy microgrids. Electronics **11**(23), 3962 (2022)
24. P. Skaloumpakas, E. Sarmas, M. Rachmanidis, V. Marinakis, Reshaping the energy landscape of create through renewable energy valleys. Sci. Rep. **14**(1), 8038 (2024)
25. V. Sze, Y.-H. Chen, J. Emer, A. Suleiman, Z. Zhang, Hardware for machine learning: Challenges and opportunities, in *2017 IEEE Custom Integrated Circuits Conference (CICC)*. (IEEE, 2017), pp. 1–8
26. M. Tamke, P. Nicholas, M. Zwierzycki, Machine learning for architectural design: practices and infrastructure. Int. J. Archit. Comput. **16**(2), 123–143 (2018)
27. J. Tollefson, K.R. Weiss, Nations adopt historic global climate accord: agreement commits world to holding warming'well below'2 [degrees] c. Nature **582**(7582), 315–317 (2015)
28. P. Weigel, M. Fischedick, Review and categorization of digital applications in the energy sector. Appl. Sci. **9**(24), 5350 (2019)

Chapter 2
The Role of Artificial Intelligence in Transforming the Energy Sector: A Comprehensive Review

Abstract This chapter explores the transformative role of Artificial Intelligence (AI) in addressing significant challenges within the energy sector, including the integration of renewable energy, the management of distributed systems, the optimization of electrical grids, and the enhancement of building energy efficiency. By leveraging large volumes of data, AI provides innovative solutions to improve sustainability, flexibility, and operational efficiency. The discussion emphasizes the critical need for high-quality, standardized datasets to ensure the accuracy and scalability of AI applications. Advancements are categorized into three primary areas: distributed energy generation, microgrid flexibility, and building energy optimization. For each area, this chapter examines the latest methodologies, breakthroughs, and practical implementations that are shaping the future of energy management. Through a comprehensive analysis of the intersection between advanced technologies and energy systems, this chapter emphasizes the potential of AI to facilitate the energy transition while addressing essential challenges such as data standardization and accessibility, which are vital for realizing its full potential.

Keywords Artificial intelligence · Machine learning · Optimization · Energy sector · Renewable energy sources · Literature review

2.1 Introduction

The rapid development of sensor technology, wireless transmission technologies, and network communications, coupled with the swift evolution of cloud computing and smart portable measurement devices, has resulted in the accumulation of big data across various aspects of our lives. It is not an exaggeration to assert that we are currently living in the era of big data, as noted in McKinsey's article nearly a decade ago [75]—the volume of available information sources is increasing rapidly, creating unprecedented opportunities for a new industrial revolution. Concurrently, the complexity of data is escalating, necessitating modern methods for managing, storing, and processing information.

E. Sarmas et al., *Artificial Intelligence for Energy Systems*, Learning and Analytics in Intelligent Systems 46, https://doi.org/10.1007/978-3-031-85209-1_2

The energy sector cannot remain unaffected by these developments. Innovation driven by big data is reshaping the landscape of the energy industry. The availability of extensive volumes of energy production and consumption data facilitates the integration of innovative IT applications, accelerating the digitization of energy systems [226]. Modern energy systems effectively address significant contemporary challenges in the energy sector, such as operational efficiency, optimal management of renewable energy sources (RES), system stability, reliability, as well as the design of buildings, smart grids, and integrated energy systems, among others [15, 124]. All of these efforts aim to achieve both a seamless green transition and cost control and reduction through the implementation of intelligent energy management systems.

Furthermore, the technological revolution has introduced new terminology and innovations that were previously unfeasible within traditional energy systems. One of the most significant innovations is the smart grid, which can be defined as an electricity network based on digital technology. Smart grids offer benefits such as more efficient electricity transmission, faster recovery after power outages, optimal utilization of RES, and reduced operating and management costs [16, 55]. Smart grids, along with innovative technical solutions for energy storage and the integration of distributed resources, can play a key role in the transition to green energy. At the same time, smart grids incorporate the concept of 'information flow', which involves storing large volumes of data that can later be leveraged using AI methods to support optimization decisions at the network level [227] or the energy production or consumption level at the building level [58]. Therefore, it is evident that big data analysis and the application of AI methods enable more systematic and efficient decision support for various stakeholders, including producers, consumers, and regulatory authorities.

This chapter provides a detailed review of the literature on the main problems in the energy sector that can be addressed using AI methods and computational optimization. As mentioned in the Introduction chapter, these problems are grouped into three main components: distributed generation **(AI-4-RES)**, flexibility at the microgrid level **(AI-4-FLEX)**, and building energy efficiency **(AI-4-ENEF)**. These three components are further illustrated in detail in Fig. 2.1. For each of the aforementioned categories, the principal issues identified in the international literature are documented, and notable AI applications developed in recent decades are presented for each category of challenges.

The structure of the chapter is organized as follows. First, a comprehensive description of the role and importance of big data in the energy sector is presented. Subsequently, the three main categories of problems are analyzed, as outlined above. The emphasis is placed on identifying the most significant specific issues within each category, as well as on the ways, methods, and techniques through which AI has contributed to solving these problems. Finally, useful conclusions and future perspectives are summarized from the literature review.

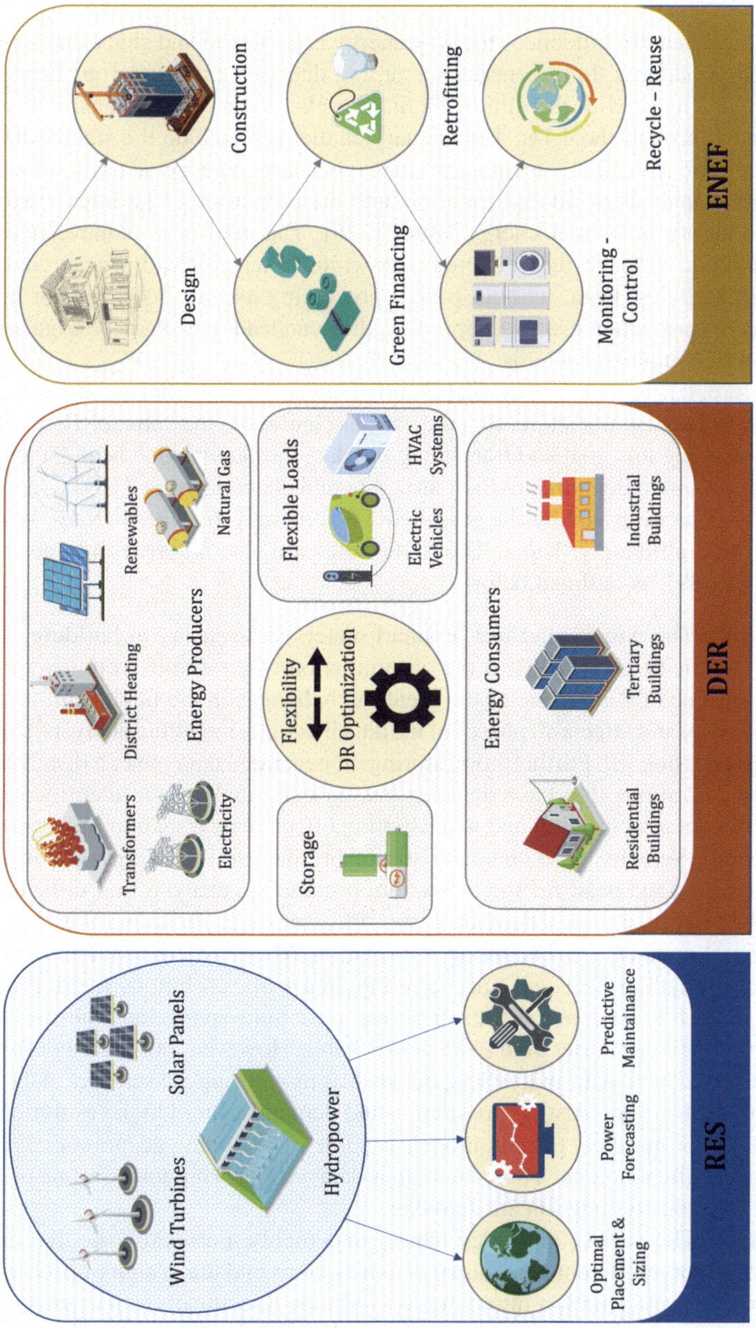

Fig. 2.1 The three subcategories of AI applications in the energy sector

2.2 Big Data in the Energy Sector

Big data significantly influences the way energy is produced and shapes new energy consumption patterns. In the energy sector, big data presents numerous opportunities and challenges, with the aim of facilitating the green energy transition, which is expressed through the 4 Ds. The primary challenges include the standardization of methods for collecting, storing, and managing large energy datasets efficiently, the effective analysis of this information, and the utilization of findings to support decision-making in critical energy issues [226]. The goal is to achieve real-time insight and add value to the collected data while adhering to all privacy and confidentiality policies. Data, without proper processing and analysis, remain merely digits. In contrast, when used appropriately, they can lead to new knowledge and the resolution of complex problems.

Therefore, it is particularly critical to fully harness the potential and opportunities offered by big data while addressing the various challenges they create. The first step toward achieving this goal and translating big data into knowledge is understanding their sources and understanding their structure and characteristics [153]. Generally, big data are characterized by the so-called 5 Vs, namely *Volume*, *Variety*, *Velocity*, *Veracity*, and, ultimately, *Value*. These attributes are also inherent in large energy datasets [81, 89], as outlined below:

➢ **Volume:** The widespread use of smart meters and sensors in buildings, RES, and electricity grids leads to the accumulation of enormous volumes of data. The volume of data is one of the greatest challenges in the field of information technology and affects all phases of the data lifecycle, including analysis, storage, and processing [8]. Furthermore, storing all generated data poses a disadvantage for organizations that base their actions on data, making the identification of specific datasets difficult and necessitating effective aggregation processes.

➢ **Velocity:** Velocity, as a characteristic of big data in the energy sector, refers primarily to the need for the collection, processing, analysis and utilization of available energy data. There are many different cases regarding the speed and temporal resolution of incoming information. Many meters provide the ability to record data in near real time, offering data with resolutions ranging from a few seconds to a minute, while others operate at more sparse intervals, providing one value every quarter hour or an hourly value. However, most energy problems require real-time data analysis. Issues such as monitoring the stability of the electricity grid and forecasting production and demand require high-resolution data (seconds or minutes) to train models capable of providing accurate predictions and recommendations. Thus, managing data with varying temporal resolutions and velocities is a significant challenge.

➢ **Variety:** The variety of energy data is undeniably a critical issue that creates the need for intelligent data management systems and the design of processes to standardize information into common formats. Specifically, smart energy systems encompass different types of data, such as relationally structured data (e.g., load time series), semi-structured data (e.g., weather data), and fully unstructured data (e.g., data in audio, text, or image format), which together constitute the overall mix of available information [205].

- **Veracity:** The reliability or veracity of the data encompasses concepts such as data protection, cleaning limitations, and the need for backups. The reliability of big data is a significant challenge both in research and in practice, as the distinction between useful and "dirty" data directly impacts the validity of AI models [133].
- **Value:** As previously mentioned, large energy datasets have little intrinsic meaning unless their true value is "unlocked" through data analysis models and techniques aimed at supporting decision making throughout the energy management process. Furthermore, according to [117], the value of large energy datasets is described as sparse, which means that the value derived from big data may be limited. Therefore, the need to develop intelligent systems that make optimal use of the available data and transform it into actionable knowledge for the relevant problem is undeniable.

From the above, it becomes clear that big data alone cannot provide solutions to complex modern energy problems. They serve as a crucial tool that, when properly combined with data analysis methods, can offer reliable applications and decision support systems. One of the most modern categories of data analysis methods is AI and its subfields, which will be explored in detail in the next chapter. The combination of big data and AI methods results in the creation of intelligent models and energy management applications. The roles of these two fields are inextricably linked, as AI relies on the existence of big data for models to learn from historical observations and improve their performance. In the following sections, a wide range of examples is presented in which big data and AI are successfully combined to develop intelligent decision support systems.

2.3 Artificial Intelligence and Distributed Energy Production

The penetration of RES, particularly wind and solar energy, is steadily increasing. The global capacity for photovoltaic electricity generation, wind turbines and other renewable energy technologies is expected to accelerate in the coming years, accounting for nearly 95% of the increase in global power capacity by 2026 [86]. The *IEA* projects that the RES electricity capacity will increase by over 60% globally between 2021 and 2026, reaching 4,800 GW, equivalent to the current global capacity of fossil fuels and nuclear power. Simultaneously, the addition of onshore wind farms by 2026 is expected to be nearly 25% higher on average compared to the 2015–2020 period. In 2020 alone, global wind farm installations doubled, reaching a remarkable level of nearly 110 GW [86].

Moreover, the trend observed in the increase of RES capacity during the 2021–2026 period suggests that RES growth in the European Union as a whole will exceed the projections outlined in the current National Energy and Climate Plans (NECPs) for 2030 [206]. This ever-growing momentum for RES stems from the emphasis placed on energy and climate policies both in the EU and globally. In this context, the European Union has designed and proposed the *Fit for 55* package, aimed at reducing the EU's greenhouse gas emissions by 55% by 2030 [38].

Thus, the coming years are critical for the successful transition to RES, both in terms of developing new technologies to optimize production from available natural resources and improving the management and distribution of the energy produced [172]. These goals can be achieved with the help of Artificial Intelligence (AI), which is already widely used in numerous studies to help achieve future RES goals [87] (Table 2.1).

2.3.1 Predictive Models for RES

The penetration of RES-particularly solar and wind energy-is increasing, and accurate forecasts of their production contribute to their smooth integration into existing electricity grids [178]. Consequently, predicting energy production from RES systems is one of the most pressing research issues in the energy sector. The primary RES systems that attract research interest are energy forecasting for photovoltaic (PV) systems and wind systems. The following paragraphs provide an in-depth description of these two issues and an extensive review of the most significant and recent literature.

Energy production forecasting in PV systems Accurate prediction of the energy production of PV systems is a complex computational problem. The challenges associated with this problem stem from the inherent variability of weather conditions, specifically solar radiation, temperature, and cloud cover. Fluctuations in meteorological phenomena result in uncertainty in the PV output, leading to problematic intermittent penetrations, overvoltages, reverse power flows, fluctuations in harmonic distortion in current and voltage waveforms, and various other consequences that increase the need for robust predictive models [146]. The unpredictable nature of production significantly impacts the reliability, stability, and operational planning of the system, also resulting in economic implications [169]. In this context, reliable forecasting of PV production can substantially mitigate uncertainty, enhance stability, and improve economic viability, thereby rendering it a critical area of research [48, 195].

Perhaps the most critical element to consider when developing a forecasting model for PV energy production is the forecast horizon. The forecast horizon is defined as the future time period for which the production forecast is made or the time interval between the actual time and the forecast time [42]. Based on the forecast horizon, some researchers [45, 145] have categorized the forecasting models into three main categories: short-, medium- and long-term forecasts. However, there are no universally agreed criteria for distinguishing models across these horizons [130, 173]. For

Table 2.1 Indicative AI studies for renewable energy sources

Study	Year	Category	Subcategory
[113]	2021	Production forecasting	PV systems
[200]	2019	Production forecasting	PV systems
[156]	2022	Production forecasting	PV systems
[161]	2023	Production forecasting	PV systems
[39]	2020	Production forecasting	PV systems
[120]	2024	Production forecasting	PV systems
[102]	2020	Production forecasting	PV systems
[122]	2020	Production forecasting	PV systems
[37]	2022	Production forecasting	PV systems
[2]	2021	Production forecasting	PV systems
[12]	2020	Production forecasting	PV systems
[118]	2021	Production forecasting	PV systems
[4]	2022	Production forecasting	PV systems
[40]	2001	Production forecasting	Wind systems
[79]	2019	Production forecasting	Wind systems
[47]	2019	Production forecasting	Wind systems
[199]	2017	Production forecasting	Wind systems
[104]	2020	Production forecasting	Wind systems
[103]	2021	Production forecasting	Wind systems
[79]	2019	Production forecasting	Wind systems
[19]	2019	Production forecasting	Geothermal systems
[67]	2020	Production forecasting	Geothermal systems
[155]	2020	Production forecasting	Hydroelectric systems
[11]	2020	Fault detection	PV systems
[192]	2021	Fault detection	PV systems
[54]	2020	Fault detection	PV systems
[136]	2020	Fault detection	PV systems
[60]	2019	Fault detection	PV systems
[128]	2023	Fault detection	PV systems
[85]	2022	Fault detection	PV systems
[190]	2022	Fault detection	PV systems
[76]	2018	Fault detection	Wind systems
[220]	2018	Fault detection	Wind systems
[80]	2020	Fault detection	Wind systems
[33]	2019	Fault detection	Wind systems

(continued)

Table 2.1 (continued)

Study	Year	Category	Subcategory
[225]	2021	Optimal siting	PV systems
[90]	2022	Optimal siting	PV systems
[49]	2024	Optimal siting	PV systems
[150]	2022	Optimal siting	PV systems
[164]	2021	Optimal siting	On-shore Wind systems
[141]	2015	Optimal siting	On-shore Wind systems
[50]	2025	Optimal siting	Off-shore Wind systems
[100]	2010	Optimal siting	Off-shore Wind systems
[208]	2013	Optimal siting	Off-shore Wind systems

example, [17] an additional category is identified: ultra-short- or extremely short-term forecasts, which are utilized for energy management in microgrids or smart grids [146].

Taking these studies into account, four categories of forecasting models can be identified based on the forecast horizon, namely: Ultra-Short-Term Forecasts, Short-Term Forecasts, Medium-Term Forecasts, and Long-Term Forecasts. Each category serves different purposes and faces unique challenges that will be analyzed in the following sections.

➢ **Ultra/very short-term forecasting** which refers to predictions made seconds to minutes ahead and is critical for immediate energy management in microgrids or smart grids. Ultra/very short-term forecasting is primarily used in the design of power systems within smart microgrids, with a forecasting horizon ranging from a few seconds to less than 30 min, according to Ren et al. [151]. The utility of these forecasts lies in areas such as electricity pricing, real-time monitoring of electricity dispatch, power smoothing processes in microgrids, and the control of energy storage in PV systems [42]. It is worth noting that some researchers also include forecasts spanning several hours within this category, though this is not the standard practice.

➢ **Short-term forecasting** which refers to predictions spanning a few hours to a day and is important for daily operational planning. Short-term forecasting is primarily applied in economic load dispatch decisions and power system scheduling. The forecasting horizon in this category ranges from 1 h to up to 6 h ahead, according to Ren et al. [151].

➢ **Medium-term forecasting** which refers to predictions over a few days to weeks and is useful for maintenance and scheduling. Medium-term forecasting spans from 6 h to up to one day ahead. The primary application of this category is in the maintenance scheduling of conventional or integrated solar energy systems, which include high-tech transformers and various types of electrical machinery [42].

➢ **Long-term forecasting** which refers to predictions extending from months to years and is relevant for policy and strategic planning. Long-term forecasting encompasses predictions for more than 24 h ahead. This category includes cases where the long-term production, transmission, and distribution of energy generated by PV systems need to be estimated [21]. Naturally, long-term models have reduced accuracy since short-term weather fluctuations spanning one or a few days cannot be accurately predicted over such extended horizons. Thus, these models are better suited for estimating production trends based on seasonality. Notable examples of such applications include studies by Vehviläinen and Pyykkönen [193], Hong et al. [78], who have developed medium-term and year-ahead models for monthly scheduling of power systems, electricity pricing, and load forecasting.

Several factors influence the accuracy of the PV production forecasting, making the development of predictive models a complex process. Key considerations include the forecasting horizon, inputs to the predictive model, and metrics to evaluate performance. To improve accuracy, it is essential to optimize correlation analysis and estimate uncertainty in predictive models through probabilistic forecasting techniques.

Another element that varies between studies is the selection of input features for models [198]. The choice of appropriate inputs for predictive models directly affects forecast accuracy, serving as a critical factor in determining a model's success. Critical evaluation and systematic filtering are necessary when selecting input features, as arbitrary inclusion of features can diminish model performance while increasing latency, cost, and computational complexity.

For PV system forecasting models, inputs are primarily meteorological parameters such as solar radiation, ambient temperature, PV module temperature, wind speed, and humidity [41, 197]. The correlation between PV output and these meteorological inputs varies, demonstrating either positive or negative relationships; the strength of these correlations is contingent upon the specific variable. Notably, solar radiation displays the strongest correlation with PV output compared to the other variables [91]. In the following paragraphs, the key input features commonly used in the literature are presented in greater detail.

➢ **Solar Radiation**: This is the most critical input variable in the problem of solar energy forecasting. It represents the radiant energy emitted by the sun in the form of electromagnetic radiation. Solar radiation has the highest correlation with the power output of a PV system and is directly proportional to the amount of solar energy that can be collected by it. In the study by [44], a neural network was employed with solar radiation as an input, resulting in highly accurate forecasts.

➢ **Temperature**: This measures the average heat or thermal energy in a substance and is primarily determined by the amount of sunlight absorbed by the Earth's surface and the amount of heat emitted into the atmosphere by greenhouse gases. Ambient temperature is another significant parameter affecting the performance of PV systems, as studies have shown that the correlation between ambient temperature and PV output power is strong enough to warrant its use as an input variable.

- **Cloud Cover**: Cloud cover (or cloudiness) is one of the most critical factors determining the intensity of solar radiation reaching the Earth's surface. The portion of solar energy passing through clouds is not constant and depends on the type and density of cloud formations [123]. According to Wang et al. [177, 202], the movement, dispersion, and deformation of clouds cause fluctuations in sunlight intensity, thereby impacting the output of PV systems. Typically, information on cloud cover is collected from satellite imagery [115], though this method loses accuracy for specific locations. Consequently, ground-based imaging techniques are now extensively used for improved precision.
- **Wind Speed**: Wind speed is one of the most critical input variables in predictive models for PV systems, acting as an additional source of uncertainty [148]. Its primary role lies in how it facilitates heat dissipation from the PV system and reduces the temperature of PV cells [149].
- **Temporal Variables**: Temporal or periodic variables include the time of day, month of the year, and season. These variables are crucial for model accuracy, as energy production is directly related to the sun's position, which changes periodically throughout the day and year. A characteristic example is the study by Al-Dahidi et al. [7], where the current timestamp was used in the form of hours starting from the beginning of each year.

As highlighted in the preceding paragraphs, forecasting the energy output of a PV system is a highly complex problem that continues to receive extensive attention from the research community. According to the review by Ahmed et al. [5], various modeling approaches have been employed for PV output predictions, categorized primarily into physical models, purely statistical models, and AI-based models (including machine learning, deep learning, and neural networks). Additionally, hybrid models, which combine these approaches, have also been proposed. Many studies compare the performance of different models to identify the optimal one; however, no evidence has yet emerged to suggest that any specific model consistently outperforms others across a sufficiently large number of experiments. For instance, Almonacid et al. [13] compared a neural network-based model with three conventional statistical models, concluding that the predictive ability of the neural network was superior. Similarly, in the study by Oudjana et al. [138], a neural network was compared with a simple regression model, yielding similar results. It is evident that many factors influence a model's accuracy, making direct comparisons challenging. The three primary categories of predictive techniques-physical, statistical, and AI-based-are discussed below.

Physical forecasting methods use a theoretical simulation model to calculate the output power of a PV system based on its key design parameters [221]. These parameters include air pressure, surface roughness, temperature, disturbances, and other obstacles [101, 217]. Physical methods have generally been shown to be more suitable for long-term forecasting horizons. Further details on physical methods can be found in the studies by Yang et al. [212] and Mayer and Gróf [116].

Statistical methods rely on historical production data, using time series of historical data and real-time data. These models apply mathematical equations to extract data patterns and relationships based on historical input data. Common algorithms used include moving average models, autoregressive models, and combinations such as ARIMA models. While these methods require less input data compared to machine learning models, they often exhibit lower forecasting accuracy when compared to modern neural network architectures.

Methods based on AI models represent a rapidly growing category due to their superior predictive capabilities. These approaches leverage AI's ability to learn from historical data and enhance predictive performance through complex training processes. A notable drawback is the frequent need for powerful computing resources to execute multiple iterations before reaching a final prediction, owing to the increased computational complexity. The main AI models will be discussed in greater detail in the next chapter.

Energy production forecasting in wind systems The prediction of energy production from wind systems poses a critical challenge for the scientific community, representing a significant category of forecasting problems. Since wind is inherently variable, wind energy is a source of electricity with significant fluctuations. Forecasting in wind farms shares many similarities with PV systems because the methods used have parallels between the two systems, and the prediction horizon is similarly defined for both problems. The primary difference lies in the input variables, as in the case of wind systems, the most critical factors are wind speed and direction.

According to the review by Foley et al. [63], short-term forecasts (defined by the authors as ranging from 1 h to 72 h ahead) are useful for power system planning, including unit commitment and dispatch, as well as for electricity trading in markets where wind energy and storage can be exchanged [127]. Medium-term forecasts, covering a range from 3 d to 1 week ahead, are utilized for scheduling maintenance of wind farms, unit commitment and dispatch, and for shutting down thermal generators for control and maintenance purposes within the grid. Lastly, long-term forecasts, extending weeks or months ahead, are particularly instrumental for strategic planning and site selection for wind farms. It is important to note that forecast error typically increases as the prediction horizon expands.

Wind energy forecasting can be defined as the application of meteorological and climatological theories and practices [140]. More specifically, wind forecasting for energy production focuses primarily on short-term periods (up to several hours ahead) and medium-term periods (up to one week ahead). A detailed review of the short-term characteristics of wind has been developed by Landberg [98]. The emphasis on these prediction horizons arises from the significant value of such forecasts in various applications, most notably load regulation and monitoring, balancing and unit commitment, and planning at the microgrid level.

In general, as with PV systems, wind energy forecasting models can be divided into two categories based on the philosophy of the input features they use. The first category relies exclusively on the analysis of historical wind time series, while the second category of methods uses wind speed forecasts derived from numerical weather prediction models (NWP). In the first category, statistical models are used to forecast the average hourly wind speed or directly predict electricity production [181]. In contrast, models in the second category use forecasts for hourly average wind speed and direction obtained from meteorological models. In fact, the study by Nielsen et al. [132] demonstrated that using multiple meteorological forecasts reduces prediction error. In the second category of models, the selection of the meteorological model-considering factors such as geographical area, spatial and temporal resolution, and prediction horizon-is critical, as it significantly influences both accuracy and computational complexity.

Finally, analogous to solar energy forecasting models, wind energy forecasting can generally be conducted using physical methods, traditional statistical methods, or machine learning (ML)-based methods. Over time, hybrid methods that combine two or more approaches from these primary categories have also been proposed. However, the selection of the most appropriate model or method category is determined by the prediction horizon. A review of the three main categories is provided in the following paragraph.

Physical models are based on using weather data and are applied both to forecast wind speed and predict wind energy [97]. According to Landberg et al. [99], the implementation of physical modeling necessitates the integration of meteorological forecasts and depends on substantial computational systems to yield accurate results, as the estimation of wind power is conducted through comprehensive descriptions of the lower atmosphere. On the other hand, statistical methods use historical data to make predictions. Unlike physical methods, statistical methods involve a single step in converting input variables into power output forecasts, acting as a "black box." A characteristic example of a statistical method is the study by Torres et al. [184], which proposed an ARMA model trained on nine years of historical wind farm data to predict wind energy production up to several hours ahead. In recent years, however, classical statistical methods have been largely replaced by ML methods, with neural systems, deep learning networks, and fuzzy models being the most commonly used. These models identify the relationship between predicted wind and expected power output by analyzing historical data in a time series format. For instance, the research conducted by Damousis and Dokopoulos [40] introduced a fuzzy system for forecasting wind speed, whereas Hong and Rioflorido [79] presented a hybrid deep learning model for day-ahead predictions, which demonstrated excellent forecasting capability.

2.3.2 Fault/Anomaly Detection and Predictive Maintenance of RES

The second most significant category of challenges in renewable energy, in which AI has made notable contributions, is anomaly detection and predictive maintenance. Specifically for PV systems, AI has expanded the capabilities to detect many of the critical faults observed either on their DC or AC sides. This section provides a brief overview of the current state of fault detection methods in PV systems and the role of AI.

The reduction in the cost of PV systems, along with performance improvements resulting from better cell efficiency and lower energy conversion losses, has increased interest in the maintenance of PV systems [20, 53]. Maintaining these systems encompasses a range of activities, from minimal inspections to high-precision monitoring, which enable plant owners to identify potential issues or determine when cleaning is necessary [204]. Interest is further heightened because maintenance directly affects the performance of PV systems in terms of power generation and, consequently, overall revenue [43].

Over the years, numerous anomaly detection techniques have been proposed for diagnosing faults in PV systems. A brief review of various monitoring systems available for performance tracking is presented by Hare et al. [73], and Triki-Lahiani et al. [185]. However, in the past decade, numerous advanced fault detection algorithms have emerged that can accurately and quickly detect faults [9]. According to Pillai and Rajasekar [143], there are three conventional systems for protecting a PV system against faults: (i) the use of standard protective devices, (ii) the combined use of multiple protective devices in a single PV array, and (iii) real-time control algorithms for fault detection. The latter category includes contributions from machine learning models, which have significantly aided fault detection through data analysis.

The most common data analysis methods for anomaly detection include techniques such as Model-based Difference Measurement (MBDM), Real-time Difference Measurement (RDM), and Output Signal Analysis (OSA) [143]. MBDM techniques compare real-time theoretical parameters with model-predicted data to detect system faults [9]. RDM techniques, on the other hand, compare real-time parameters against threshold limits determined from experiments under real-world conditions [167]. OSA techniques typically apply appropriate analysis to the output signal to identify anomalies, such as detecting anomalies in PV system arcing through Fourier transformations of the output signal waveform [222].

Recently, numerous machine learning (ML) methods have been developed for fault detection in PV systems, offering better results than the previously mentioned methods [22]. ML algorithms are trained to learn the relationship between the input parameters and the PV system output using data from smart meters. However, data from fault cases are still required for training models on non-standard operating conditions. Once trained, these models can easily identify defective states by evaluating faults. Fault detection through ML algorithms also aids in accurately identifying and classifying faults. Nevertheless, like all data-driven techniques, ML-based anomaly detection algorithms have limitations, such as dependence on data quality, challenges in obtaining data during faulty conditions, and difficulty in applying the same model to various PV systems, which prevents the creation of generalized solutions.

AI algorithms are now widely used to detect different types of faults. For instance, Ali et al. [11] proposed a support vector machine (SVM)-based model utilizing hybrid PV features for fault detection and classification using infrared thermography techniques. Similarly, Veerasamy et al. [192] developed a model for detecting high impedance faults in integrated PV power systems using recurrent neural networks. Similar research has been conducted for line-to-line faults, such as the work by Eskandari et al. [54], who proposed an autonomous method for detecting and classifying line faults based on a combination of ML methods and genetic algorithms, leveraging PV current-voltage (I-V) curves. Another common category of faults includes series arc faults, which are among the most prevalent in PV systems, prompting significant research efforts for their detection. For example, Omran et al. [136] developed a model based on convolutional neural networks (CNNs) for identifying operational errors from the DC side of PV systems.

2.3.3 Optimum Placement of Renewable Energy Sources RES

One of the most common research challenges in the field of renewable energy (RE) is identifying the optimal location for installing RE systems, particularly PV systems and wind farms. The first step in ensuring an efficient site for energy production-and consequently economic benefits-is conducting a comprehensive analysis of the factors that directly affect production levels. In the case of PV systems, the primary factor is solar radiation, whereas for wind farms, the key criterion is wind potential. Furthermore, the process must also take into account additional decision-making criteria and evaluate all restrictive factors [163].

AI, combined with other supportive information systems, enables the identification of optimal locations for RE installations, playing a crucial role in successful planning at both national and global levels. The following paragraphs outline the advancements in the fields of PV systems and wind farms.

Placement of PV Systems One of the most fundamental computational systems used for identifying the most suitable location for PV systems is the Geographical Information System (GIS). GIS is a tool for analyzing and processing data, maps, and spatial information, and it has become highly popular in energy planning problems [51]. A common approach involves developing decision-support models that integrate GIS with multi-criteria problems to determine the ideal location for solar energy production. For instance, in the study by Zambrano–Asanza et al. [216], GIS is combined with multi-criteria analysis for optimal PV placement, using high-resolution data for spatial characterization of energy demand and conducting density analysis.

In addition to classical methods of multi-criteria decision analysis, GIS has been successfully combined with AI models, particularly deep learning, in recent years, yielding remarkable results. In the study by Zhong et al. [225], a methodological framework is proposed for estimating rooftop PV potential at a city level using high-resolution satellite images and a deep learning model that automatically extracts rooftop areas by segmenting the image. Similarly, the primary objective in the study by Kalyan and Sun [90] is to detect rooftops without PV systems but with moderate to high solar irradiance potential, employing GIS and deep learning techniques. Lastly, the challenge of analyzing shading and rooftop availability was the focus of the study by Ren et al. [150], which proposed a novel 3D-GIS system and a deep learning approach. This approach analyzes solar irradiance to predict rooftop PV production while accounting for the shading effects of surrounding buildings.

Placement of Wind Farms As discussed at the beginning of this chapter, wind energy is a rapidly evolving field that has garnered significant attention and substantial investment over recent decades. Consequently, as a highly competitive sector, it is crucial to minimize the cost of installing new wind farms and maximize energy production profits starting from the design phase. Designing a wind farm is a complex task requiring expertise from various disciplines and involving numerous optimization variables [62]. The problem of wind farm placement has not yet been fully automated, and AI offers numerous opportunities to optimize design decision-making.

Research efforts in the field of determining the optimal location for wind farm installations can be divided into two subcategories: onshore and offshore wind farms. Each category involves different constraints, making it logical to focus efforts within a specific framework. Regarding onshore wind farms, a methodology has been developed to determine the optimal placement of a wind farm within an electricity transmission network, with a primary focus on assessing the network's security [164]. Additionally, Petrov and Wessling [141] propose an alternative methodology for the optimal placement of wind turbines based on environmental and ecological factors, combining multi-criteria analysis with machine learning.

For offshore wind farms, Lee et al. [100] developed an optimization methodology based on a genetic algorithm. Initially, neural network models are employed to construct wind and bathymetry maps. Subsequently, a genetic algorithm is implemented with the objective of maximizing energy density while adhering to constraints related to maximum water depth and maximum distance from the shoreline. Similarly, the study by Wu et al. [208] proposes a methodology for optimizing the placement of offshore wind turbines using AI techniques and genetic algorithms.

2.4 Artificial Intelligence and Smart Grids

As detailed in previous sections, RES have proliferated in recent years, becoming a major driver for the decentralization and democratization of energy [18, 186]. Small production units located on the consumer side, such as rooftop PV systems and small wind generation units, are referred to as distributed energy resources (DERs). On the other hand, electrical loads that can be shifted during the day or week are called flexible loads. The integration of DERs, combined with the increasing use of storage systems and the optimal scheduling of flexible loads, can form a low-voltage distribution network known as a microgrid, which can operate either in connection with the main grid or entirely autonomously [166].

Essentially, any advanced power system that incorporates information from smart metering systems and can optimally manage the production and consumption of its constituent nodes can be considered an intelligent grid. The main challenges in smart grids include the optimal utilization of grid resources, achieving the highest possible energy efficiency, ensuring maximum system reliability and security, and delivering electricity to consumers at the lowest cost. In the field of smart grids, the generated data, along with AI algorithms and optimization techniques, play a critical role in solving these challenges and achieving objectives at the microgrid level. This section examines issues related to DERs that can be addressed using optimization algorithms and AI. These issues can be divided into four main components:

1. **Demand Forecasting at the Grid Level**: Predicting energy demand to ensure efficient resource allocation.
2. **Demand Response Programs and Energy Market Correlations**: Exploring strategies for aligning demand response initiatives with market dynamics.
3. **Optimization of Flexible Load Scheduling**: Developing models to schedule flexible loads for maximum efficiency and cost savings.
4. **Energy Storage System Challenges**: Investigating storage-related problems, which are expected to reshape energy management practices.

By leveraging AI and optimization techniques, these components address key aspects of distributed energy resource management, contributing to the development of smarter, more efficient energy systems (Table 2.2).

2.4.1 Load Prediction Models

Load forecasting plays a vital role in the energy transition through modern power systems, serving as the foundation for many innovative solutions such as smart grids, optimal energy management, and energy system design [145]. Consequently, in recent years, much of the energy research has focused on accurate short-term load forecasting due to its importance for the reliable operation of power systems and, therefore, the economy. Moreover, significant functions of the electrical system, such as scheduling, maintenance, and tariff adjustments, rely heavily on accurate load forecasts. Thus, it is clear that load forecasting can directly influence energy policy, as numerous decisions in energy management systems take energy demand predictions into account.

Load forecasting is generally divided into three categories by most researchers based on the prediction time horizon. Short-term forecasts cover periods from a few minutes to a day or a week ahead. Short-term load demand plays a crucial role in optimal unit commitment, energy reserve control, and the evaluation of purchase and sales contracts between different companies. Medium-term forecasts focus on periods ranging from a week to a year ahead and are used for the efficient operation and maintenance of the power system. Finally, long-term forecasts are associated

Table 2.2 Studies of AI applications for distributed energy resources

Study	Year	Category
[119]	2003	Load forecasting
[14]	2017	Load forecasting
[224]	2017	Load forecasting
[162]	2022	Load forecasting
[10]	2020	Load forecasting
[147]	2017	Load forecasting
[218]	2010	Load forecasting
[158]	2024	Load profiling
[203]	2015	Demand response
[196]	2020	Demand response
[108]	2019	Demand response
[121]	2024	Demand response
[88]	2016	Dynamic pricing
[109]	2018	Dynamic pricing
[188]	2015	Flexibility management
[3]	2019	Flexibility management
[59]	2014	Flexibility management
[74]	2022	Flexibility management
[160]	2024	Flexibility management
[31]	2019	Flexibility management
[168]	2017	Flexibility management
[209]	2018	Storage system
[52]	2019	Storage system
[230]	2021	Storage system
[135]	2020	Storage system
[134]	2020	Storage system
[35]	2018	Storage system
[201]	2023	Storage system

with policy-making and long-term planning, with a forecasting horizon of several years. Notably, long-term load forecasting is used for the long-term planning of the power system in line with future energy demand and the energy policy of each state. For instance, the research by Rahman and Hazim [144] distinguishes four categories, though this is not the common classification.

As mentioned above, the primary tasks supported by load forecasting include production planning, the safe and consistent operation of power generation units, economic dispatch, and reliability maintenance. More specifically, the reliability of the power system fluctuates significantly due to abrupt variations in load demand. To better understand the problem, consider the two potential failure scenarios in demand forecasting. If the load demand is underestimated, the system may face a supply shortage, while if it is overestimated, many available resources may be wasted to meet the overestimated demand.

Generally, it has been observed that there is a strong correlation between weather variables and load demand [77]. In most cases, demand increases during the summer due to rising temperatures, while it is lower during the winter period. In any case, weather variables are among the most critical inputs in load forecasting models. Additionally, load demand varies cyclically throughout the day due to consumers' daily routines and regular activities performed at specific times. Therefore, temporal variables also play an essential role in improving forecasting accuracy.

In the literature, many studies can be found that have addressed accurate load demand forecasting. Specifically, in the study by Amarasinghe et al. [14], the authors present a load forecasting methodology based on deep learning, using convolutional neural networks to predict the demand of a residential building, demonstrating that this model exhibits better predictive ability than most machine learning algorithms. Another study by Zheng et al. [224] employed recurrent neural networks to predict demand at a microgrid level with a forecasting horizon of up to several days ahead. This research showed that such neural networks exhibit excellent ability to forecast load time series despite the inherent difficulty of the problem. However, load forecasting remains a challenging task. For this reason, it is considered that many research efforts will focus on the development of predictive models in the coming years.

2.4.2 Demand Response and the Energy Market

Demand response can be defined as the change in electricity consumption by residential, commercial, and industrial consumers to maintain the balance between electricity supply and demand [183]. Demand response events occur at specifically designated times when consumers are pre-notified to be mindful of their energy consumption due to high grid congestion. The benefits of demand response encompass energy savings, diminished reliance on fossil fuel power plants, and the facilitation of renewable energy sources into the grid, thereby providing enhanced stability and management [23]. Consequently, demand response enables consumers to decrease energy consumption through load shedding, shifting energy use to different periods, or generating and storing energy at specific times, thereby offering greater flexibility to the grid. In return, consumers usually receive reduced energy bills or other forms of incentives [170].

There are two main categories of demand response programs according to Vázquez-Canteli and Nagy [191]. The first category includes incentive-based programs, in which consumers voluntarily participate in a system where the system operator can directly turn off certain devices to reduce energy consumption during peak demand periods. In contrast, the second category comprises time-based programs, which emphasize dynamic pricing and aim to flatten demand curves by offering electricity prices that fluctuate over time in response to grid congestion [165]. Unlike the first category, in time-based programs, the operator cannot directly intervene in specific loads.

Although demand response programs represent a promising strategy for enhancing flexibility in contemporary microgrids, research conducted in the U.S. indicates that these programs have not yet fully realized their potential [36]. One primary reason for this underachievement is that the value of electricity often exceeds its price for consumers, who are generally unwilling to alter their typical electricity consumption patterns. Additionally, consumers face challenges in accessing information regarding electricity prices and comprehending the financial benefits associated with participation in demand response programs [191]. These issues highlight the need for modern technical frameworks to implement effective demand response programs. Key features of such programs should include the automation of decision-making, the monitoring of environmental conditions, and the active engagement of consumers in the process, all while minimizing their level of discomfort.

AI achieves these objectives to a large extent and plays a significant role in the integration and improvement of demand response programs. Among AI algorithms, reinforcement learning plays a pivotal role, with many studies emerging in recent years. For instance, O'Neill et al. [137] presented a novel energy management system designed for residential demand response, which utilizes an algorithm that indirectly estimates the impact of future energy prices and consumer decisions on long-term costs while scheduling the usage of home appliances. A similar study by Wen et al. [203] introduced an energy management system for residential and small commercial buildings. This system is based on a well-known reinforcement learning algorithm (Q-learning), allowing users to input their preferences and self-initiate tasks through flexible requests.

As discussed in previous sections, demand response presents a challenge with clear behavioral implications, as providing appropriate incentives is crucial for the successful implementation of such programs. Consequently, there is a direct connection with the energy market and the energy pricing policy as an incentive to achieve the goal. Specifically, the identified challenge involves dynamic energy pricing, which entails the timely adjustment of energy prices to encourage consumers to adapt their energy consumption patterns appropriately. A notable example is the study by Lu et al. [109], which employs a reinforcement learning algorithm to facilitate dynamic energy pricing within demand response programs. The simulation results showed that the proposed dynamic pricing algorithm could increase the profitability of the provider, reduce energy costs for customers, and balance energy supply and demand in the electricity market, thus improving the reliability of the power system.

2.4.3 Flexible Load Management and Demand Management

In the previous section, demand response programs were introduced, with the aim of incentivizing consumers to shift part of their energy consumption from peak hours to off-peak periods. However, the response to demand represents only one dimension of the challenge of flexibility in the grid. The broader issue can be formulated as an optimization problem for the optimal allocation of flexible loads, which is discussed in more detail in this section.

Energy systems require flexibility to meet the energy demand, which fluctuates over time. There are various approaches to increasing the flexibility of the energy system, either on the supply side or on the demand side. On the supply side, flexibility is achieved through power generation plants with different response times. However, the use of variable energy sources, such as wind and solar power, increases the need for flexibility in the energy system [111]. Of course, there are different precise definitions of flexibility, resulting in varying metrics depending on the context [84].

Demand-side management (DSM) is perhaps the most intriguing issue in the field of grid flexibility, which encompasses a wide range of means to alter electricity consumption. According to Gellings and Smith [69], there are three main categories: peak shaving, valley filling, and load shifting. The shifting of the load requires some form of intermediate storage and a utilization rate less than 100%, as it involves both increasing and decreasing the load [72]. Although the functionality of load shifting is similar in philosophy to energy storage, the difference lies in that demand-side management does not require energy conversion to or from another form (Fig. 2.2).

Although the concept of demand management was introduced decades ago, its successful implementation involves several challenges [176]. AI and mathematical optimization, in general, aim to address some of these challenges through algorithms that optimally allocate flexible loads. For this reason, numerous studies have explored the issues of peak shaving, valley filling, and load shifting [189].

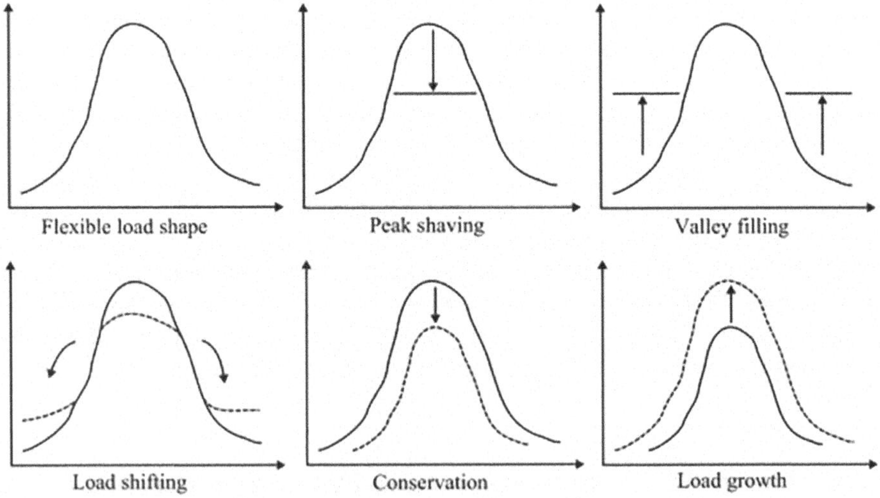

Fig. 2.2 Different types of flexibility provision through optimal shifting of flexible loads

In the study by Chapaloglou et al. [31], an optimization algorithm was developed to manage power flows in an islanded power system. This algorithm achieves a more stable system operation while simultaneously reducing peak demand values for days of the year that exhibit nighttime peaks in their load curves. The proposed algorithm combines deep learning models for load prediction, a pattern recognition process, and a custom optimized power flow programming algorithm.

A different application was developed by Shirazi and Jadid [168], who proposed a smart home energy management model that considers both energy consumption and production. The proposed model regulates household electrical and thermal appliances to operate in a way that minimizes the monetary cost of the customer based on a time-varying pricing model.

2.4.4 Energy Storage Systems

Energy storage technologies play an active role in achieving greater flexibility in the energy grid, improving energy consumption capabilities, ensuring stable and cost-effective operation of power systems, and promoting the widespread application of renewable energy technologies. Numerous innovative solutions, ideas, and technologies have recently been introduced in the field of energy storage from various perspectives, including materials science, system management, and control through AI algorithms aimed at optimizing grid operation [1]. In this context, many research efforts have introduced different algorithms and methods for integrating renewable energy and storage systems, offering solutions to several challenges, such as system optimization, energy control strategies, and energy storage system sizing.

Most AI applications focus on energy management in hybrid systems, consisting of energy storage systems and energy generation sources such as PV systems or wind farms. For example, Xiong et al. [209] developed a reinforcement learning algorithm for real-time energy management. The algorithm can learn the current system consumption data as well as the battery's health status and subsequently update the strategy. The study results showed that the proposed energy management strategy can limit the maximum discharge current, reduce the frequency of battery charging, minimize energy losses, and optimize system performance.

Similarly, Duan et al. [52] proposed a reinforcement learning-based control method for a hybrid energy storage system in AC microgrids that include PV systems and diesel generators, demonstrating the effectiveness of algorithms in complex systems. This method was successfully tested on grid-connected and off-grid island networks. These two examples are just a few of the many successful applications of AI in hybrid energy storage systems, with deep learning gaining significant traction and showing outstanding results compared to rule-based methods Zsembinszki et al. [230].

2.5 Artificial Intelligence and Energy Efficiency in Buildings

The revolution in the field of big data, the development of powerful, advanced algorithms, and access to extensive computational resources have been significant factors in scaling applications based on machine learning for building sector research over the past decades, showcasing the potential of AI to improve building performance. This section presents a range of challenges in the building sector where machine learning algorithms have been applied at various stages of the building life cycle. The primary conclusion drawn from the subsequent analysis is that data analysis using AI methods opens new horizons in the design, operation, and control of buildings. An increasing number of studies are becoming available and growing significantly year by year.

Among the applications developed in recent years, there are several successful studies aimed at fault detection and diagnosis in various building systems (e.g. HVAC systems), predicting or estimating building energy consumption, classifying buildings based on their energy efficiency, learning the behavior of building occupants, and optimizing the use of building subsystems, among others. However, only a limited number of these studies have been widely adopted by the construction industry due to numerous prevalent challenges. These challenges include the lack of large-scale, standardized data for training and validating machine learning models, the difficulty in transferring a machine learning model from one building to another-which limits the applicability of a model across a range of buildings-and the often justified skepticism toward machine learning models, which may not demonstrate reliable performance in numerous instances. This unreliability can lead to erroneous estimates with negative repercussions for both building occupants and energy efficiency. Consequently, it is imperative for future studies to concentrate on specific key pillars, such as: (a) enhancing the thermal comfort of building occupants, (b) achieving energy efficiency in buildings, (c) facilitating demand flexibility, and (d) ensuring building resilience. For all of the above, it is essential to have standardized data available so that the developed models can be generalized and reusable.

The problems related to energy efficiency in buildings and intersecting with the field of AI can be categorized into four main groups. A common goal across all categories is the advancement of green buildings that minimize the carbon footprint, or, in certain cases, attain a positive carbon footprint where feasible (i.e., positive-energy buildings). These categories are as follows (Table 2.3):

- ➢ Building Design
- ➢ Building Construction
- ➢ Building Operation
- ➢ Building Renovation

Table 2.3 Indicative AI studies for energy efficiency in the building sector

Study	Year	Category	Subcategory
[219]	2018	Design	Parametric design
[64]	2018	Design	Parametric design
[27]	2019	Design	Parametric design
[179]	2018	Design	Parametric design
[129]	2017	Design	Genetic design
[30]	2022	Design	Genetic design
[83]	2018	Design	Genetic design
[105]	2017	Design	Genetic design
[95]	2010	Design	Design assessment
[70]	2018	Design	Design assessment
[142]	2017	Design	Design assessment
[46]	2018	Design	Design assessment
[32]	2016	Design	Design assessment
[107]	2018	Design	Design assessment
[106]	2017	Design	Design assessment
[34]	2010	Construction	Cost analysis
[94]	2013	Construction	Cost analysis
[182]	2009	Construction	Cost analysis
[82]	2021	Construction	BIM
[180]	2010	Construction	BIM
[24]	2019	Construction	BIM
[96]	2021	Construction	BIM
[26]	2008	Construction	Construction management
[25]	2005	Construction	Construction management
[228]	2010	Construction	Construction management
[6]	2006	Construction	Defect detection
[229]	2010	Construction	Defect detection
[213]	2011	Operation	Energy efficiency
[93]	2016	Operation	Energy efficiency
[57]	2018	Operation	Energy efficiency
[28]	2013	Operation	Thermal comfort
[57]	2018	Operation	Thermal comfort
[110]	2019	Operation	Thermal comfort
[61]	2012	Operation	Thermal comfort
[126]	2013	Operation	Thermal comfort
[187]	2023	Operation	Thermal comfort
[171]	2023	Operation	Thermal comfort

(continued)

Table 2.3 (continued)

Study	Year	Category	Subcategory
[125]	2016	Operation	Thermal comfort
[92]	2015	Operation	Maintenance
[211]	2018	Operation	Maintenance
[56]	2010	Operation	Maintenance
[207]	2011	Operation	Maintenance
[112]	2017	Operation	Intelligent control
[215]	2014	Operation	Intelligent control
[152]	2015	Operation	Intelligent control
[139]	2019	Operation	Intelligent control
[29]	2019	Renovation	Renovation planning
[68]	2014	Renovation	Renovation planning
[174]	2024	Renovation	Renovation planning
[210]	2006	Renovation	Renovation planning
[114]	2016	Renovation	Renovation planning
[71]	2017	Renovation	Savings estimation
[66]	2018	Renovation	Savings estimation
[157]	2024	Renovation	Savings estimation
[65]	2018	Renovation	Savings estimation

2.5.1 Building Design

The design stage is one of the most critical phases in the life cycle of a building. Advances in AI have fundamentally transformed the way architects design modern buildings, significantly reshaping the landscape of the construction sector. Building data is becoming increasingly accessible, offering opportunities to optimize and transform traditional workflows related to building design. According to a comprehensive review recently published, machine learning has improved building design in three distinct components: parametric design emulation, generative design, and design evaluation.

In the parametric design category, the application of machine learning models combined with statistical methods facilitates the processes carried out during design workflows. Traditional parametric design requires defining design rules, and the advent of AI introduces the need to encode these rules into the software. As a result, exploring the design space has often been limited by the designers' capabilities. However, in recent years, several parametric design tools have been successfully tested in architectural and structural design, with the aim of optimizing design workflows. These tools enable designers not only to specify a final geometric solution but also to describe entire systems and specific parameters that guide variations in the design. Machine learning techniques have been used to simulate strictly encoded rules and parameters, automating related processes.

A notable example is the study by Zhang et al. [219], which presents a framework to assess post-seismic structural safety. In this research, machine learning algorithms map response and damage patterns to the structural safety status of a building based on an acceptable residual collapse capacity threshold. Predictive models are then used to probabilistically determine the structural safety of earthquake-damaged buildings. Another study by Fuhrimann et al. [64] aims to introduce a novel structural design process that enables architects and engineers to expand their typical design horizons. Specifically, they propose a hybrid method that combines structural form-finding algorithms (Combinatorial Equilibrium Modeling) with modern machine learning algorithms (Self-Organizing Maps) to analyze the design space and control interactions between input parameters.

In the field of generative design, numerous studies in the last decade have focused on Generative Adversarial Networks (GANs), a category of ML systems that automate the time-consuming manual application of design rules. These models have been successfully applied to tasks such as floor plan generation and style transfer. For example, Huang et al. [82] developed a variant of GAN to recognize architectural plans and identify rooms with different colors, ultimately aiming to generate apartment layouts. Moreover, machine learning aids in learning and transforming architectural styles, thereby contributing to a profound understanding of the functional rules that govern the internal organization of the design process. Liu et al. [104] tackled the challenge of converting a raster floor plan image into a vector graphic representation, facilitating the visualization of interior spaces, direct manipulation of models for architectural remodeling, and computational applications such as data analysis during the design stage.

Finally, the use of AI in the design evaluation process serves as another valuable tool for designers and architects. This application emphasizes the optimal utilization of building spaces, safety, occupant well-being, and energy efficiency. Frequently, the performance of AI models surpasses the predictions made by professional designers. For example, Phelan et al. [142] proposed a neural network to predict the usage of conference rooms based on historical data from 56 other buildings. The results were highly promising, indicating that the neural network outperformed the estimates provided by designers. Additionally, a study by De Rycke et al. [46] illustrates the development of a machine learning model aimed at classifying the least desirable offices through simulations conducted within a Building Information Modeling (BIM) tool.

Design evaluation also involves assessing user comfort and the utilization of natural light. AI-based techniques have also been developed for these domains. For example, Chatzikonstantinou and Sariyildiz [32] compared the precision and computational cost of three machine learning methods to approximate daylight autonomy and the likelihood of daylight reflection. The models were trained inductively using visual comfort indices derived from simulations in office spaces. Similarly, Lorenz et al. [107] proposed a neural network to predict levels of daylight autonomy based on indoor climate conditions. The results were highly competitive when compared to computationally intensive daylight simulations.

2.5.2 Building Construction

The construction process represents the second stage of its life cycle and follows the design process analyzed in the previous section. AI finds broad applications in the construction stage as well, with its benefits classified into four main problem areas: construction cost analysis, building information modeling, construction management, and defect detection.

Firstly, with regard to construction cost analysis, AI has numerous applications, primarily in estimating total costs using predictive methods. Accurate construction cost prediction is essential for any new project as it helps project managers identify potential problems and control project expenses. However, a precise calculation is challenging in the initial stages when the project information is limited. During the past decade, numerous studies have focused on this issue. For example, Cheng et al. [34] proposed an evolutionary inference model for cost estimation based on the support vector machines algorithm and the "fast messy genetic algorithm". Similarly, in the research of Kim et al. [94], various machine learning algorithms for cost estimation were compared, revealing that the neural network-based algorithms provided the highest precision.

Building Information Modeling (BIM) represents the most contemporary approach to technical design and construction management. BIM is an advanced technology aimed at digitizing the construction industry through a standardized framework for storing and sharing building-related information. AI plays a pivotal role in this domain through fields such as image processing and pattern recognition Huang et al. [82]. One of the major challenges in this area is that BIM cannot accurately capture all details of an installation as it exists in reality from a digital CAD design. To address this issue, AI provides models for the automatic conversion of 3D point cloud data into BIM. Notable findings have been presented by Tang et al. [180], who compared automation techniques to generate BIM from raw data. According to this study, there are three core functionalities: building geometry modeling, object recognition, and object relationship modeling. Similar studies combining BIM with AI have been conducted by Braun and Borrmann [24] and Koo et al. [96].

The third category of problems focuses on construction management, a crucial process for the successful and timely completion of construction projects. The role of AI in this domain emphasizes computer vision techniques for processing, storing, interpreting, and analyzing images captured from construction sites, which are increasingly replacing traditional written reports. The information detected in such images is used for more effective classification, retrieval, and management of site images to automate the monitoring of the construction process. Although this category of problems is less common than others, several studies have employed computer vision techniques to identify structural objects such as beams, columns, and walls [26] or structural materials [25]. A more recent study segments image analysis into two stages: dividing images into smaller subframes and classifying the visual characteristics of each subframe to identify the presence of structural materials [229].

Finally, machine learning has numerous applications in defect detection, again utilizing computer vision algorithms. An example of defect detection is the study by Akinci et al. [6], which proposed a method for quality control through automated construction defect detection. The core idea of this approach involves applying a deep learning algorithm to analyze images from a laser scanner and comparing the identified objects to reference standards to pinpoint deviations and assess whether these are due to construction defects. Another study by Zhu and Brilakis [228] focuses on using computer vision techniques to automatically inspect defects such as air pockets and discoloration on concrete surfaces.

2.5.3 Building Operation

Having completed the stages of design and construction, the life cycle of the building continues with its operation. Buildings must operate efficiently and undergo maintenance when necessary to maintain high performance. Daily operations and maintenance (known as operation and maintenance or O&M) affect the building's energy efficiency, the comfort levels of its users, and its operational costs. Modern buildings, equipped with advanced infrastructure and technologies, intelligent metering systems, and sophisticated communication protocols, pave the way for the development of intelligent systems [154, 175, 223].

During the operational phase, technological advances promise significant improvements in central building operation and management, building maintenance through fault detection, and optimizing user comfort through smart control of specific systems. In addition, smart meters (Internet of Things - IoT) have become a critical tool in modern buildings, enabling the collection of historical data in real-time, either for the entire building or specific subsystems. As a result, a large volume of building data is gathered and stored, providing rich information for investigating and evaluating the building's actual energy efficiency [214]. Recent research on building operation and control focuses on applying machine learning techniques to support daily decision making and optimize building system controls. In the following sections, we analyze these methods, categorizing AI applications into the following problem areas: improving building energy efficiency, ensuring thermal comfort for users, diagnosing anomalies for building maintenance, and intelligent control of building devices.

Energy Efficiency: In terms of building energy efficiency, significant progress has been made using AI to design tools that improve energy efficiency globally [159]. AI studies in the construction sector focus on identifying the precise energy consumption of buildings and optimizing energy usage rates. For example, the research by Yıldız and Arsan [213] examines the most critical factors affecting energy heating and cooling loads in buildings located in hot and humid climates, using experimental analysis of the design of an existing apartment. Similarly, Khayatian et al. [93] uses artificial neural networks to predict the thermal energy demand of buildings based on historical data on energy consumption. Another application by Fan et al. [57]

proposed a pattern recognition method to identify useful patterns from operational data. This method was applied to uncover correlations between the opening of cold- and hot-water valves in an air handling unit, the opening of the hot-water valve, and the supply-return temperature difference of hot water, aiming to improve the building's energy efficiency.

Thermal Comfort for Users: In the area of thermal comfort for building users (workers or residents), several AI algorithms have been used to develop predictive models, with neural networks and tree-structured models being prominent [131]. These methodologies aim to directly estimate the levels of thermal comfort or improve existing thermal comfort indices, such as the Predicted Mean Vote (PMV) [61]. For example, many studies focus on predicting indoor air temperature using variables such as internal temperature and humidity, weather data, and deep learning techniques [125, 126]. Beyond temperature and humidity, advanced studies use machine learning to estimate PMV as a function of four physical factors (air temperature, relative humidity, air velocity, mean radiant temperature) and two personal factors (clothing and activity) [28, 110, 194].

Building Maintenance: AI has also been widely applied in building maintenance during the operational phase. Various methods have been developed to create profiles for electrical devices or detect anomalies. Energy disaggregation calculates the electricity consumption of individual devices using a single meter that measures the total household energy demand. Recent advances in deep neural networks have significantly improved the classification performance of device profiles. For example, the research by Kelly and Knottenbelt [92] used LSTM networks and autoencoders. Predictive models have also been developed to minimize manual checks and reduce overall maintenance costs. Machine learning has been applied for fault detection and diagnosis in cooling systems [211], adaptive fault detection in air handling units [56], and defect identification in HVAC units at various levels [207].

Intelligent Building Control: Significant progress has also been made in the intelligent control of buildings using ambient intelligence combined with reinforcement learning algorithms. Traditionally, building control relies on feedback rules and preset schedules to select optimal set-points (e.g., air and water supply temperatures, thermostat settings), followed by classic control techniques to maintain these set-points. However, with advances in machine learning, control processes can now be largely automated. Automation applies to various building subsystems, such as HVAC, window operation, lighting systems, hot water control, and appliance scheduling [112]. For example, Zakula et al. [215] developed a predictive control model for dynamic HVAC optimization in two buildings with encouraging results. Similarly, Ruelens et al. [152] proposed a learning algorithm for thermostat control using a setback strategy, which lowers the temperature when residents are away. Another application, LightLearn, developed by Park et al. [139], focuses on optimal lighting control. This library learns personalized occupant behavior and indoor conditions to adjust lighting control parameters accordingly. Furthermore, a classification algorithm of Marasco and Kontokosta [114] demonstrated excellent results in predicting the suitability of energy-saving measures for specific characteristics of the building.

2.5.4 Building Renovation

The renovation of buildings is a critical factor for reducing energy consumption and greenhouse gas emissions in the building sector. However, the building renovation process is complex, as it involves a series of actions that must be carried out before, during, and after the renovation process. The most important of these actions include monitoring the current efficiency of the building prior to renovation, identifying the most suitable energy conservation measures (ECM) and evaluating the energy savings after the renovation actions are completed. Generally, machine learning methods have been used in these categories to facilitate the entire renovation process. The evaluation of a renovation action involves two stages: (a) the a priori evaluation, which takes place before the renovation, and (b) the ex ante analysis, which requires the renovation action to have been completed. In the first category, the primary focus is the techno-economic evaluation of the investment, taking into account the relationship between the cost of the actions and the estimated energy savings. In the second category, research efforts focus on evaluating the energy savings achieved by the action, comparing pre- and post-action consumption while normalizing for weather conditions.

The typical process for retrofitting buildings requires detailed energy audits. However, these audits are often time-consuming and quite expensive. As the volume of data from such energy audits in buildings grows and the information collected becomes more enriched, AI offers valuable methods to analyze and explore the underlying patterns in the data to support renovation planning in the building sector. An example of such applications is the study by Cecconi et al. [29], which evaluated the energy savings potential of public school buildings. The research relied on Italy's open building energy certification database. Schools were grouped according to their age and characteristics and then suitable strategies and renovation plans were designed for each group. In the study, Gao and Malkawi [68] follows a similar approach, which classified the buildings according to their structural characteristics to propose appropriate renovation actions in a second stage. A slightly different approach is taken by Marasco and Kontokosta [114], who trained a classification model to identify renovation opportunities. Specifically, the model was trained on energy audit data from more than 1,100 buildings in New York City and was used to predict renovation opportunities in various categories, such as cooling systems, hot water distribution systems, and lighting systems.

The process of Measurement & Verification (M&V) is one of the most important processes used to evaluate the effectiveness of energy conservation measures. Specifically, the term M&V is used to describe automated approaches and methodologies that rely on large datasets and statistical computational models to estimate energy savings. It is evident that with advances in machine learning, simple statistical models are gradually being replaced by more complex predictive machine learning models that offer increased accuracy [71]. An analytical methodology for this problem was developed in the research by Gallagher et al. [66], who used various AI models, such as artificial neural networks, support vector machines, the k-nearest neighbors algorithm, and multiple regression, to model energy consumption during the pre-renovation period. Another significant application was presented by Gallagher et al. [65], where it was demonstrated that machine learning-based approaches are particularly beneficial in cases where a lack of data limits the training period of the model.

2.6 Conclusions

The analysis presented in this chapter highlighted the potential of AI to model complex systems and offer solutions to a wide range of energy-related problems, including renewable energy management, distributed energy resources, and grid issues, as well as building energy efficiency challenges. However, the adoption of new technologies also presents several challenges, such as the lack of reliable datasets, which often result in inaccurate model predictions. Consequently, the quality and quantity of data are critical features that must be met in the data sets used to ensure the generalization of various machine learning and deep learning models.

Another common concern related to data is the lack of standardization, which would allow the generalization of applications. This critical aspect has been identified as a significant barrier to the implementation of AI in the energy sector. Secure, organized, and standardized data production in the energy sector could offer benefits including the exchange of information between companies, government agencies, and academic institutions; access to high-quality data for research; and the development of new specialized AI tools. Achieving this overarching goal will require the development of protective policies that ensure the secure and fair management of information.

References

1. A.N. Abdalla, M.S. Nazir, H. Tao, S. Cao, R. Ji, M. Jiang, L. Yao, Integration of energy storage system and renewable energy sources based on artificial intelligence: an overview. J. Energy Storage **40**, 102811 (2021)
2. M. Abdel-Basset, H. Hawash, R.K. Chakrabortty, M. Ryan, Pv-net: an innovative deep learning approach for efficient forecasting of short-term photovoltaic energy production. J. Clean. Prod. **303**, 127037 (2021)
3. M. Afzalan, F. Jazizadeh, Residential loads flexibility potential for demand response using energy consumption patterns and user segments. Appl. Energy **254**, 113693 (2019)

4. A. Agga, A. Abbou, M. Labbadi, Y. El Houm, I.H.O. Ali, Cnn-lstm: an efficient hybrid deep learning architecture for predicting short-term photovoltaic power production. Electric Power Syst. Res. **208**, 107908 (2022)
5. R. Ahmed, V. Sreeram, Y. Mishra, M. Arif, A review and evaluation of the state-of-the-art in pv solar power forecasting: techniques and optimization. Renew. Sustain. Energy Rev. **124**, 109792 (2020)
6. B. Akinci, F. Boukamp, C. Gordon, D. Huber, C. Lyons, K. Park, A formalism for utilization of sensor systems and integrated project models for active construction quality control. Autom. Constr. **15**(2), 124–138 (2006)
7. S. Al-Dahidi, O. Ayadi, M. Alrbai, J. Adeeb, Ensemble approach of optimized artificial neural networks for solar photovoltaic power prediction. IEEE Access **7**, 81741–81758 (2019)
8. A.M. Al-Salim, A.Q. Lawey, T.E. El-Gorashi, J.M. Elmirghani, Energy efficient big data networks: impact of volume and variety. IEEE Trans. Netw. Serv. Manage. **15**(1), 458–474 (2017)
9. M.K. Alam, F. Khan, J. Johnson, J. Flicker, A comprehensive review of catastrophic faults in PV arrays: types, detection, and mitigation techniques. IEEE J. Photovolt. **5**(3), 982–997 (2015)
10. M. Alhussein, K. Aurangzeb, S.I. Haider, Hybrid CNN-LSTM model for short-term individual household load forecasting. IEEE Access **8**, 180544–180557 (2020)
11. M.U. Ali, H.F. Khan, M. Masud, K.D. Kallu, A. Zafar, A machine learning framework to identify the hotspot in photovoltaic module using infrared thermography. Sol. Energy **208**, 643–651 (2020)
12. M. AlKandari, I. Ahmad, Solar power generation forecasting using ensemble approach based on deep learning and statistical methods. Appl. Comput. Inf. (2020)
13. F. Almonacid, C. Rus, L. Hontoria, F. Munoz, Characterisation of PV cis module by artificial neural networks. a comparative study with other methods. Renew. Energy **35**(5), 973–980 (2010)
14. K. Amarasinghe, D.L. Marino, M. Manic, Deep neural networks for energy load forecasting, in *2017 IEEE 26th International Symposium on Industrial Electronics (ISIE)*. (IEEE, 2017), pp. 1483–1488
15. M. Amin, Challenges in reliability, security, efficiency, and resilience of energy infrastructure: Toward smart self-healing electric power grid, in *2008 IEEE Power And Energy Society General Meeting-conversion and Delivery of Electrical Energy in the 21st Century*. (IEEE, 2008), pp. 1–5
16. S.M. Amin, B.F. Wollenberg, Toward a smart grid: power delivery for the 21st century. IEEE Power Energ. Mag. **3**(5), 34–41 (2005)
17. N. Amral, C. Ozveren, D. King, Short term load forecasting using multiple linear regression, in *2007 42nd International Universities Power Engineering Conference*. (IEEE, 2007), pp. 1192–1198
18. Arsenopoulos, A., Sarmas, E., Stavrakaki, A., Giannouli, I., and Psarras, J. (2021). A data-driven decision support tool at the service of energy suppliers and utilities for tackling energy poverty: A case study in greece. In *2021 12th International Conference on Information, Intelligence, Systems & Applications (IISA)*, pages 1–6. IEEE
19. B. Baruque, S. Porras, E. Jove, J.L. Calvo-Rolle, Geothermal heat exchanger energy prediction based on time series and monitoring sensors optimization. Energy **171**, 49–60 (2019)
20. M. Bazilian, I. Onyeji, M. Liebreich, I. MacGill, J. Chase, J. Shah, D. Gielen, D. Arent, D. Landfear, S. Zhengrong, Re-considering the economics of photovoltaic power. Renewable Energy **53**, 329–338 (2013)
21. M.K. Behera, I. Majumder, N. Nayak, Solar photovoltaic power forecasting using optimized modified extreme learning machine technique. Engineering Science and Technology, an International Journal **21**(3), 428–438 (2018)
22. T. Berghout, M. Benbouzid, T. Bentrcia, X. Ma, S. Djurović, L.-H. Mouss, Machine learning-based condition monitoring for pv systems: State of the art and future prospects. Energies **14**(19), 6316 (2021)

23. P. Bradley, M. Leach, J. Torriti, A review of the costs and benefits of demand response for electricity in the uk. Energy Policy **52**, 312–327 (2013)
24. A. Braun, A. Borrmann, Combining inverse photogrammetry and bim for automated labeling of construction site images for machine learning. Autom. Constr. **106**, 102879 (2019)
25. I. Brilakis, L. Soibelman, Y. Shinagawa, Material-based construction site image retrieval. J. Comput. Civ. Eng. **19**(4), 341–355 (2005)
26. I.K. Brilakis, L. Soibelman, Shape-based retrieval of construction site photographs. J. Comput. Civ. Eng. **22**(1), 14–20 (2008)
27. N.C. Brown, C.T. Mueller, Design variable analysis and generation for performance-based parametric modeling in architecture. Int. J. Archit. Comput. **17**(1), 36–52 (2019)
28. M. Castilla, J. Álvarez, M. Ortega, M. Arahal, Neural network and polynomial approximated thermal comfort models for hvac systems. Build. Environ. **59**, 107–115 (2013)
29. F.R. Cecconi, N. Moretti, L.C. Tagliabue, Application of artificial neutral network and geographic information system to evaluate retrofit potential in public school buildings. Renew. Sustain. Energy Rev. **110**, 266–277 (2019)
30. S. Chaillou, *Artificial Intelligence and Architecture: From Research to Practice.* (Birkhäuser, 2022)
31. S. Chapaloglou, A. Nesiadis, P. Iliadis, K. Atsonios, N. Nikolopoulos, P. Grammelis, C. Yiakopoulos, I. Antoniadis, E. Kakaras, Smart energy management algorithm for load smoothing and peak shaving based on load forecasting of an island's power system. Appl. Energy **238**, 627–642 (2019)
32. I. Chatzikonstantinou, S. Sariyildiz, Approximation of simulation-derived visual comfort indicators in office spaces: a comparative study in machine learning. Archit. Sci. Rev. **59**(4), 307–322 (2016)
33. F. Chen, Z. Fu, Z. Yang, Wind power generation fault diagnosis based on deep learning model in internet of things (iot) with clusters. Clust. Comput. **22**(6), 14013–14025 (2019)
34. M.-Y. Cheng, H.-S. Peng, Y.-W. Wu, T.-L. Chen, Estimate at completion for construction projects using evolutionary support vector machine inference model. Autom. Constr. **19**(5), 619–629 (2010)
35. L.W. Chong, Y.W. Wong, R.K. Rajkumar, D. Isa, An adaptive learning control strategy for standalone pv system with battery-supercapacitor hybrid energy storage system. J. Power Sources **394**, 35–49 (2018)
36. F.E.R. Commission et al., Assessment of demand response and advanced metering (2008)
37. M. Cordeiro-Costas, D. Villanueva, P. Eguía-Oller, E. Granada-Álvarez, Machine learning and deep learning models applied to photovoltaic production forecasting. Appl. Sci. **12**(17), 8769 (2022)
38. E. Council, Fit for 55: The eu's plan for a green transition (2020)
39. A. Dairi, F. Harrou, Y. Sun, S. Khadraoui, Short-term forecasting of photovoltaic solar power production using variational auto-encoder driven deep learning approach. Appl. Sci. **10**(23), 8400 (2020)
40. I.G. Damousis, P. Dokopoulos, A fuzzy expert system for the forecasting of wind speed and power generation in wind farms, in *PICA 2001. Innovative Computing for Power-Electric Energy Meets the Market. 22nd IEEE Power Engineering Society. International Conference on Power Industry Computer Applications (Cat. No. 01CH37195)* (IEEE, 2001), pp. 63–69
41. U.K. Das, K.S. Tey, M. Seyedmahmoudian, M.Y. Idna Idris, S. Mekhilef, B. Horan, A. Stojcevski, Svr-based model to forecast pv power generation under different weather conditions. Energies **10**(7), 876 (2017)
42. U.K. Das, K.S. Tey, M. Seyedmahmoudian, S. Mekhilef, M.Y.I. Idris, W. Van Deventer, B. Horan, A. Stojcevski, Forecasting of photovoltaic power generation and model optimization: A review. Renew. Sustain. Energy Rev. **81**, 912–928 (2018)
43. M. De Benedetti, F. Leonardi, F. Messina, C. Santoro, A. Vasilakos, Anomaly detection and predictive maintenance for photovoltaic systems. Neurocomputing **310**, 59–68 (2018)
44. M.G. De Giorgi, P.M. Congedo, M. Malvoni, Photovoltaic power forecasting using statistical methods: impact of weather data. IET Sci. Meas. Technol. **8**(3), 90–97 (2014)

45. R.A. de Marcos, A. Bello, J. Reneses, Electricity price forecasting in the short term hybridising fundamental and econometric modelling. Electric Power Syst. Res. **167**, 240–251 (2019)
46. K. De Rycke, C. Gengnagel, O. Baverel, J. Burry, C. Mueller, N.M., Nguyen, P. Rahm, M.R. Thomsen, Humanizing digital reality. *Stone Morphologies: Erosion-Based Digital Fabrication Through Event-Driven Control, Steve De Micoli, Katja Rinderspacher, Achim Menges* (2018), pp. 113–124
47. H. Demolli, A.S. Dokuz, A. Ecemis, M. Gokcek, Wind power forecasting based on daily wind speed data using machine learning algorithms. Energy Convers. Manage. **198**, 111823 (2019)
48. M. Diagne, M. David, P. Lauret, J. Boland, N. Schmutz, Review of solar irradiance forecasting methods and a proposition for small-scale insular grids. Renew. Sustain. Energy Rev. **27**, 65–76 (2013)
49. I.C. Dimitriou, A. Arsenopoulos, G.P. Trachanas, E. Sarmas, V. Marinakis, Assessing the adoption level of solar pv installations, on district scale of urban environment. Renew. Energy **237**, 121676 (2024)
50. I.C. Dimitriou, E. Sarmas, G.P. Trachanas, V. Marinakis, H. Doukas, Multi-criteria gis-based offshore wind farm site selection: Case study in Greece. Renew. Sustain. Energy Rev. **207**, 114962 (2025)
51. I.C. Dimitriou, G.P. Trachanas, A. Arsenopoulos, E. Sarmas, Future heat risk index: The case of Attica regions, Greece, in *2024 15th International Conference on Information, Intelligence, Systems & Applications (IISA)* (IEEE, 2024), pp. 1–7
52. J. Duan, Z. Yi, D. Shi, C. Lin, X. Lu, Z. Wang, Reinforcement-learning-based optimal control of hybrid energy storage systems in hybrid ac-dc microgrids. IEEE Trans. Ind. Inf. **15**(9), 5355–5364 (2019)
53. S. Eftekharnejad, V. Vittal, G.T. Heydt, B. Keel, J. Loehr, Impact of increased penetration of photovoltaic generation on power systems. IEEE Trans. Power Syst. **28**(2), 893–901 (2012)
54. A. Eskandari, J. Milimonfared, M. Aghaei, A.H. Reinders, Autonomous monitoring of line-to-line faults in photovoltaic systems by feature selection and parameter optimization of support vector machine using genetic algorithms. Appl. Sci. **10**(16), 5527 (2020)
55. Z.M. Fadlullah, M.M. Fouda, N. Kato, A. Takeuchi, N. Iwasaki, Y. Nozaki, Toward intelligent machine-to-machine communications in smart grid. IEEE Commun. Mag. **49**(4), 60–65 (2011)
56. B. Fan, Z. Du, X. Jin, X. Yang, Y. Guo, A hybrid fdd strategy for local system of ahu based on artificial neural network and wavelet analysis. Build. Environ. **45**(12), 2698–2708 (2010)
57. C. Fan, Y. Sun, K. Shan, F. Xiao, J. Wang, Discovering gradual patterns in building operations for improving building energy efficiency. Appl. Energy **224**, 116–123 (2018)
58. S. Fan, R.J. Hyndman, Short-term load forecasting based on a semi-parametric additive model. IEEE Trans. Power Syst. **27**(1), 134–141 (2011)
59. B. Favre, B. Peuportier, Application of dynamic programming to study load shifting in buildings. Energy Build. **82**, 57–64 (2014)
60. R. Fazai, K. Abodayeh, M. Mansouri, M. Trabelsi, H. Nounou, M. Nounou, G.E. Georghiou, Machine learning-based statistical testing hypothesis for fault detection in photovoltaic systems. Sol. Energy **190**, 405–413 (2019)
61. P.M. Ferreira, S.M. Silva, A.E. Ruano, A.T. Negrier, E.Z. Conceicao, Neural network PMV estimation for model-based predictive control of HVAC systems, in *The 2012 International Joint Conference on Neural Networks (IJCNN)* (IEEE, 2012), pp. 1–8
62. M. Fischetti, D. Pisinger, Mathematical optimization and algorithms for offshore wind farm design: An overview. Bus. Inf. Syst. Eng. **61**(4), 469–485 (2019)
63. A.M. Foley, P.G. Leahy, A. Marvuglia, E.J. McKeogh, Current methods and advances in forecasting of wind power generation. Renew. Energy **37**(1), 1–8 (2012)
64. L. Fuhrimann, V. Moosavi, P.O. Ohlbrock, P. D'acunto, Data-driven design: Exploring new structural forms using machine learning and graphic statics, in *Proceedings of IASS Annual Symposia*, number 2 in 1. International Association for Shell and Spatial Structures (IASS) (2018), pp. 1–8

65. C.V. Gallagher, K. Bruton, K. Leahy, D.T. O'Sullivan, The suitability of machine learning to minimise uncertainty in the measurement and verification of energy savings. Energy Build. **158**, 647–655 (2018)
66. C.V. Gallagher, K. Leahy, P. O'Donovan, K. Bruton, D.T. O'Sullivan, Development and application of a machine learning supported methodology for measurement and verification (m&v) 2.0. Energy Build. **167**, 8–22 (2018)
67. P. Gangwani, J. Soni, H. Upadhyay, S. Joshi, A deep learning approach for modeling of geothermal energy prediction. Int. J. Comput. Sci. Inf. Sec. (IJCSIS) **18**(1) (2020)
68. X. Gao, A. Malkawi, A new methodology for building energy performance benchmarking: An approach based on intelligent clustering algorithm. Energy Build. **84**, 607–616 (2014)
69. C.W. Gellings, W.M. Smith, Integrating demand-side management into utility planning. Proc. IEEE **77**(6), 908–918 (1989)
70. P. Geyer, S. Singaravel, Component-based machine learning for performance prediction in building design. Appl. Energy **228**, 1439–1453 (2018)
71. J. Granderson, S. Touzani, S. Fernandes, C. Taylor, Application of automated measurement and verification to utility energy efficiency program data. Energy Build. **142**, 191–199 (2017)
72. X. Han, T. Ji, Z. Zhao, H. Zhang, Economic evaluation of batteries planning in energy storage power stations for load shifting. Renew. Energy **78**, 643–647 (2015)
73. J. Hare, X. Shi, S. Gupta, A. Bazzi, Fault diagnostics in smart micro-grids: A survey. Renew. Sustain. Energy Rev. **60**, 1114–1124 (2016)
74. S. He, H. Gao, J. Liu, X. Zhang, Z. Chen, Distribution system planning considering peak shaving of energy station. Appl. Energy **312**, 118692 (2022)
75. S. Heck, M. Rogers, Are you ready for the resource revolution. McKinsey Quart. **2**, 32–45 (2014)
76. G. Helbing, M. Ritter, Deep learning for fault detection in wind turbines. Renew. Sustain. Energy Rev. **98**, 189–198 (2018)
77. L. Hernández, C. Baladrón, J.M. Aguiar, L. Calavia, B. Carro, A. Sánchez-Esguevillas, D.J. Cook, D. Chinarro, J. Gómez, A study of the relationship between weather variables and electric power demand inside a smart grid/smart world framework. Sensors **12**(9), 11571–11591 (2012)
78. T. Hong, J. Wilson, J. Xie, Long term probabilistic load forecasting and normalization with hourly information. IEEE Trans. Smart Grid **5**(1), 456–462 (2013)
79. Y.-Y. Hong, C.L.P.P. Rioflorido, A hybrid deep learning-based neural network for 24-h ahead wind power forecasting. Appl. Energy **250**, 530–539 (2019)
80. J.-Y. Hsu, Y.-F. Wang, K.-C. Lin, M.-Y. Chen, J.H.-Y. Hsu, Wind turbine fault diagnosis and predictive maintenance through statistical process control and machine learning. IEEE Access **8**, 23427–23439 (2020)
81. J. Hu, A.V. Vasilakos, Energy big data analytics and security: challenges and opportunities. IEEE Trans. Smart Grid **7**(5), 2423–2436 (2016)
82. M. Huang, J. Ninić, Q. Zhang, Bim, machine learning and computer vision techniques in underground construction: Current status and future perspectives. Tunn. Undergr. Space Technol. **108**, 103677 (2021)
83. W. Huang, H. Zheng, Architectural drawings recognition and generation through machine learning (2018)
84. M. Huber, D. Dimkova, T. Hamacher, Integration of wind and solar power in Europe: Assessment of flexibility requirements. Energy **69**, 236–246 (2014)
85. M. Hussain, H. Al-Aqrabi, R. Hill, Statistical analysis and development of an ensemble-based machine learning model for photovoltaic fault detection. Energies **15**(15), 5492 (2022)
86. IEA, Renewables 2021: Analysis and forecasts to 2026 (2021)
87. S.K. Jha, J. Bilalovic, A. Jha, N. Patel, H. Zhang, Renewable energy: Present research and future scope of artificial intelligence. Renew. Sustain. Energy Rev. **77**, 297–317 (2017)
88. L. Jia, L. Tong, Dynamic pricing and distributed energy management for demand response. IEEE Trans. Smart Grid **7**(2), 1128–1136 (2016)

89. H. Jiang, K. Wang, Y. Wang, M. Gao, Y. Zhang, Energy big data: A survey. IEEE Access **4**, 3844–3861 (2016)
90. S. Kalyan, Q. Sun, Interrogating the installation gap and potential of solar photovoltaic systems using gis and deep learning. Energies **15**(10), 3740 (2022)
91. F. Kasten, G. Czeplak, Solar and terrestrial radiation dependent on the amount and type of cloud. Sol. Energy **24**(2), 177–189 (1980)
92. J. Kelly, W. Knottenbelt, Neural nilm: Deep neural networks applied to energy disaggregation, in *Proceedings of the 2nd ACM International Conference On Embedded Systems For Energy-efficient Built Environments* (2015), pp. 55–64
93. F. Khayatian, L. Sarto et al., Application of neural networks for evaluating energy performance certificates of residential buildings. Energy Build. **125**, 45–54 (2016)
94. G.-H. Kim, J.-M. Shin, S. Kim, Y. Shin, Comparison of school building construction costs estimation methods using regression analysis, neural network, and support vector machine. J. Build. Constr. Plan. Res. (2013)
95. H. Kim, A.L. Stumpf, R.L. Schneider, *Developing energy efficient building design in machine learning*. Tribun EU (2010)
96. B. Koo, R. Jung, Y. Yu, Automatic classification of wall and door bim element subtypes using 3d geometric deep neural networks. Adv. Eng. Inform. **47**, 101200 (2021)
97. L. Landberg, A mathematical look at a physical power prediction model. Wind Energy: Int. J. Progr. Appl. Wind Power Convers. Technol. **1**(1), 23–28 (1998)
98. L. Landberg, Short-term prediction of the power production from wind farms. J. Wind Eng. Ind. Aerodyn. **80**(1–2), 207–220 (1999)
99. L. Landberg, L. Myllerup, O. Rathmann, E.L. Petersen, B.H. Jørgensen, J. Badger, N.G. Mortensen, Wind resource estimation-an overview. Wind Energy: Int. J. Progr. Appl. Wind Power Convers. Technol. **6**(3), 261–271 (2003)
100. K.-H. Lee, S.-O. Jun, K.-H. Pak, D.-H. Lee, K.-W. Lee, J.-P. Park, Numerical optimization of site selection for offshore wind turbine installation using genetic algorithm. Curr. Appl. Phys. **10**(2), S302–S306 (2010)
101. M. Lei, L. Shiyan, J. Chuanwen, L. Hongling, Z. Yan, A review on the forecasting of wind speed and generated power. Renew. Sustain. Energy Rev. **13**(4), 915–920 (2009)
102. P. Li, K. Zhou, X. Lu, S. Yang, A hybrid deep learning model for short-term pv power forecasting. Appl. Energy **259**, 114216 (2020)
103. W.-H. Lin, P. Wang, K.-M. Chao, H.-C. Lin, Z.-Y. Yang, Y.-H. Lai, Wind power forecasting with deep learning networks: Time-series forecasting. Appl. Sci. **11**(21), 10335 (2021)
104. B. Liu, S. Zhao, X. Yu, L. Zhang, Q. Wang, A novel deep learning approach for wind power forecasting based on wd-lstm model. Energies **13**(18), 4964 (2020)
105. C. Liu, J. Wu, P. Kohli, Y. Furukawa, Raster-to-vector: Revisiting floorplan transformation, in *Proceedings of the IEEE International Conference on Computer Vision* (2017), pp. 2195–2203
106. L. Liu, E.A. Silva, C. Wu, H. Wang, A machine learning-based method for the large-scale evaluation of the qualities of the urban environment. Comput. Environ. Urban Syst. **65**, 113–125 (2017)
107. C.-L. Lorenz, M. Packianather, A. Spaeth, C. Bleil De Souza, Artificial neural network-based modelling for daylight evaluations. Society for Modeling & Simulation International (2018)
108. R. Lu, S.H. Hong, M. Yu, Demand response for home energy management using reinforcement learning and artificial neural network. IEEE Trans. Smart Grid **10**(6), 6629–6639 (2019)
109. R. Lu, S.H. Hong, X. Zhang, A dynamic pricing demand response algorithm for smart grid: reinforcement learning approach. Appl. Energy **220**, 220–230 (2018)
110. S. Lu, W. Wang, C. Lin, E.C. Hameen, Data-driven simulation of a thermal comfort-based temperature set-point control with ashrae rp884. Build. Environ. **156**, 137–146 (2019)
111. P.D. Lund, J. Lindgren, J. Mikkola, J. Salpakari, Review of energy system flexibility measures to enable high levels of variable renewable electricity. Renew. Sustain. Energy Rev. **45**, 785–807 (2015)
112. N. Luo, T. Hong, H. Li, R. Jia, W. Weng, Data analytics and optimization of an ice-based energy storage system for commercial buildings. Appl. Energy **204**, 459–475 (2017)

113. X. Luo, D. Zhang, X. Zhu, Deep learning based forecasting of photovoltaic power generation by incorporating domain knowledge. Energy **225**, 120240 (2021)
114. D.E. Marasco, C.E. Kontokosta, Applications of machine learning methods to identifying and predicting building retrofit opportunities. Energy Build. **128**, 431–441 (2016)
115. M. Martínez-Chico, F. Batlles, J. Bosch, Cloud classification in a mediterranean location using radiation data and sky images. Energy **36**(7), 4055–4062 (2011)
116. M.J. Mayer, G. Gróf, Extensive comparison of physical models for photovoltaic power forecasting. Appl. Energy **283**, 116239 (2021)
117. V. Mayer-Schönberger, K. Cukier, *Big data: A Revolution That Will Transform How We Live, Work, And Think.* (Houghton Mifflin Harcourt, 2013)
118. A. Mellit, A.M. Pavan, V. Lughi, Deep learning neural networks for short-term photovoltaic power forecasting. Renew. Energy **172**, 276–288 (2021)
119. K. Methaprayoon, W. Lee, P. Didsayabutra, J. Liao, R. Ross, Neural network-based short term load forecasting for unit commitment scheduling, in *IEEE Technical Conference on Industrial and Commercial Power Systems, 2003.* (IEEE, 2003), pp. 138–143
120. V. Michalakopoulos, E. Sarantinopoulos, E. Sarmas, V. Marinakis, Empowering federated learning techniques for privacy-preserving pv forecasting. Energy Rep. **12**, 2244–2256 (2024)
121. V. Michalakopoulos, E. Sarmas, I. Papias, P. Skaloumpakas, V. Marinakis, H. Doukas, A machine learning-based framework for clustering residential electricity load profiles to enhance demand response programs. Appl. Energy **361**, 122943 (2024)
122. M. Mishra, P.B. Dash, J. Nayak, B. Naik, S.K. Swain, Deep learning and wavelet transform integrated approach for short-term solar pv power prediction. Measurement **166**, 108250 (2020)
123. K. Mohammadi, N. Goudarzi, Study of inter-correlations of solar radiation, wind speed and precipitation under the influence of el niño southern oscillation (enso) in california. Renew. Energy **120**, 190–200 (2018)
124. J.A. Momoh, Smart grid design for efficient and flexible power networks operation and control, in *2009 IEEE/PES Power Systems Conference and Exposition* (IEEE, 2009), pp. 1–8
125. J.W. Moon, S.K. Jung, Algorithm for optimal application of the setback moment in the heating season using an artificial neural network model. Energy Build. **127**, 859–869 (2016)
126. J.W. Moon, S.-H. Yoon, S. Kim, Development of an artificial neural network model based thermal control logic for double skin envelopes in winter. Build. Environ. **61**, 149–159 (2013)
127. C.F. Moyano, J.A.P. Lopes, An optimization approach for wind turbine commitment and dispatch in a wind park. Electric Power Syst. Res. **79**(1), 71–79 (2009)
128. Z. Mustafa, A.S. Awad, M. Azzouz, A. Azab, Fault identification for photovoltaic systems using a multi-output deep learning approach. Expert Syst. Appl. **211**, 118551 (2023)
129. D. Nagy, D. Lau, J. Locke, J. Stoddart, L. Villaggi, R. Wang, D. Zhao, D. Benjamin, Project discover: An application of generative design for architectural space planning, in *Proceedings of the Symposium on Simulation for Architecture and Urban Design* (2017), pp. 1–8
130. A. Nespoli, E. Ogliari, S. Leva, A. Massi Pavan, A. Mellit, V. Lughi, A. Dolara, Day-ahead photovoltaic forecasting: A comparison of the most effective techniques. Energies **12**(9), 1621 (2019)
131. J. Ngarambe, G.Y. Yun, M. Santamouris, The use of artificial intelligence (AI) methods in the prediction of thermal comfort in buildings: Energy implications of ai-based thermal comfort controls. Energy Build. **211**, 109807 (2020)
132. H.A. Nielsen, T.S. Nielsen, H. Madsen, M.J.S.I. Pindado, I. Marti, Optimal combination of wind power forecasts. Wind Energy: Int. J. Progr. Appl. Wind Power Convers. Technol. **10**(5), 471–482 (2007)
133. K. Normandeau, Beyond volume, variety and velocity is the issue of big data veracity. *Inside big data* (2013)
134. B.E. Nyong-Bassey, D. Giaouris, C. Patsios, S. Papadopoulou, A.I. Papadopoulos, S. Walker, S. Voutetakis, P. Seferlis, S. Gadoue, Reinforcement learning based adaptive power pinch analysis for energy management of stand-alone hybrid energy storage systems considering uncertainty. Energy **193**, 116622 (2020)

135. E. Oh, H. Wang, Reinforcement-learning-based energy storage system operation strategies to manage wind power forecast uncertainty. IEEE Access **8**, 20965–20976 (2020)
136. A.H. Omran, D.M. Said, S.M. Hussin, N. Ahmad, H. Samet, A novel intelligent detection schema of series arc fault in photovoltaic (pv) system based convolutional neural network. Period. Eng. Nat. Sci. (PEN) **8**(3), 1641–1653 (2020)
137. D. O'Neill, M. Levorato, A. Goldsmith, U. Mitra, Residential demand response using reinforcement learning, in *2010 First IEEE international conference on smart grid communications* (IEEE, 2010), pp. 409–414
138. S. Oudjana, A. Hellal, I.H. Mahammed, Power forecasting of photovoltaic generation. Int. J. Electr. Comput. Eng. **7**(6), 627–631 (2013)
139. J.Y. Park, T. Dougherty, H. Fritz, Z. Nagy, Lightlearn: An adaptive and occupant centered controller for lighting based on reinforcement learning. Build. Environ. **147**, 397–414 (2019)
140. E.L. Petersen, N.G. Mortensen, L. Landberg, J. Højstrup, H.P. Frank, *Wind power meteorology* (Risø National Laboratory, Roskilde, Denmark, 1997)
141. A.N. Petrov, J.M. Wessling, Utilization of machine-learning algorithms for wind turbine site suitability modeling in Iowa, USA. Wind Energy **18**(4), 713–727 (2015)
142. N. Phelan, D. Davis, C. Anderson, Evaluating architectural layouts with neural networks, in *Proceedings of the Symposium on Simulation for Architecture and Urban Design* (2017), pp. 1–7
143. D.S. Pillai, N. Rajasekar, A comprehensive review on protection challenges and fault diagnosis in pv systems. Renew. Sustain. Energy Rev. **91**, 18–40 (2018)
144. S. Rahman, O. Hazim, A generalized knowledge-based short-term load-forecasting technique. IEEE Trans. Power Syst. **8**(2), 508–514 (1993)
145. M.Q. Raza, A. Khosravi, A review on artificial intelligence based load demand forecasting techniques for smart grid and buildings. Renew. Sustain. Energy Rev. **50**, 1352–1372 (2015)
146. M.Q. Raza, M. Nadarajah, C. Ekanayake, On recent advances in pv output power forecast. Sol. Energy **136**, 125–144 (2016)
147. M.Q. Raza, M. Nadarajah, D.Q. Hung, Z. Baharudin, An intelligent hybrid short-term load forecasting model for smart power grids. Sustain. Urban Areas **31**, 264–275 (2017)
148. S.S. Reddy, Optimal scheduling of thermal-wind-solar power system with storage. Renew. Energy **101**, 1357–1368 (2017)
149. S.S. Reddy, P. Bijwe, Real time economic dispatch considering renewable energy resources. Renew. Energy **83**, 1215–1226 (2015)
150. H. Ren, C. Xu, Z. Ma, Y. Sun, A novel 3d-geographic information system and deep learning integrated approach for high-accuracy building rooftop solar energy potential characterization of high-density cities. Appl. Energy **306**, 117985 (2022)
151. Y. Ren, P. Suganthan, N. Srikanth, Ensemble methods for wind and solar power forecasting-a state-of-the-art review. Renew. Sustain. Energy Rev. **50**, 82–91 (2015)
152. F. Ruelens, S. Iacovella, B.J. Claessens, R. Belmans, Learning agent for a heat-pump thermostat with a set-back strategy using model-free reinforcement learning. Energies **8**(8), 8300–8318 (2015)
153. S. Sagiroglu, D. Sinanc, Big data: A review, in *2013 International Conference on Collaboration Technologies and Systems (CTS)* (IEEE, 2013), pp. 42–47
154. P. Samaras, E. Stamatopoulos, A. Arsenopoulos, E. Sarmas, E. Marinakis, Readiness to adopt the smart readiness indicator scheme across europe: A multi-criteria decision analysis approach, in *2024 IEEE International Workshop on Metrology for Living Environment (MetroLivEnv)* (IEEE, 2024), pp. 268–273
155. M. Sapitang, M. Ridwan, W., Faizal Kushiar, K., Najah Ahmed, A., and El-Shafie, A., Machine learning application in reservoir water level forecasting for sustainable hydropower generation strategy. Sustainability **12**(15), 6121 (2020)
156. E. Sarmas, N. Dimitropoulos, V. Marinakis, Z. Mylona, H. Doukas, Transfer learning strategies for solar power forecasting under data scarcity. Sci. Rep. **12**(1), 1–13 (2022)
157. E. Sarmas, A. Forouli, V. Marinakis, H. Doukas, Baseline energy modeling for improved measurement and verification through the use of ensemble artificial intelligence models. Inf. Sci. **654**, 119879 (2024)

158. E. Sarmas, A. Fragkiadaki, V. Marinakis, Explainable ai-based ensemble clustering for load profiling and demand response. Energies **17**(22), 5559 (2024)
159. E. Sarmas, E. Spiliotis, N. Dimitropoulos, V. Marinakis, H. Doukas, Estimating the energy savings of energy efficiency actions with ensemble machine learning models. Appl. Sci. **13**(4), 2749 (2023)
160. E. Sarmas, E. Spiliotis, V. Marinakis, M.A. Bucarelli, F. Santori, H. Doukas, Revving up energy autonomy: A forecast-driven framework for reducing reverse power flow in microgrids. Sustain. Energy, Grids Netw. **38**, 101376 (2024)
161. E. Sarmas, E. Spiliotis, E. Stamatopoulos, V. Marinakis, H. Doukas, Short-term photovoltaic power forecasting using meta-learning and numerical weather prediction independent long short-term memory models. Renew. Energy **216**, 118997 (2023)
162. E. Sarmas, S. Strompolas, V. Marinakis, F. Santori, M.A. Bucarelli, H. Doukas, An incremental learning framework for photovoltaic production and load forecasting in energy microgrids. Electronics **11**(23), 3962 (2022)
163. M. Sengupta, A. Habte, S. Wilbert, C. Gueymard, J. Remund, Best practices handbook for the collection and use of solar resource data for solar energy applications. Technical report, National Renewable Energy Lab.(NREL), Golden, CO (United States) (2021)
164. U. Shahzad, Application of machine learning for optimal wind farm location. J. Electr. Eng. Electron. Control Comput. Sci. **8**(3), 9–20 (2021)
165. F. Shariatzadeh, P. Mandal, A.K. Srivastava, Demand response for sustainable energy systems: A review, application and implementation strategy. Renew. Sustain. Energy Rev. **45**, 343–350 (2015)
166. H. Shayeghi, E. Shahryari, M. Moradzadeh, P. Siano, A survey on microgrid energy management considering flexible energy sources. Energies **12**(11), 2156 (2019)
167. T. Shimakage, K. Nishioka, H. Yamane, M. Nagura, M. Kudo, Development of fault detection system in pv system, in *2011 IEEE 33rd International Telecommunications Energy Conference (INTELEC)*, (IEEE, 2011) pp. 1–5
168. E. Shirazi, S. Jadid, Cost reduction and peak shaving through domestic load shifting and ders. Energy **124**, 146–159 (2017)
169. S. Shivashankar, S. Mekhilef, H. Mokhlis, M. Karimi, Mitigating methods of power fluctuation of photovoltaic (pv) sources-a review. Renew. Sustain. Energy Rev. **59**, 1170–1184 (2016)
170. P. Siano, Demand response and smart grids-a survey. Renew. Sustain. Energy Rev. **30**, 461–478 (2014)
171. P. Skaloumpakas, E. Sarmas, Z. Mylona, A. Cavadenti, F. Santori, V. Marinakis, Predicting thermal comfort in buildings with machine learning and occupant feedback, in *2023 IEEE International Workshop on Metrology for Living Environment (MetroLivEnv)* (IEEE, 2023), pp. 34–39
172. P. Skaloumpakas, E. Sarmas, M. Rachmanidis, V. Marinakis, Reshaping the energy landscape of create through renewable energy valleys. Sci. Rep. **14**(1), 8038 (2024)
173. S. Sobri, S. Koohi-Kamali, N.A. Rahim, Solar photovoltaic generation forecasting methods: A review. Energy Convers. Manage. **156**, 459–497 (2018)
174. E. Stamatopoulos, A. Forouli, D. Stoian, P. Kouloukakis, E. Sarmas, V. Marinakis, An adaptive framework for assessing climate resilience in buildings. Build. Environ. **264**, 111869 (2024)
175. D. Stoian, E. Spiliotis, E. Stamatopoulos, E. Sarmas, P. Rekomaa, V. Marinakis, A neural network-based approach for hierarchical electricity consumption forecasting in buildings, in *2024 IEEE International Workshop on Metrology for Living Environment (MetroLivEnv)* (IEEE, 2024), pp. 274–279
176. G. Strbac, Demand side management: Benefits and challenges. Energy Policy **36**(12), 4419–4426 (2008)
177. Y. Sun, F. Wang, Z. Zhen, Z. Mi, C. Liu, B. Wang, J. Lu, Research on short-term module temperature prediction model based on bp neural network for photovoltaic power forecasting, in *2015 IEEE Power & Energy Society General Meeting* (IEEE, 2015), pp. 1–5
178. C. Sweeney, R.J. Bessa, J. Browell, P. Pinson, The future of forecasting for renewable energy. Wiley Interdiscip. Rev.: Energy Environ. **9**(2), e365 (2020)

179. M. Tamke, P. Nicholas, M. Zwierzycki, Machine learning for architectural design: Practices and infrastructure. Int. J. Archit. Comput. **16**(2), 123–143 (2018)
180. P. Tang, D. Huber, B. Akinci, R. Lipman, A. Lytle, Automatic reconstruction of as-built building information models from laser-scanned point clouds: A review of related techniques. Autom. Constr. **19**(7), 829–843 (2010)
181. J.W. Taylor, R. Buizza, Using weather ensemble predictions in electricity demand forecasting. Int. J. Forecast. **19**(1), 57–70 (2003)
182. J. Tian, M. Gao, S. Zhou, The research of building logistics cost forecast based on regression support vector machine, in *2009 International Conference on Computational Intelligence and Security*, volume 1 (IEEE, 2009), pp. 648–652
183. L. Toderean, T. Cioara, I. Anghel, E. Sarmas, V. Michalakopoulos, V. Marinakis, Demand response optimization for smart grid integrated buildings: Review of technology enablers landscape and innovation challenges. Energy Build. 115067 (2024)
184. J.L. Torres, A. Garcia, M. De Blas, A. De Francisco, Forecast of hourly average wind speed with Arma models in Navarre (Spain). Sol. Energy **79**(1), 65–77 (2005)
185. A. Triki-Lahiani, A.B.-B. Abdelghani, I. Slama-Belkhodja, Fault detection and monitoring systems for photovoltaic installations: A review. Renew. Sustain. Energy Rev. **82**, 2680–2692 (2018)
186. G.A. Tsihrintzis, E. Sarmas, V. Marinakis, D. Panagoulias, E.A. Tsihrintzi, M. Virvou, Evaluating non-expert stakeholder interaction with artificial intelligence on energy urban domain using virtsi: The case of Chatgpt, in *2024 15th International Conference on Information, Intelligence, Systems & Applications (IISA)* (IEEE, 2024), pp. 1–10
187. C. Tsolkas, E. Spiliotis, E. Sarmas, V. Marinakis, H. Doukas, Dynamic energy management with thermal comfort forecasting. Build. Environ. **237**, 110341 (2023)
188. W. Turner, I. Walker, J. Roux, Peak load reductions: Electric load shifting with mechanical pre-cooling of residential buildings with low thermal mass. Energy **82**, 1057–1067 (2015)
189. M. Uddin, M.F. Romlie, M.F. Abdullah, S. Abd Halim, T.C. Kwang et al., A review on peak load shaving strategies. Renew. Sustain. Energy Rev. **82**, 3323–3332 (2018)
190. J. Van Gompel, D. Spina, C. Develder, Satellite based fault diagnosis of photovoltaic systems using recurrent neural networks. Appl. Energy **305**, 117874 (2022)
191. J.R. Vázquez-Canteli, Z. Nagy, Reinforcement learning for demand response: A review of algorithms and modeling techniques. Appl. Energy **235**, 1072–1089 (2019)
192. V. Veerasamy, N.I.A. Wahab, M.L. Othman, S. Padmanaban, K. Sekar, R. Ramachandran, H. Hizam, A. Vinayagam, M.Z. Islam, Lstm recurrent neural network classifier for high impedance fault detection in solar pv integrated power system. IEEE Access **9**, 32672–32687 (2021)
193. I. Vehviläinen, T. Pyykkönen, Stochastic factor model for electricity spot price-the case of the Nordic market. Energy Econ. **27**(2), 351–367 (2005)
194. J. von Grabe, Potential of artificial neural networks to predict thermal sensation votes. Appl. Energy **161**, 412–424 (2016)
195. C. Voyant, G. Notton, S. Kalogirou, M.-L. Nivet, C. Paoli, F. Motte, A. Fouilloy, Machine learning methods for solar radiation forecasting: A review. Renew. Energy **105**, 569–582 (2017)
196. B. Wang, Y. Li, W. Ming, S. Wang, Deep reinforcement learning method for demand response management of interruptible load. IEEE Trans. Smart Grid **11**(4), 3146–3155 (2020)
197. F. Wang, Z. Zhen, Z. Mi, H. Sun, S. Su, G. Yang, Solar irradiance feature extraction and support vector machines based weather status pattern recognition model for short-term photovoltaic power forecasting. Energy Build. **86**, 427–438 (2015)
198. F. Wang, Z. Zhen, B. Wang, Z. Mi, Comparative study on knn and svm based weather classification models for day ahead short term solar pv power forecasting. Appl. Sci. **8**(1), 28 (2017)
199. H.-Z. Wang, G.-Q. Li, G.-B. Wang, J.-C. Peng, H. Jiang, Y.-T. Liu, Deep learning based ensemble approach for probabilistic wind power forecasting. Appl. Energy **188**, 56–70 (2017)

200. K. Wang, X. Qi, H. Liu, A comparison of day-ahead photovoltaic power forecasting models based on deep learning neural network. Appl. Energy **251**, 113315 (2019)
201. X. Wang, J. Zhou, B. Qin, L. Guo, Coordinated control of wind turbine and hybrid energy storage system based on multi-agent deep reinforcement learning for wind power smoothing. J. Energy Storage **57**, 106297 (2023)
202. Z. Wang, F. Wang, S. Su, Solar irradiance short-term prediction model based on bp neural network. Energy Proc. **12**, 488–494 (2011)
203. Z. Wen, D. O'Neill, H. Maei, Optimal demand response using device-based reinforcement learning. IEEE Trans. Smart Grid **6**(5), 2312–2324 (2015)
204. C. Whaley, Best practices in photovoltaic system operations and maintenance. Technical report, National Renewable Energy Lab.(NREL), Golden, CO (United States) (2016)
205. T. Wilcox, N. Jin, P. Flach, J. Thumim, A big data platform for smart meter data analytics. Comput. Ind. **105**, 250–259 (2019)
206. K. Williges, W. Van der Gaast, K. de Bruyn-Szendrei, A. Tuerk, G. Bachner, The potential for successful climate policy in national energy and climate plans: highlighting key gaps and ways forward. Sustain. Earth **5**(1), 1–17 (2022)
207. S. Wu, J. Sun, A top-down strategy with temporal and spatial partition for fault detection and diagnosis of building hvac systems. Energy Build. **43**(9), 2134–2139 (2011)
208. Y.-K. Wu, C.-Y. Lee, C.-R. Chen, K.-W. Hsu, H.-T. Tseng, Optimization of the wind turbine layout and transmission system planning for a large-scale offshore windfarm by ai technology. IEEE Trans. Ind. Appl. **50**(3), 2071–2080 (2013)
209. R. Xiong, J. Cao, Q. Yu, Reinforcement learning-based real-time power management for hybrid energy storage system in the plug-in hybrid electric vehicle. Appl. Energy **211**, 538–548 (2018)
210. M. Yalcintas, An energy benchmarking model based on artificial neural network method with a case example for tropical climates. Int. J. Energy Res. **30**(14), 1158–1174 (2006)
211. K. Yan, L. Ma, Y. Dai, W. Shen, Z. Ji, D. Xie, Cost-sensitive and sequential feature selection for chiller fault detection and diagnosis. Int. J. Refrig **86**, 401–409 (2018)
212. D. Yang, E. Wu, J. Kleissl, Operational solar forecasting for the real-time market. Int. J. Forecast. **35**(4), 1499–1519 (2019)
213. Y. Yıldız, Z.D. Arsan, Identification of the building parameters that influence heating and cooling energy loads for apartment buildings in hot-humid climates. Energy **36**(7), 4287–4296 (2011)
214. Z.J. Yu, F. Haghighat, B.C. Fung, Advances and challenges in building engineering and data mining applications for energy-efficient communities. Sustain. Urban Areas **25**, 33–38 (2016)
215. T. Zakula, P.R. Armstrong, L. Norford, Modeling environment for model predictive control of buildings. Energy Build. **85**, 549–559 (2014)
216. S. Zambrano-Asanza, J. Quiros-Tortos, J.F. Franco, Optimal site selection for photovoltaic power plants using a gis-based multi-criteria decision making and spatial overlay with electric load. Renew. Sustain. Energy Rev. **143**, 110853 (2021)
217. A. Zameer, J. Arshad, A. Khan, M.A.Z. Raja, Intelligent and robust prediction of short term wind power using genetic programming based ensemble of neural networks. Energy Convers. Manage. **134**, 361–372 (2017)
218. H-T. Zhang, F.-Y. Xu, L. Zhou Artificial neural network for load forecasting in smart grid, in *2010 international conference on machine learning and cybernetics*, vol. 6 (IEEE, 2010), pp. 3200–3205
219. Y. Zhang, H.V. Burton, H. Sun, M. Shokrabadi, A machine learning framework for assessing post-earthquake structural safety. Struct. Saf. **72**, 1–16 (2018)
220. H. Zhao, H. Liu, W. Hu, X. Yan, Anomaly detection and fault analysis of wind turbine components based on deep learning network. Renew. Energy **127**, 825–834 (2018)
221. J. Zhao, Z.-H. Guo, Z.-Y. Su, Z.-Y. Zhao, X. Xiao, F. Liu, An improved multi-step forecasting model based on wrf ensembles and creative fuzzy systems for wind speed. Appl. Energy **162**, 808–826 (2016)

222. Y. Zhao, B. Lehman, R. Ball, J. Mosesian, J.-F. de Palma, Outlier detection rules for fault detection in solar photovoltaic arrays, in *2013 twenty-eighth annual IEEE applied power electronics conference and exposition (APEC)* (IEEE, 2013), pp. 2913–2920
223. J. Zheng, D.W. Gao, L. Lin, Smart meters in smart grid: An overview, in *2013 IEEE Green Technologies Conference (GreenTech)* (IEEE, 2013), pp. 57–64
224. J. Zheng, C. Xu, Z. Zhang, X. Li, Electric load forecasting in smart grids using long-short-term-memory based recurrent neural network, in *2017 51st Annual Conference on Information Sciences and Systems (CISS)* (IEEE, 2017), pp. 1–6
225. T. Zhong, Z. Zhang, M. Chen, K. Zhang, Z. Zhou, R. Zhu, Y. Wang, G. Lü, J. Yan, A city-scale estimation of rooftop solar photovoltaic potential based on deep learning. Appl. Energy **298**, 117132 (2021)
226. K. Zhou, C. Fu, S. Yang, Big data driven smart energy management: From big data to big insights. Renew. Sustain. Energy Rev. **56**, 215–225 (2016)
227. K. Zhou, S. Yang, Z. Chen, S. Ding, Optimal load distribution model of microgrid in the smart grid environment. Renew. Sustain. Energy Rev. **35**, 304–310 (2014)
228. Z. Zhu, I. Brilakis, Machine vision-based concrete surface quality assessment. J. Constr. Eng. Manag. **136**(2), 210–218 (2010)
229. Z. Zhu, I. Brilakis, Parameter optimization for automated concrete detection in image data. Autom. Constr. **19**(7), 944–953 (2010)
230. G. Zsembinszki, C. Fernández, D. Vérez, L.F. Cabeza, Deep learning optimal control for a complex hybrid energy storage system. Buildings **11**(5), 194 (2021)

Chapter 3
Scalable Framework for Intelligent System Architecture to Address Challenges in the Energy Sector

Abstract This chapter examines the intersection of artificial intelligence and energy systems, addressing challenges such as distributed energy generation, grid flexibility, and building energy efficiency. A central focus is on a comprehensive and scalable architecture designed to facilitate the development and deployment of AI-driven models and solutions tailored to the energy sector. The proposed architecture comprises three interconnected layers: a data layer for robust acquisition and preprocessing, a model layer for training and optimizing intelligent models, and an application layer for delivering actionable insights through user-friendly interfaces. This architecture ensures streamlined data management, efficient model training, and accessible interpretation of results, addressing the diverse needs of stakeholders, including energy producers, grid operators, residential users, and policymakers. Additionally, the chapter introduces a methodological framework that spans all phases of AI model development, from data preprocessing to monitoring and deployment, emphasizing the significance of interpretability and scalability in delivering impactful energy solutions. By integrating advanced AI techniques with domain-specific knowledge, the framework and architecture collectively provide a roadmap for addressing current energy challenges and fostering innovation in energy management systems. This work serves as a valuable guide for researchers and practitioners seeking to leverage AI for sustainable and efficient energy solutions.

Keywords Artificial intelligence · System architecture · Optimization · Energy sector · Renewable energy sources

3.1 Introduction

The energy sector is a critical domain of modern society, facilitating the seamless operation of various activities by supplying energy to residential, commercial, and infrastructural entities. Nonetheless, it faces numerous challenges, including the necessity to reduce carbon dioxide emissions, improve efficiency, and ensure a reliable energy supply [19]. Consequently, innovative solutions to these complex issues are being actively pursued.

E. Sarmas et al., *Artificial Intelligence for Energy Systems*, Learning and Analytics in Intelligent Systems 46, https://doi.org/10.1007/978-3-031-85209-1_3

Furthermore, Artificial Intelligence (AI) has emerged as a promising and innovative field for addressing various challenges. AI-based solutions can contribute to optimizing energy production, reducing energy consumption, and enhancing the reliability of energy systems [2]. Additionally, AI can be employed for forecasting energy demand and supply, detecting anomalies and faults in energy systems, and managing energy storage systems, among numerous other issues [1]. In light of the preceding discussion, this monograph identifies a collection of issues that necessitate further investigation through AI-based solutions. These issues represent a subsehose examined in this chapter, associate them with relevant stakeholders for each issue, and delineate the corresponding data necessary for their resolution.

Additionally, this chapter delineates the stages involved in the development of models and learning algorithms designed to address the challenges presented in this monograph. The proposed methodological framework for algorithm development aims to assist various stakeholders within the energy sector throughout the process of creating effective algorithms and models. It encompasses both the stages of model and learning algorithm development as well as the architecture of artificial intelligence models. This framework incorporates all necessary levels, including the management of data, models, and results.

The chapter is structured as follows. Section 3.2 provides an overview of the issues addressed by the AI methods developed in this monograph, including the primary stakeholders involved and the relevant data. Section 3.3 outlines the stages involved in the development of AI models. Section 3.4 describes the architecture of the proposed solution, which comprises three distinct levels: the data level, encompassing data management aspects such as ingestion, storage, and visualization; the model level, focusing on the development, monitoring, and management of machine learning models; and the application level, which includes the interpretation, explainability, and scalability of the results produced by the models, as well as the presentation of these results to the end stakeholders. Finally, Sect. 3.5 presents the chapter's conclusions and suggests directions for future research.

3.2 Definition of Stakeholders, Problems and Data

AI has emerged as a potent tool for addressing a diverse array of energy-related challenges, including forecasting electricity demand, predicting renewable energy production, and optimizing the operation of energy storage and distribution systems. Consequently, an increasing number of stakeholders-including utilities, regulators, and households-are adopting AI-based solutions to enhance energy management and reduce their carbon footprints [3, 4].

Moreover, developing effective AI solutions necessitates the identification of relevant data sources and the specific problems to be addressed. The availability and quality of data vary significantly depending on the application domain [17]. This data may encompass time-series information on energy consumption and production, weather forecasts, characteristics of buildings and infrastructure, socioeconomic indicators, and other pertinent factors [10].

In this section, we delineate the set of problems related to energy management that are addressed by the AI models presented in this monograph, along with the associated stakeholders and data sources. Our emphasis is on problems pertinent to the design of AI algorithms and models that can be approached using a common methodology, albeit with minor variations among them.

3.2.1 Stakeholders

According to the general definition, stakeholders are individuals or groups with interests in or who are affected by a particular issue or system [9]. In the context of energy systems, stakeholders may include individuals, businesses, organizations, and governmental entities involved in the production, distribution, and consumption of energy. These stakeholders often have differing perspectives, interests, and goals, which can lead to conflicts and challenges during the development and implementation of energy policies and projects [16].

Identifying and understanding the various stakeholders involved in an energy system is essential for effective decision-making and problem-solving. Stakeholder analysis serves as a valuable tool for identifying and assessing the interests, needs, and concerns of different stakeholders, as well as their potential impacts on an energy project or policy. By involving stakeholders in the decision-making process, their concerns and interests can be addressed, thereby fostering support and ownership of energy projects and policies [28].

Relying on existing studies and synthesizing their findings [8, 12, 14, 30], we will define the stakeholders involved in a modern energy system and present their roles and interests in the development and implementation of energy policies and projects. The ultimate goal is to link these stakeholders to the challenges addressed in this monograph. The stakeholders, referred to as Key Stakeholders (KS), are outlined below and summarized in Fig. 3.1.Please provide high-resolution source file for Fig. 3.2.

KS01—Producers (PD): Energy producers are responsible for generating the electricity consumed by households and businesses. They can use various methods to produce electricity, such as nuclear power, coal-fired power plants, natural gas power plants, wind farms, and solar parks. Their activities have a significant environmental impact, as they may contribute to greenhouse gas emissions and other pollutants. This category also includes individual owners of energy parks.

KS02—Suppliers (SP): Energy suppliers are responsible for providing energy to households and businesses. They purchase energy from producers and sell it to consumers. The liberalization of the energy market has enabled consumers to choose from a variety of suppliers. This has led to increased competition and potentially lower prices for consumers.

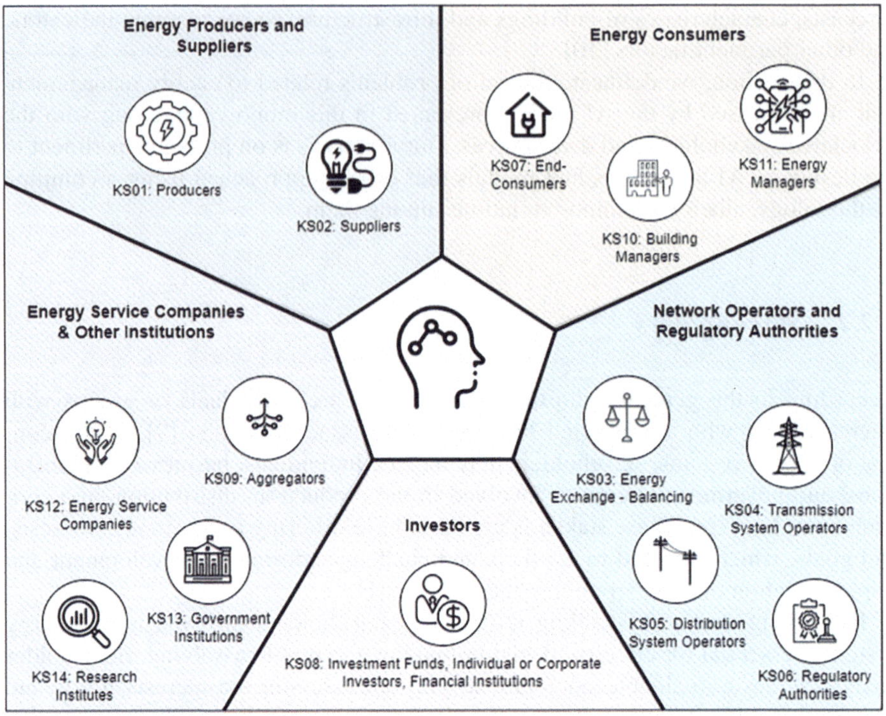

Fig. 3.1 Stakeholders

KS03—Balance Responsible Party (BRP): Balancing entities are responsible for ensuring the stability of the energy grid. They oversee one or more access points to the transmission network, and their role involves the continuous balancing of production and consumption. This helps ensure that there is always enough energy available to meet demand.

KS04—Transmission System Operator (TSO): Transmission system operators are responsible for transporting electricity from power generation stations to industrial customers or the distribution network. They are also responsible for maintaining the overall balance of the grid by coordinating production and consumption. Due to the capital-intensive nature of these businesses, it is challenging to introduce competition within the sector.

KS05—Distribution System Operators (DSO): Distribution system operators are responsible for constructing and managing medium- and low-voltage networks that connect transmission networks to private residences. They also assist in managing power outages and energy meter issues. Their role is crucial in ensuring the safe and efficient delivery of energy to consumers.

KS06—Regulatory Bodies (REG): Regulatory authorities are tasked with ensuring the transparency and competitiveness of the energy market. They oversee the activities of producers, suppliers, and other stakeholders to ensure they act in the best interests of consumers. They also certify the operation of energy markets and advise authorities on energy-related matters.

KS07—Electricity Customers, Consumers, Residents of Buildings, Occupants (BUI): These stakeholders are the end users of energy. They consume energy at various scales, ranging from individual households to large industrial customers. They have a significant impact on energy demand and can contribute to reducing energy consumption through energy efficiency measures.

KS08—Project Developers, Investing Funds, Governmental Institutions (INV): These stakeholders seek funding opportunities to invest in energy efficiency projects. They may invest in building renovations at the building, district, or city level, playing a key role in promoting renewable energy sources and reducing energy consumption. Investors are individuals or organizations that provide financing for energy-related projects and initiatives. This can include individual investors, corporate investors, and financial institutions such as banks and venture capital firms. This category includes project developers, investment funds, private investors, corporate investors, as well as financial institutions.

KS09—Aggregators (AGG): Aggregators are a new type of energy service provider that can increase or decrease the electricity consumption of a group of consumers depending on the overall electricity demand on the grid. They aggregate flexibility from the devices of prosumers (producer-consumers) and sell it to DSOs (Distribution System Operators), BRPs (Balancing Responsible Parties), and TSOs (Transmission System Operators).

KS10—Facility Managers, Building Operators (FMB): Building managers are responsible for ensuring the efficient operation of a building's heating, cooling, mechanical, and electrical systems. They ensure that the facility operates successfully and in the most efficient manner, which can help reduce energy consumption and costs.

KS11—Energy Managers (ENM): Energy managers are professionals tasked with overseeing energy usage and enhancing energy efficiency within organizations. Their primary goal is to reduce energy consumption and costs while maintaining or improving productivity. They may develop and implement energy management plans, conduct energy audits, and identify and apply energy-saving measures.

KS12—Energy Service Companies (ESCOs): Energy service companies (ESCOs) are businesses that provide energy-efficient solutions and services to clients. These solutions may include energy audits, energy-saving measures, and the implementation of energy-efficient technologies. Typically, they operate under a performance-based contract model, where compensation is contingent upon the amount of energy saved through their initiatives.

KS13—Government Agencies (GA): Government agencies are entities responsible for overseeing and regulating energy and environmental policies. These agencies may include energy departments, environmental agencies, and other governmental bodies operating at local, state, and national levels. They are tasked with the development and enforcement of regulations, the provision of funding for research and development, and the implementation of programs designed to encourage the adoption of renewable energy sources and energy-efficient technologies. This category includes energy departments, environmental organizations, and other governmental entities.

KS14—Research Institutions (RI): Research institutions are organizations dedicated to advancing scientific knowledge through research and development. In the realm of energy, these institutions conduct studies on energy efficiency, renewable energy technologies, energy policy, and other related subjects. Examples of such institutions include universities, national laboratories, and other research organizations.

The fourteen stakeholder categories outlined above offer a comprehensive overview of the stakeholders. However, organizing them into six broader categories enhances the clarity of the analysis. The six broader categories are presented below.

- **Energy producers and suppliers:** This category includes producers (KS01) and suppliers (KS02), who are responsible for generating and delivering energy to end users.
- **Network managers and regulatory authorities:** This category includes balancing entities (KS03), transmission system operators (KS04), distribution system operators (KS05), and regulatory authorities (KS06), who are responsible for managing the energy network and ensuring its efficient and fair operation.
- **Consumers and energy managers:** This category includes electricity consumers (KS07), facility managers (KS10), and energy managers (KS11), who are responsible for energy consumption and its efficient management.
- **Energy Investors:** This category includes investment funds and financial institutions (KS08), which invest in energy-related projects and initiatives to promote renewable energy sources and reduce energy consumption.
- **Energy service providers and researchers:** This category includes aggregators (KS09), energy service companies (KS12), governmental organizations (KS13), and research institutions (KS14). All these entities provide efficient solutions and services, conduct research on energy-related issues, and implement policies and programs to promote sustainable energy practices.

3.2.2 Problems

The AI algorithms and models discussed in this chapter can address various energy-related challenges, including energy demand and production forecasting, energy efficiency, and energy management. For instance, energy demand forecasting entails predicting the energy required at a specific time and location, whereas energy efficiency emphasizes identifying methods to reduce energy consumption without compromising performance.

These problems can be classified into three primary categories: (a) issues pertaining to distributed energy production (RES), (b) challenges associated with achieving microgrid flexibility (FLEX), and (c) concerns related to enhancing energy efficiency in buildings (ENEF).

The first category, RES, addresses the variability and uncertainty inherent in renewable energy sources, particularly solar and wind energy. Integrating these sources into the electricity grid presents challenges, including accurate production forecasting, optimization of utilization, and management of variability to ensure grid stability and reliability. The second category, FLEX, encompasses distributed energy resources, including energy storage systems, flexible loads, and demand response programs. These resources play a crucial role in mitigating electricity demand during peak periods and enhancing grid stability. However, optimal scheduling and sizing of these resources necessitate precise predictions of energy demand, energy prices, and weather conditions. The third category, ENEF, emphasizes the reduction of energy consumption in buildings while maintaining comfort and efficiency. This includes the implementation of advanced controls, building automation systems, and data analytics to optimize energy use and minimize waste. Challenges in this category may include forecasting energy consumption, optimizing building HVAC systems, and estimating energy savings resulting from retrofits and renovations.

This monograph addresses the following problems, which are approached through the development of microservices (Analytics Services—AS) based on artificial intelligence models and optimization algorithms:

- **AS01—Photovoltaic (PV) Production Forecasting:** The problem of estimating the energy production of a photovoltaic system in the short term (e.g., one hour ahead) or medium term (e.g., one week ahead) using photovoltaic production time series, historical weather data, and numerical weather forecasts [21].
- **AS02—Consumption Forecasting at the Building Level:** The problem of estimating the future energy consumption of a building using building consumption time series [22].
- **AS03—Load Forecasting of the Grid:** The problem of estimating the future load of the grid using consumption time series, historical weather data, and numerical weather forecasts [27].
- **AS04—Estimation of Energy Savings as a Result of Renovations:** The problem of estimating the actual energy savings from renovations using historical building consumption time series, historical weather data, and post-renovation measurements [23].
- **AS05—Financing of Energy Conservation Measures:** The problem of determining the renovation category (A, B, or C) based on the renovation's potential for energy savings relative to cost, using historical records of financing projects and building characteristics [25].
- **AS06—Optimization of DERs in micro-grids:** The problem of finding the optimal scheduling of flexible loads and the optimal sizing of the storage system, using forecasts of photovoltaic production, building consumption, and other loads, as well as profiles of flexible loads and storage systems. This service is directly linked to services AS01 and AS02 [24].

- **AS07—Scheduling of flexible loads for peak shaving at the grid level:** The problem of finding the optimal scheduling of flexible loads to minimize peak loads, using forecasts of photovoltaic production, grid load forecasts, and profiles of flexible loads. This service is directly linked to services AS01 and AS03 [26].
- **AS08—Thermal Comfort with Dynamic Energy Management:** The problem of scheduling the HVAC system to ensure the thermal comfort of occupants, using room-specific building consumption, humidity, and temperature data. This service is directly linked to service AS02 [29].

Table 3.1 presents a mapping of services to the stakeholders defined in the previous section. The services encompass a broad spectrum of issues, including forecasting, load management, and optimization, aiming to enhance energy efficiency and reduce energy costs. The stakeholders are categorized into fourteen groups, each with a unique set of interests, and Table 3.1 emphasizes which services are most pertinent to each group.

For instance, the load forecasting service (AS03) is particularly advantageous for transmission system operators and distribution system operators, as it enhances their capability to manage the grid more effectively. Similarly, the financing service (AS05) holds particular significance for building owners and energy service companies, which bear the responsibility of implementing energy-saving measures. By mapping services to stakeholders, the table offers a valuable overview of the services that are most beneficial to various groups within the energy sector.

Table 3.1 Stakeholders and Analytics Services

Stakeholders	AS01	AS02	AS03	AS04	AS05	AS06	AS07	AS08
KS01 - PD	X					X	X	
KS02 - SP	X		X	X				
KS03 - BRP		X	X					
KS04 - TSO	X		X			X		
KS05 - DSO		X	X			X	X	
KS06 - REG					X			
KS07 - BUI		X		X				X
KS08 - INV	X			X	X			
KS09 - AGG	X							
KS10 - FMB		X		X				
KS11 - ENM		X		X				X
KS12 - ESCOs		X		X	X	X		
KS13 - GA						X		X
KS14 - RI	X	X	X	X	X	X	X	X

3.2.3 Data

The data used in energy applications, such as those described above, can be collected from a wide range of heterogeneous sources, including sensors, smart meters, and online platforms [13]. These sources may provide data in different formats and structures, which can pose challenges for data integration and processing [18]. Additionally, data may come from various domains, such as weather data, energy consumption data, and building data, further increasing data heterogeneity. Therefore, careful preprocessing and curation of the data are essential to ensure their quality, completeness, and consistency before using them for the development of AI algorithms and models.

In this monograph, various data sources were used to develop AI models for energy-related problems. The data sources included publicly available datasets, such as weather data, as well as energy consumption and energy production data. Furthermore, data were collected from various stakeholders, such as building owners, energy companies, and energy management firms. The data collected from stakeholders included building characteristics, energy consumption patterns, energy production data, and weather data specific to the building's location. All data were anonymized and stored securely to ensure privacy and data security. Preprocessing and cleaning were performed on the data to remove outliers, missing values, or errors that could impact the performance of the algorithms and models. The final dataset used for model development was a combination of publicly available datasets and stakeholder data, providing a rich and diverse data pool for training the algorithms and models.

Regarding the problems studied in this monograph, the most relevant data for each problem are presented below:

➢ **AS01—Photovoltaic (PV) Production Forecasting:** The data required for forecasting photovoltaic production include historical production data, meteorological data, and information about the installation (e.g., panel type, orientation).
➢ **AS02—Building-Level Consumption Forecasting:** The data required for forecasting consumption at the building level include historical consumption data, meteorological data, occupancy data, and building characteristics (e.g., area, number of occupants).
➢ **AS03—Grid Load Forecasting:** The data required for forecasting grid load include historical load data, weather data, and potentially information about the electrical grid (e.g., topology, power flow data, etc.).
➢ **AS04—Estimation of Energy Savings from Renovations:** The data required to estimate energy savings from renovations include historical consumption data before and after the renovations, as well as information about the renovations themselves (e.g., type of insulation installed, efficiency of new appliances).
➢ **AS05—Financing Energy-Saving Measures:** The data required for financing energy-saving measures include information about the cost of the measures, available financing options, and the expected return on investment.
➢ **AS06—Optimization of DERs in Microgrids:** The data required for optimizing energy resources in microgrids include information about local energy demand,

the capacity of available resources (e.g., solar panels, batteries, etc.), and the characteristics of the microgrid itself (e.g., network topology, communication infrastructure, etc.).

➢ **AS07—Scheduling Flexible Loads for Peak Saving at the Grid Level:** The data required for scheduling flexible loads for peak saving include historical load data, weather data, information about the electrical grid (e.g., topology, power flow data, etc.), and data on the flexible loads themselves (e.g., their capacity, flexibility, etc.).

➢ **AS08—Thermal Comfort with Dynamic Energy Management:** The data required for dynamic management of thermal comfort include information about the building's heating and cooling systems, meteorological data, usage data, and user preferences. It may also include data from sensors monitoring temperature, humidity, and other environmental factors.

3.3 Development of AI Models and Optimization Algorithms

The development of a Machine Learning (ML) model involves various stages, from problem selection to model deployment and monitoring. Each stage is essential for ensuring the accuracy, reliability, and effectiveness of the model in addressing the specific problem at hand. This section presents the four stages of ML model development: problem selection and data collection, data preprocessing and feature engineering, model training, evaluation and selection, and model deployment and monitoring. A comprehensive diagram illustrating these four stages is presented in Fig. 3.2.

3.3.1 Problem Selection and Data Collection

The first step in developing a ML model is selecting a problem to solve and collecting relevant data (Problem Selection and Data Collection). Initially, the user must choose one of these problems and input the necessary data related to it. Problem selection should include defining the problem, identifying the business or research objectives, and specifying the target variable. A comprehensive understanding of the problem is critical for identifying relevant data sources and gathering the necessary data. This phase may necessitate collaboration with domain experts or stakeholders to ensure that both the problem and the target variable are well-defined and accurately represent the real-world issue. In this instance, the user is required to select one of the eight problems presented in the preceding section.

The problems addressed through the models of this monograph may belong to one or more of the following categories: classification, regression, forecasting, and

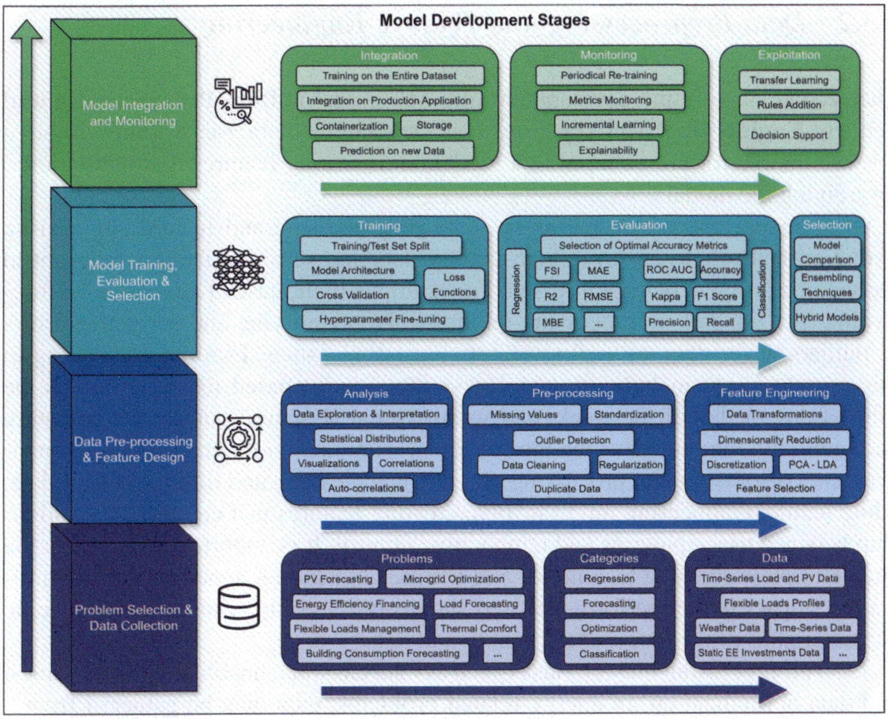

Fig. 3.2 Methodology

optimization. For example, network load forecasting may require regression analysis, while flexible load scheduling for peak shaving at the network level may require optimization techniques. Each problem category may require a different approach to data collection, preparation, and model development. These categories possess unique characteristics and necessitate specific considerations, including the availability and quality of data sources, the complexity of the system being modeled, and the desired level of accuracy and precision. Therefore, it is essential for users to critically assess the category of the problem they aim to address and select the appropriate methodologies for collecting, preparing, and developing the corresponding models.

Once the problem is defined, data collection involves identifying data sources, gathering and preparing the data, and ensuring the data adequately represents both the problem and the target variable. The user must input the requisite data pertaining to the selected problem. Data can be collected from various sources, including sensors, smart meters, and online platforms. It is vital to recognize that the quality and quantity of the data can substantially affect the accuracy and reliability of the model, rendering the data collection and preparation phase critical for successful model development.

3.3.2 Data Preprocessing and Feature Engineering

The second stage in the development of a ML model is **Data Preprocessing and Feature Engineering**. This stage involves cleaning the data, transforming it into a format suitable for use as input to a model, and creating features that will be used for training the model.

In the energy sector, in particular, data preprocessing and feature engineering play a critical role in ensuring high-quality data and selecting features relevant to the specific problem being addressed. For example, in the case of network load forecasting (AS03), data preprocessing might involve identifying and correcting errors or outliers in the data, as well as imputing missing values. Feature engineering in this context might include creating lag variables, time-based features such as the time of day or day of the week, and weather-related features such as temperature and humidity. Similarly, for the thermal comfort model with dynamic energy management (AS08), data preprocessing may include cleaning and filtering sensor data, while feature engineering might involve creating features that capture the relationship between usage patterns and temperature control. It is important to note that the choice of preprocessing techniques and feature engineering approaches can significantly impact model performance and should be carefully tailored to the specific characteristics of the data and the problem at hand.

Data preprocessing includes tasks such as data cleaning, handling missing values, and data normalization. In energy-related problems, data may be gathered from a variety of sources, including sensors or smart meters. Consequently, data cleaning may involve the removal or correction of erroneous or irrelevant data. For instance, data collected from sensors may contain noise or outliers due to hardware malfunctions or external factors, including weather conditions. Techniques like smoothing or filtering can be used to remove noise or outliers in such cases. Addressing missing data represents a critical aspect of the preprocessing phase. In energy-related issues, missing data may arise from equipment malfunctions or communication failures. Various methods, including mean imputation, regression-based techniques, data augmentation, or inputting values from the previous day, can be utilized to manage missing data. Data normalization involves transforming data into a standardized scale, thereby preventing biases that may arise from variations in units or scales. This process is particularly essential when data is collected from diverse sources or across different time periods. Common techniques for data normalization include min-max scaling, Z-score normalization, and logarithmic transformation.

Feature engineering plays a crucial role in building accurate and effective ML models for energy-related problems. It involves creating new features from raw data, selecting the most relevant features, and reducing data dimensionality. In the context of energy-related issues, feature engineering may entail the extraction of pertinent information from time-series data, encompassing patterns, trends, seasonality, and weather-related data. For instance, in photovoltaic production forecasting, features such as the time of day, season, weather conditions, and historical production data can be used to train the model. Similarly, in forecasting building-level consumption, relevant features may include usage patterns, historical consumption data, and weather conditions. Thus, the design of appropriate features can significantly influence the performance and accuracy of ML models.

Various techniques can be used for feature engineering, including domain knowledge, statistical methods, and ML-based methods. Domain knowledge involves understanding the problem domain and leveraging prior knowledge to select features. Statistical methods utilize techniques such as correlation analysis, Principal Component Analysis (PCA), and feature selection algorithms to identify the most relevant features. Finally, ML-based methods use algorithms like decision trees and random forests to identify the most important features. Combining these approaches helps identify the most relevant features for a given problem.

3.3.3 Model Training, Evaluation, and Selection

The next stage in the development of a ML model involves selecting an algorithm, training the model on data, and evaluating its performance (Model Training, Evaluation, and Selection). The primary objective is to create a model that generalizes effectively to new, unseen data while maintaining accuracy and interpretability. This phase is generally iterative, involving modifications to the model as required, informed by the results of the evaluation.

Model training involves adapting the chosen algorithm to the data to learn the underlying patterns and relationships. This stage encompasses the division of the available data into training, validation, and testing sets. The training set is employed to fit the model, while the validation set aids in hyperparameter tuning and minimizes the risk of overfitting. Once the model is trained, its performance is assessed on the test set.

Model evaluation measures the performance of the trained model on the test set. The choice of evaluation metrics depends on the type of problem, such as classification or regression.

For classification problems, common metrics include accuracy, precision, recall, F1-Score, and the Area Under the Curve—Receiver Operating Characteristic (AUC-ROC). Accuracy evaluates the proportion of correct predictions, while precision and recall quantify the ratios of true positives among all positive predictions and true positives among all actual positive samples, respectively. The AUC-ROC curve illustrates the true positive rate against the false positive rate, providing a measure of the model's capability to differentiate between positive and negative samples.

For regression problems, metrics such as Mean Absolute Error (MAE), Mean Squared Error (MSE), and R-squared (R^2) are commonly used to assess performance. MAE and MSE quantify the average absolute and squared differences between predicted and actual values, respectively, whereas R^2 quantifies the proportion of variance in the target variable that can be explained by the model. Detailed descriptions of the metrics used in the experimental applications are provided in subsequent chapters.

Model selection entails the comparison of various algorithms and the optimization of hyperparameters to identify the model that demonstrates optimal performance. Hyperparameters are adjusted to enhance the performance of the chosen model. Typically, model selection occurs during or after training and evaluation. This process involves comparing the performance of distinct algorithms or variations of the same algorithm, culminating in the selection of the model that exhibits superior performance according to specific metrics. Following the identification of the optimal model, further refinement and optimization may be pursued. It is essential to recognize that the model selection process is iterative and may include feedback loops among training, evaluation, and selection until a satisfactory model is achieved.

3.3.4 Model Deployment and Monitoring

The final stage in building a ML model is Model Deployment and Monitoring. Once the model has been selected, trained, and evaluated, it is ready to be deployed in a production environment. This involves integrating the model into existing systems and infrastructure, as well as establishing processes for data input, preprocessing, and prediction. In addition to deploying the model, it is important to create a monitoring framework to track the model's performance over time. This includes implementing automated processes to monitor the model's accuracy, detect any anomalies or errors in predictions, and trigger alerts when necessary.

Finally, it is essential to continuously retrain the model with new data to ensure it remains up-to-date and continues to provide accurate predictions. This requires creating a continuous data pipeline to feed new data into the model, retraining it regularly, and deploying the updated model to production. Overall, the objective of model deployment and monitoring is to ensure that the model remains accurate and reliable over time, thereby contributing to the success of the business or research project.

3.4 Proposed Architecture

In the field of AI, the development of models that accurately predict corresponding energy metrics constitutes only one aspect of the overall process and does not inherently ensure the delivery of valuable services to stakeholders. To effectively utilize these models, it is essential that they be integrated into a comprehensive architecture capable of managing data acquisition, preprocessing, and storage, as well as disseminating results to end users through various channels. At this juncture, a robust architecture is required for the management of data, machine learning (ML) models, and results.

There are various approaches to structuring such an architecture, contingent on the specific requirements of the application and the available resources [11, 13, 15]. This

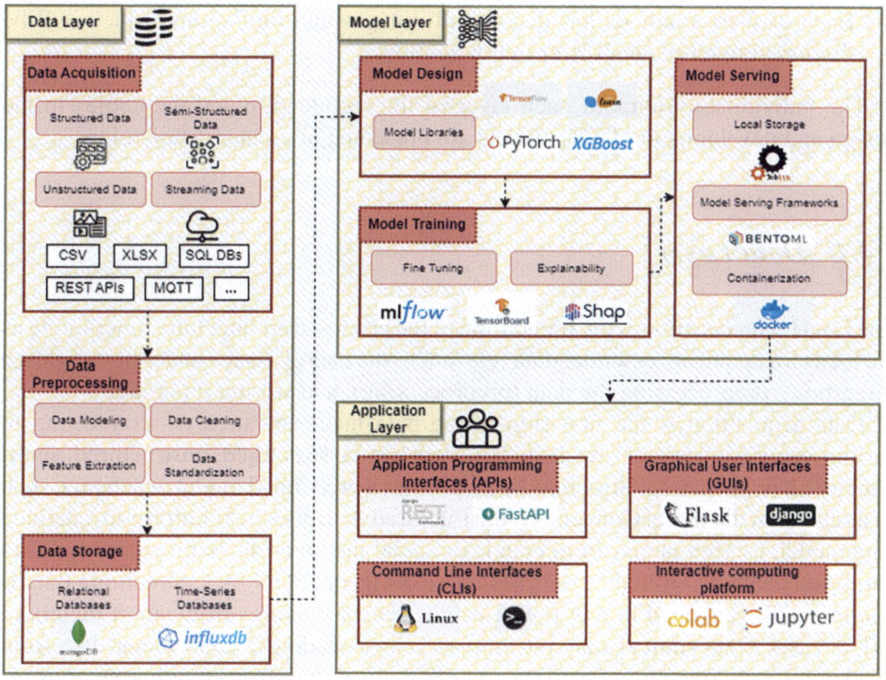

Fig. 3.3 Architecture

section presents a three-tier architecture that includes the data layer, the model layer, and the application layer, and explores some of the common tools and techniques used at each layer. The proposed architecture is illustrated in Fig. 3.3.

The various individual levels of this architecture are briefly presented below:

➢ **Data Layer:** The data layer provides a robust and scalable solution for data management, which is a critical component of any MM system. Its main functionality is to handle different types of data-structured, unstructured, semi-structured, or streaming-supporting their cleaning, feature engineering, and normalization. Furthermore, it facilitates the integration of data from different sources and enables efficient storage in various types of databases.

➢ **Model Layer:** The model layer is an essential part of any MM system, as it is responsible for the development, training, and storage of models that address the problems outlined at the beginning of the chapter. The model layer enables the creation of complex models capable of identifying patterns in large datasets, making accurate predictions, and adapting to changing conditions over time.

➢ **Application Layer:** This layer enables users to access and interpret the decisions made by the trained models and gain insights into the factors influencing the models' predictions. It can be implemented in various ways, such as through web applications, reporting tools, or application programming interfaces (APIs).

Each of these methods has distinct advantages and limitations, which will be discussed in more detail later in this section.

The individual systems of each layer of the architecture and their roles in the design and development of ML models are presented in greater detail below.

3.4.1 Data Layer

The Data Layer constitutes the foundation of the proposed architecture, functioning as the locus for acquiring essential data relevant to the energy issues under investigation (ingestion), preprocessing the data (preprocessing), and storing it (storage).

Data acquisition is the first step in the data layer and involves connecting to various sources to collect raw data. This data can be structured, unstructured, semi-structured, or streaming. Structured data follows a predefined model or schema, such as a relational database. Conversely, unstructured data does not conform to a defined data model, as seen in text from emails or social media posts. Semi-structured data combines elements of both, examples being XML or JSON files. Streaming data is continuously generated in real time, such as sensor readings from smart meters or IoT devices. Data sources may encompass sensor readings, smart meter data from residential and commercial entities, meteorological data from weather stations, or energy consumption data sourced from energy providers.

After data acquisition, the preprocessing stage ensures data quality and accuracy, a critical step. This stage includes data modeling, data cleaning, feature engineering, and data normalization:

- Data modeling defines the schema, types, and structures of the data.
- Data cleaning involves identifying and removing outliers, handling missing values, and resolving inconsistencies.
- Feature engineering entails selecting and transforming features to enhance model performance.
- Data normalization scales data to a common range, improving the efficiency and accuracy of models.

Following preprocessing, data is stored in various database types depending on its characteristics and the requirements of ML models. For example, structured data might use an SQL database, while unstructured data might benefit from a NoSQL database. The stored data can then be processed further in the model layer for training and evaluating ML models.

Data acquisition and preprocessing pose significant challenges due to the high volume of data, the diversity of data sources, and the inherent complexity of the data itself. For instance, energy consumption data can vary significantly between households, and energy providers may use different data formats and schemas. Therefore, robust acquisition and preprocessing processes are critical to ensure the accuracy and reliability of the models.

3.4.2 Models Layer

The Models Layer constitutes the second tier of the architecture, encompassing the design, training, and storage of models. This layer is responsible for creating models that address the proposed energy problems using the appropriate data standardized in the data layer.

Model design encompasses all actions required to create ML models. This stage includes selecting appropriate libraries, algorithms, and techniques for the specific problem. In this study, various libraries such as SkLearn, PyTorch, TensorFlow Keras, XGBoost, and others have been used and supported [5–7, 20]. Depending on the nature of the problem, different algorithms, such as decision trees, random forests, support vector machines, and neural networks, can be employed. The selection of libraries, algorithms, and techniques should be informed by the characteristics of the problem, the available data, and the desired performance criteria. In the energy sector, various problems can be addressed using ML techniques, such as:

- Fault classification in power systems.
- Regression for energy consumption forecasting.
- Prediction of renewable energy production.
- Clustering of energy consumers based on consumption patterns.

For instance, a predictive model can estimate energy production from photovoltaic systems, wind turbines, or other renewable sources. Conversely, a clustering model can categorize consumers based on their consumption patterns to promote energy-saving behaviors.

After development, models must be trained using available data. Model training involves tuning the model's hyperparameters to optimize its performance on training data. Hyperparameters are configuration parameters that govern the training process, such as learning rate, the number of hidden layers, and the number of neurons. This study supports various tools for hyperparameter tuning, such as TensorBoard, and training environments like MLFlow. Additionally, model interpretability can be evaluated using tools such as Shap.

The quality and quantity of data significantly affect model accuracy and reliability, making hyperparameter tuning and interpretability evaluation crucial for successful model development. For instance, a regression model designed to predict energy consumption should be trained on a large and diverse dataset that includes inputs such as weather conditions, historical consumption, and building characteristics to enhance accuracy.

Once trained, models need to be stored for future use. The proposed architecture supports various storage methods, including:

- Local storage using tools like joblib, suitable for small models easily loaded and used in the same environment.
- Cloud storage using tools like BentoML, ideal for storing larger models accessible from anywhere and usable across different environments.

- Containerization using Docker, enabling models to be deployed as standalone packages and used across diverse environments without dependencies.

These storage options ensure flexibility and accessibility, catering to different deployment needs.

3.4.3 Applications Layer

The Applications Layer is responsible for delivering a comprehensive view of the results to end users in a user-friendly, comprehensible, and accessible manner. This can be achieved through various means, including graphical user interfaces (GUIs), application programming interfaces (APIs), command-line interfaces (CLIs), and reporting tools.

One of the key functions of this layer is the development of applications that enhance the interpretability and explainability of model predictions. This is particularly valuable in energy management problems, where understanding the underlying factors contributing to a prediction or recommendation is crucial for making informed decisions. Moreover, the applications layer should enable users to discern which features significantly impact the model's predictions and how they influence the outcomes.

In the context of this monograph, the results can be made available to the end user through one or more of the following options:

- ➢ **Application Programming Interfaces—APIs:** This category includes interfaces that allow software applications to communicate with each other. Models can be integrated into applications via APIs, enabling the model's results to be accessed and utilized programmatically. In general, this method of accessing a model's results is commonly used by large organizations or companies to handle bulk predictions involving a large number of variables to be forecasted.
- ➢ **Graphical User Interfaces—GUIs:** These are interactive applications that provide a visual interface for users to interact with the results of the models. Graphical interfaces can be designed to display results in various formats, such as charts, graphs, tables, or maps. These interfaces may take the form of a web application that presents predictions, key performance metrics, and other relevant information. In such cases, the application should be user-friendly, allowing users quick access to the information they need and enabling them to delve into more detailed data as required.
- ➢ **Command-Line Interfaces—CLIs:** Command-line interfaces (CLIs) allow users to interact with MM models by entering commands into a terminal window. One of the advantages of using a CLI is the speed and flexibility they offer. CLIs enable users to quickly execute a prediction with a simple command, without the need to navigate through a graphical interface. Additionally, they can be easily integrated into scripts and automated workflows, facilitating seamless integration with other tools and processes.

➢ **Reporting Applications:** This category includes web-based applications like Jupyter Notebooks and Google Colab, which allow users to create reports that incorporate model results and their analysis. One of their main advantages is their ability to produce interactive reports, enabling stakeholders to interact with the models and data. For example, users can adjust a model's input parameters and observe how the output changes. Additionally, they can include rich visualizations that help make the results more accessible and comprehensible to stakeholders.

3.5 Conclusions

In this chapter, the issues addressed in this monograph are presented concisely, along with the stages of model development aimed at resolving these issues. Specifically, the chapter covered the stages of ML model development, the architecture of the proposed models, and the management of data, models, and results. The proposed framework can serve as a guideline for developing AI models in other domains.

The quality of data significantly impacts model performance, making data management a critical aspect of ML model development. In this context, a three-layer architecture was introduced, which comprises a layer focused on data management, an intermediate layer dedicated to model development, and a layer centered on result handling.

Future research directions for developing intelligent models in the energy sector could include exploring new ML algorithms for various energy management and efficiency problems. Key areas for investigation might include:

- Optimization of energy systems, such as specialized energy storage systems.
- Demand response programs.
- Smart grid management using reinforcement learning algorithms.
- Additional research could focus on improving the interpretability and explainability of ML models, which is vital for building stakeholder trust in the models.

The development of artificial intelligence models for energy-related challenges represents a rapidly evolving field. The proposed framework serves as an initial foundation for the creation of such models. Ongoing research in this area is anticipated to yield new insights and innovations, facilitating more efficient and effective energy management through intelligent, user-centered applications.

References

1. A.N. Abdalla, M.S. Nazir, H. Tao, S. Cao, R. Ji, M. Jiang, L. Yao, Integration of energy storage system and renewable energy sources based on artificial intelligence: an overview. J. Energy Storage **40**, 102811 (2021)
2. T. Ahmad, D. Zhang, C. Huang, H. Zhang, N. Dai, Y. Song, H. Chen, Artificial intelligence in sustainable energy industry: status quo, challenges and opportunities. J. Clean. Prod. **289**, 125834 (2021)
3. T. Ahmad, H. Zhu, D. Zhang, R. Tariq, A. Bassam, F. Ullah, A.S. AlGhamdi, S.S. Alshamrani, Energetics systems and artificial intelligence: applications of industry 4.0. Energy Rep. **8**, 334–361 (2022)
4. A. Arsenopoulos, E. Sarmas, A. Stavrakaki, I. Giannouli, J. Psarras, A data-driven decision support tool at the service of energy suppliers and utilities for tackling energy poverty: A case study in Greece, in *2021 12th International Conference on Information, Intelligence, Systems & Applications (IISA)* (IEEE, 2021), pp. 1–6
5. L. Buitinck, G. Louppe, M. Blondel, F. Pedregosa, A. Mueller, O. Grisel, V. Niculae, P. Prettenhofer, A. Gramfort, J. Grobler, R. Layton, J. VanderPlas, A. Joly, B. Holt, G. Varoquaux, *API design for machine learning software: experiences from the scikit-learn project* (Languages for Data Mining and Machine Learning, In ECML PKDD Workshop, 2013), pp.108–122
6. T. Chen, C. Guestrin, XGBoost: a scalable tree boosting system, in *Proceedings of the 22nd ACM SIGKDD International Conference on Knowledge Discovery and Data Mining*, KDD '16. (ACM, , New York, NY, USA), pp. 785–794
7. F. Chollet et al., Keras (2015). https://keras.io
8. S. Franzò, F. Frattini, E. Cagno, A. Trianni, A multi-stakeholder analysis of the economic efficiency of industrial energy efficiency policies: empirical evidence from ten years of the italian white certificate scheme. Appl. Energy **240**, 424–435 (2019)
9. A.L. Friedman, S. Miles, *Stakeholders: Theory and Practice* (OUP Oxford, 2006)
10. J. Hu, A.V. Vasilakos, Energy big data analytics and security: challenges and opportunities. IEEE Trans. Smart Grid **7**(5), 2423–2436 (2016)
11. L. Linder, D. Vionnet, J.-P. Bacher, J. Hennebert, Big building data-a big data platform for smart buildings. Energy Proc. **122**, 589–594 (2017)
12. P. Liu, Y. Zhou, D.K. Zhou, L. Xue, Energy performance contract models for the diffusion of green-manufacturing technologies in China: a stakeholder analysis from smes' perspective. Energy Policy **106**, 59–67 (2017)
13. V. Marinakis, H. Doukas, J. Tsapelas, S. Mouzakitis, Á. Sicilia, L. Madrazo, S. Sgouridis, From big data to smart energy services: an application for intelligent energy management. Futur. Gener. Comput. Syst. **110**, 572–586 (2020)
14. N. Martin, J. Rice, Improving Australia's renewable energy project policy and planning: a multiple stakeholder analysis. Energy Policy **84**, 128–141 (2015)
15. M.U. Mehmood, D. Chun, H. Han, G. Jeon, K. Chen et al., A review of the applications of artificial intelligence and big data to buildings for energy-efficiency and a comfortable indoor living environment. Energy and Build. **202**, 109383 (2019)
16. A.K. Nasr, M.K. Kashan, A. Maleki, N. Jafari, H. Hashemi, Assessment of barriers to renewable energy development using stakeholders approach. Entrepreneurship Sustain. Issues **7**(3), 2526 (2020)
17. B. Nastasi, M. Manfren, M. Noussan, Open data and energy analytics (2020)
18. M. Obitko, V. Jirkovský, J. Bezdíček, Big data challenges in industrial automation, in *Industrial Applications of Holonic and Multi-Agent Systems: 6th International Conference, HoloMAS 2013, Prague, Czech Republic, August 26-28, 2013. Proceedings*. D(Springer, 2013), pp. 305–316
19. E. Papadis, G. Tsatsaronis, Challenges in the decarbonization of the energy sector. Energy **205**, 118025 (2020)

20. Paszke, A., Gross, S., Massa, F., Lerer, A., Bradbury, J., Chanan, G., Killeen, T., Lin, Z., Gimelshein, N., Antiga, L., Desmaison, A., Kopf, A., Yang, E., DeVito, Z., Raison, M., Tejani, A., Chilamkurthy, S., Steiner, B., Fang, L., Bai, J., and Chintala, S. (2019). Pytorch: An imperative style, high-performance deep learning library. In *Advances in Neural Information Processing Systems 32*, pages 8024–8035. Curran Associates, Inc
21. E. Sarmas, N. Dimitropoulos, V. Marinakis, Z. Mylona, H. Doukas, Transfer learning strategies for solar power forecasting under data scarcity. Sci. Rep. **12**(1), 14643 (2022)
22. Sarmas, E., Dimitropoulos, N., Strompolas, S., Mylona, Z., Marinakis, V., Giannadakis, A., Romaios, A., and Doukas, H. (2022b). A web-based building automation and control service. In *2022 13th International Conference on Information, Intelligence, Systems & Applications (IISA)*, pages 1–6. IEEE
23. E. Sarmas, E. Spiliotis, N. Dimitropoulos, V. Marinakis, H. Doukas, Estimating the energy savings of energy efficiency actions with ensemble machine learning models. Appl. Sci. **13**(4), 2749 (2023)
24. E. Sarmas, E. Spiliotis, V. Marinakis, M.A. Bucarelli, F. Santori, H. Doukas, Revving up energy autonomy: a forecast-driven framework for reducing reverse power flow in microgrids. Sustain. Energy,Grids Netw. **38**, 101376 (2024)
25. E. Sarmas, E. Spiliotis, V. Marinakis, T. Koutselis, H. Doukas, A meta-learning classification model for supporting decisions on energy efficiency investments. Energy Build. **258**, 111836 (2022)
26. E. Sarmas, E. Spiliotis, V. Marinakis, G. Tzanes, J.K. Kaldellis, H. Doukas, Ml-based energy management of water pumping systems for the application of peak shaving in small-scale islands. Sustain. Urban Areas **82**, 103873 (2022)
27. E. Sarmas, S. Strompolas, V. Marinakis, F. Santori, M.A. Bucarelli, H. Doukas, An incremental learning framework for photovoltaic production and load forecasting in energy microgrids. Electronics **11**(23), 3962 (2022)
28. P. Skaloumpakas, E. Sarmas, M. Rachmanidis, V. Marinakis, Reshaping the energy landscape of create through renewable energy valleys. Sci. Rep. **14**(1), 8038 (2024)
29. C. Tsolkas, E. Spiliotis, E. Sarmas, V. Marinakis, H. Doukas, Dynamic energy management with thermal comfort forecasting. Build. Environ. **237**, 110341 (2023)
30. S. Widya Yudha, B. Tjahjono, Stakeholder mapping and analysis of the renewable energy industry in Indonesia. Energies **12**(4), 602 (2019)

Part II
Advances in Energy Forecasting

Chapter 4
Deep Learning Models for Short-Term Forecasting of Photovoltaic Energy Production

Abstract This chapter examines the application of two advanced learning methodologies in addressing energy-related challenges: (i) a meta-learning methodology designed to dynamically integrate the core predictions of multiple deep learning models, and (ii) transfer learning approaches to enhance photovoltaic (PV) production forecasts. First, it is demonstrated that learning the conditions under which each model in a meta-learning scheme performs optimally enhances overall predictive accuracy. Specifically, four baseline Long Short-Term Memory (LSTM) models with distinct architectures are developed to predict short-term energy production in PV systems. The accuracy of the meta-learning model is evaluated using data from three PV systems located in Lisbon, Portugal. The results indicate that different baseline models excel with various PV systems, underscoring the potential of meta-learning to significantly improve accuracy, particularly during peak PV production periods. Conversely, the chapter illustrates that the use of transfer learning enhances PV forecasts. Specifically, the LSTM model is employed with three transfer learning strategies to improve accuracy. Transfer learning is applied for both initializing the LSTM model's weights and for feature extraction, utilizing distinct approaches for each case. These strategies are compared against a conventional learning model that does not incorporate transfer learning, as well as the widely accepted smart persistence model. The methodology is assessed in the context of forecasting hourly production for six PV systems. Results reveal that transfer learning models significantly outperform conventional models, achieving a 12.6% reduction in mean squared error and a 16.3% improvement in prediction accuracy utilizing one year of training data.

Keywords Photovoltaic power · Forecasting · Deep learning · Meta-learning · Transfer learning · Long short-term memory neural networks

E. Sarmas et al., *Artificial Intelligence for Energy Systems*, Learning and Analytics in Intelligent Systems 46, https://doi.org/10.1007/978-3-031-85209-1_4

4.1 Introduction

Solar PV power generation involves the conversion of solar energy into electricity utilizing solar collector technologies. These technologies offer a clean, silent, and cost-effective means of electricity generation, contributing to reduced carbon footprints and financial savings [35, 41, 61]. In recent years, PV energy production has experienced a surge of 22%, representing the second-largest absolute increase in renewable energy output, following wind energy [69, 82]. However, the operation of PV systems presents challenges due to significant variability in production, which is influenced by weather-dependent factors such as solar radiation and temperature [99]. This variability complicates grid management as the penetration of solar energy increases, thereby impacting electricity prices, energy market efficiency, and the operating costs of energy systems [2, 5, 15, 27, 49].

Short-term PV production forecasting, which provides estimates of production for the forthcoming hours, is essential for several reasons. Accurate predictions support optimal economic load dispatch, reducing electricity production costs [90, 105]. On a localized scale, short-term forecasts are critical for the management of smart grids and microgrids. They facilitate the efficient scheduling of energy storage systems [71], optimize load-shifting processes [92], decrease energy costs [107], manage reserves [98], and enhance the performance of PV systems through proactive measures [14]. Consequently, the quest for precise short-term PV production forecasting models has attracted considerable global research interest over the past decade [55, 57].

A notable feature of most short-term PV production forecasting models in the literature is their reliance on Numerical Weather Predictions (NWPs) [7]. These NWPs can either be obtained through advanced radiation forecasting models or purchased from meteorological services [52]. While the former approach significantly increases the computational costs associated with the development and operation of PV production forecasting models, this may be acceptable for large-scale energy producers, such as aggregators. In contrast, it presents challenges for small-scale PV owners, rooftop PV operators, who require accurate forecasts without dependency on meteorological services. In the latter scenario, where forecasting accuracy is reliant on external services, risks such as communication failures or alterations in the format and details of the provided NWPs may arise [56]. Furthermore, NWPs frequently demonstrate considerable variability over short time frames and are challenging to estimate accurately at a localized level [9]. Given these limitations, the development of short-term PV production forecasting models that solely necessitate observed energy production and weather data emerges as a viable alternative. Such models provide a more generalized, reliable, and easier-to-implement solution for energy system managers, energy traders, and other stakeholders [21].

Even without reliance on NWPs, a wide range of algorithms remains available for developing forecasting methods [2]. However, forecasting literature generally agrees that no single model consistently outperforms all others across all scenarios, and PV production forecasting is no exception [83]. To address this limitation, forecasters combine predictions from multiple models, each presenting distinct assumptions and structures, to achieve more accurate results. For instance, certain models may excel during peak production hours, while others may perform better during off-peak times, and some may more effectively handle unusual weather variations. The core principle of forecast combination is that errors from different models tend to offset each other, leading to improved overall prediction accuracy. Numerous combination schemes have been proposed, ranging from simple methods such as averaging, taking the median, or using the mode of predictions, to more advanced approaches based on forecasting errors, information criteria, Bayesian methods, and both linear and nonlinear regression.

Recently, meta-learning has gained traction as a method to incorporate external variables (e.g., time of day or weather conditions) that influence the performance of individual forecasting models and the optimal weights for combining predictions. For example, Montero-Manso et al. [59] utilized meta-learning to assign weights to the predictions of nine base forecasting models using time-series features, achieving second place in the M4 forecasting competition [50]. This chapter explores a similar method to combine the predictions of various deep learning models, which are highly effective for short-term PV production forecasting [2]. Unlike many existing meta-learning approaches that compute optimal combination weights, we propose a model-based meta-learning approach that directly computes the optimal predictions through regression, thereby advancing beyond the linear combination of base model outputs.

The key contributions of this chapter are summarized as follows:

- We design and optimize four Long Short-Term Memory (LSTM) models with distinct architectures, each capable of generating accurate deterministic forecasts for PV production over the next hour. A primary goal of this study is to establish a generalized forecasting framework suitable for scheduling in smart grid and microgrid applications. To align with this objective, the models are tailored to produce point forecasts rather than probabilistic ones, as deterministic predictions are more applicable to optimization tasks at the grid level.
- We concentrate on NWP-independent PV forecasting methods, with careful selection of input features for the LSTM models. This approach results in models that are faster to execute and more reliable than those dependent on NWPs.
- We propose a Support Vector Regressor (SVR) model that utilizes meta-learning techniques to optimally combine the predictions from the four LSTM models. This blending process leverages external variables that affect the performance of the base forecasting models. Unlike most meta-learning approaches, the meta-learner directly computes the final predictions using a nonlinear regression method, rather than relying on a weighted linear combination of the base predictions.
- We assess the forecasting accuracy of the meta-learner using representative benchmarks and a dataset from three rooftop PV systems situated in Portugal.

The structure of this chapter is organized as follows: Sect. 4.2 outlines the problem formulation and presents a brief overview of short-term PV production forecasting. Section 4.3 details the foundational forecasting models and introduces the proposed meta-learning approach. Section 4.4 provides an empirical evaluation of the method, which includes a description of the dataset, metrics and criteria employed, and a summary of the results. Finally, Sect. 4.6 concludes the chapter with a discussion of the key findings.

4.2 Problem Formulation and Related Work

This section reviews key research in the field of PV energy production forecasting. The first part categorizes PV forecasting into specific subcategories based on criteria such as forecasting horizon, employed methods, and input variables utilized by the models. The second part focuses on forecasting methods that utilize deep learning, emphasizing various representative approaches and their implementations.

4.2.1 Overview and Categorization

Methods for forecasting PV energy production can be categorized based on various criteria, including the forecasting horizon (very short-term, short-term, medium-term, long-term), the nature of the method employed (physical, data-driven, machine learning/deep learning), and their dependence on numerical weather predictions (dependent, independent).

The forecasting horizon, defined as the future time span for which predictions are generated, has been categorized in various ways within the literature. The predominant classification divides forecasting methods into short-term, medium-term, and long-term categories [13]. Nonetheless, as noted by Gupta and Singh [26] and Ahmed et al. [2], an additional category-very short-term PV energy forecasting-can also be established, leading to the following classifications:

- **Very short-term forecasting methods** generate predictions for timeframes ranging from a few seconds to several minutes ahead. These methods are commonly applied in smart grid energy management, such as for the rapid and efficient adjustment of active and reactive power [6, 99].
- **Short-term forecasting methods** produce predictions covering timeframes from 1 h to 1 week. These methods are well-suited for smart grid applications, supporting power unit commitment, scheduling, and dispatch [13], as well as for applications like dynamic pricing and predictive maintenance [85].
- **Medium-term forecasting methods** generate predictions for timeframes ranging from 1 month to 1 year. These methods are primarily used for optimal grid planning and maintenance scheduling [2].
- **Long-term forecasting methods**, on the other hand, provide predictions over periods of 1 year to several years and are valuable for planning energy transmission, distribution, and storage [18].

Another key factor distinguishing PV energy production forecasting approaches is the type of method or algorithm employed to generate forecasts and its reliance on actual measurements. Broadly, these approaches can be classified into two main categories: physical methods and data-driven methods.

Physical methods utilize detailed meteorological data and fundamental design parameters of PV panels, applying physical equations to predict PV energy production [54], as demonstrated in the work of Zhao et al. [115], Zameer et al. [113], and Lei et al. [43]. These methods are generally more appropriate for long-term forecasts. Conversely, data-driven methods analyze patterns and correlations in weather and PV production data to generate predictions [2], rendering them better suited for short-term forecasting.

Data-driven methods can be further categorized into purely statistical approaches and machine learning (ML) techniques [75]. For instance, Sharadga et al. [80] employed statistical models such as ARMA, ARIMA, and SARIMA for forecasting PV production. More recently, Haider et al. [28] compared the performance of SARIMAX and Prophet models against various neural network (NN)-based methods in solar radiation forecasting. The study concluded that statistical models are more effective for extended forecasting horizons, while neural networks demonstrate superior performance in shorter timeframes. Similarly, Akhter et al. [3] highlighted the advantages of various ML models, including neural networks, decision trees, random forests, gradient boosting, and support vector regression (SVR). In addition to physical and data-driven methods, hybrid methods integrate characteristics from both approaches. For a comprehensive comparison of physical, data-driven, and hybrid methods, refer to Mayer [53].

Another distinguishing feature of forecasting approaches is their reliance on Numerical Weather Predictions (NWPs). Alongside the development of NWP-independent methods, including the one proposed in this study, there exist models that depend on NWPs. These models incorporate meteorological forecasts to enhance prediction accuracy, given the robust relationship between PV production and weather conditions. For example, Ogliari et al. [60] and Dolara et al. [16] examined various neural network models, such as Physical Hybrid Artificial Neural Networks (PHANN), for day-ahead forecasting. A comprehensive comparison of NWP-based PV energy production forecasting methods was conducted by Markovics and Mayer [51].

Conversely, NWP-independent methods do not rely on NWPs as input features, which renders them faster and simpler to implement while maintaining a high level of accuracy for short-term forecasts. Wang et al. [101] propose an alternative terminology for this distinction, referring to NWP-dependent and NWP-independent approaches as "indirect" and "direct" forecasting methods, respectively.

4.2.2 Deep Learning-Based Methods

Deep learning (DL), a specialized branch of ML, has exhibited remarkable performance in PV energy production forecasting, frequently surpassing traditional ML and statistical models. This advantage is attributed to the capability of DL models to discern complex patterns within high-dimensional data [86, 97] and to provide more reliable results in uncertain conditions [39]. Over time, DL architectures have evolved to become increasingly sophisticated, utilizing multilayer neural networks that effectively capture nonlinear and intricate relationships between input features and output variables [58].

In addition to advancements in the types, numbers, and combinations of units within DL architectures, the learning mechanisms of these models have also evolved. For instance, novel combinations such as transfer learning with constrained long short-term memory have been applied to forecast energy production in newly deployed PV systems [48]. Contemporary DL architectures frequently incorporate convolutional neural networks (CNNs), which are proficient at detecting patterns and structures in image data, and recurrent neural networks (RNNs), which excel in analyzing time-series data characterized by temporal relationships. Furthermore, CNNs and RNNs are often integrated with feedforward neural networks (FFNNs) by stacking dense layers on their outputs, thereby enabling more precise predictions.

The following paragraphs highlight some of the most promising DL methods, along with key results and insights. However, direct quantitative comparisons between different forecasting methods remain challenging due to the substantial influence of the quality and resolution of available historical data on their accuracy [77].

LSTM networks, a specialized type of recurrent neural network (RNN), have been extensively utilized in PV energy forecasting research. LSTMs incorporate nonlinear, data-dependent controls within the RNN cell to mitigate the vanishing or exploding gradient problem during training. For example, Wen et al. [105] employed a deep LSTM model to predict hourly PV production over short-term horizons using scheduling information, weather data, and time-scale variables, with the objective of optimizing load allocation in a community microgrid. The results indicated that LSTMs outperformed other ML algorithms in terms of accuracy.

Similarly, Lee et al. [42] developed a model that integrated Convolutional Neural Networks (CNN) and LSTM networks to forecast daily PV system production based on observations from the previous day. Wang et al. [102] also proposed a hybrid CNN-LSTM model and conducted a comparative analysis of its performance against standalone CNN and LSTM models. The findings indicated no definitive superiority; CNNs exhibited greater accuracy for extended forecasting horizons, whereas LSTMs demonstrated superior performance for specific time intervals. In an alternative approach, Song et al. [84] categorized PV energy production into four distinct tasks: residential, agricultural, industrial, and commercial. Furthermore, they introduced a multitasking recurrent neural network (RNN) capable of transferring and reusing acquired knowledge across these tasks.

Gated recurrent units (GRUs) constitute a distinct category of DL models that have recently been employed for short-term PV energy forecasting. Like LSTM networks, GRUs are engineered to manage sequential data; however, they utilize fewer parameters, which leads to a decrease in training time [103]. Expanding on the capabilities of GRU models, Abdel-Basset et al. [1] and Rai et al. [67] proposed advanced versions of these networks.

Abdel-Basset et al. [1] modified the traditional gated recurrent unit (GRU) cell by incorporating expanded couplings and residual connections, which enabled the network to capture spatiotemporal patterns in PV data more effectively. This approach demonstrated enhanced accuracy relative to other DL models, including the convolutional neural network (CNN) and LSTM models from Wang et al. [102], the LSTM model from Sharadga et al. [80], and the GRU model from Lee and Kim [40]. In a similar vein, Rai et al. [67] introduced a GRU-based autoencoder and assessed its performance against LSTM, CNN, GRU, and hybrid models over forecasting horizons of 24 h, 48 h, and 15 days. The results suggested that the GRU-based autoencoder represents a promising method for PV energy forecasting.

Convolutional neural networks (CNNs) have gained prominence in PV energy forecasting, primarily due to their capacity to process data with network topology features and efficiently extract latent structures and inherent characteristics [2]. For instance, Zang et al. [114] applied widely used CNN architectures, including ResNet [residual network; 29] and DenseNet [dense convolutional network; 33], to forecast daily PV energy production. They compared these models against statistical, physical, ML, and standard CNN approaches, reporting encouraging results. Similarly, Aslam et al. [6] conducted a comprehensive comparative study, concluding that DL models, particularly hybrid DL approaches, consistently deliver superior accuracy.

Ensembling, which entails the integration of predictions from multiple independent base forecasting models, has demonstrated considerable effectiveness in short-term PV energy production forecasting [2]. This methodology can involve either the combination of diverse models or multiple instances of an identical model with varying parameter configurations. Illustrative examples include the simple mean ensemble utilized by Zhu et al. [117] and the weighted mean ensemble model advocated by Liu et al. [45].

Beyond mean-based ensembling, some studies have explored meta-learning techniques for PV forecasting. Meta-learning employs a ML or DL model to amalgamate base predictions, rather than relying on traditional operators such as the mean, median, or mode. For instance, Lateko et al. [38] introduced a recurrent neural network (RNN)-based meta-learner to integrate the predictions of five base models. Similarly, Eom et al. [18] developed a convolutional neural network (CNN)-based meta-learner that selected features and combined the predictions of two statistical models and a LSTM network, demonstrating its effectiveness in enhancing forecast accuracy.

Thus, the increasing availability of solar panel electricity production data has established DL models as the predominant approach for forecasting PV energy production. In particular, LSTM networks are well-suited for multivariate forecasting, excel at capturing complex patterns in sequential data, and are highly effective in generating multi-step predictions. However, as this review section indicates, current solutions have yet to fully harness the potential of DL. Further research is necessary to explore the integration and comparison of various DL architectures to enhance forecasting accuracy. Moreover, few studies have concentrated on developing ensemble models that leverage the strengths of different DL architectures, thus presenting an opportunity for future advancements.

4.3 Proposed Methodology

This section outlines the proposed method. It commences with an introduction to the overall methodological framework. This is followed by a comprehensive explanation of the LSTM unit, which serves as the fundamental component of all neural network architectures employed in developing the base forecasting models. Finally, the architectures of both the base forecasting models and the meta-learner are described in detail.

4.3.1 Model Architecture

The proposed meta-learning method (Meta) consists of three essential phases: data preparation, training of base forecasting models, and training of the meta-learner. In the first phase, raw data are collected and transformed into a structured format that is suitable for input into both the base forecasting models and the meta-learner. The second phase entails the training of the base forecasting models using the processed data, which generates base predictions (level-0) as output. Finally, in the third phase, the meta-learner, as described in Sect. 4.3.4, is trained utilizing the base predictions in conjunction with additional features as input to produce the final predictions (level-1). These phases are illustrated in Fig. 4.1.

The data preparation phase consists of two primary steps. The first step involves selecting the features to be utilized as predictor variables for the base forecasting models outlined in Sect. 4.3.3, as well as the inputs for the meta-learner, which aims to achieve the "optimal" combination of base predictions. In the context of short-term PV energy production forecasting, these features may include autoregressive variables, which consist of past PV production observations, weather-related features such as past solar radiation and temperature data, seasonal variables like the time of day and month, and other relevant factors that can enhance forecast accuracy. While, in theory, the features for the base models and the meta-learner could differ, they are the same in this study for the sake of simplicity.

The second step involves processing raw data to prepare it for training foundational machine learning models, as discussed in Sect. 4.3.3. This step may encompass

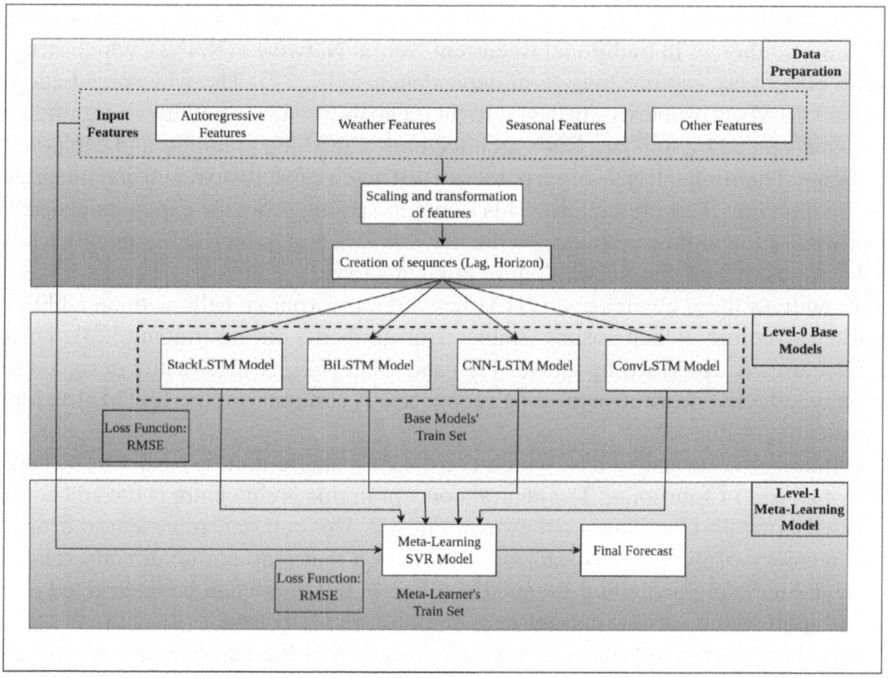

Fig. 4.1 The proposed meta-learning framework

tasks such as data cleaning, preprocessing, and the application of necessary transformations. Further details regarding the data preparation phase, tailored to the specific case study, are presented in Sect. 4.4.

The base models were implemented using the TensorFlow Keras library for Python, version 2.6.0 [11], while the meta-learner was developed using the Scikit-learn library for Python, version 1.0 [65].

4.3.2 The LSTM Model

Recurrent Neural Networks (RNNs) constitute a specialized class of neural networks designed to manage data characterized by temporal dependencies. In RNNs, the output from a specific time step is reintegrated as input for the subsequent time step, facilitating the model's ability to make predictions based on both the current input and the outputs from preceding time steps. Among the most prevalent RNN variants is the LSTM network. LSTMs incorporate internal memory units that enable the model to preserve and accumulate state information over time, rendering them particularly well-suited for time-series forecasting, where relationships of varying durations exist between observations and significant events.

The development of LSTM networks was driven by the necessity to address the limitations inherent in traditional Recurrent Neural Networks (RNNs), which struggle to effectively capture long-term dependencies [17, 72]. The widespread adoption of LSTMs is primarily attributed to their capability to surmount two significant challenges faced by standard RNN architectures: vanishing gradients and exploding gradients. Traditional RNNs often possess a restricted capacity to retain and integrate information into their predictions. This limitation results from the exponential decay or growth of the input's influence on the network's output, which subsequently leads to the issues of vanishing or exploding gradients [31].

To address these challenges, LSTM networks incorporate cells with an additive gradient structure, which ensures stable error gradients during training [23]. Their architecture, as described by Graves [22] and Olah [62], is specifically designed to maintain effective learning over long sequences. A summary of the LSTM structure is presented in Fig. 4.2.

Unlike conventional RNNs, which consist of a single unit, typical LSTM cells are composed of four units. The central concept in this architecture is the cell state, depicted as the horizontal line at the top of Fig. 4.2. This cell state represents the long-term memory of the network, storing information that may not be directly related to the immediately preceding events. Each line in the figure can be interpreted as a vector upon which various pointwise operations are performed.

In addition to the cell state, LSTM cells include a **hidden state** that serves as the network's working memory, retaining information from recent events. The hidden state is updated at each time step and interacts dynamically with the cell state to enhance learning. An LSTM cell receives three inputs and generates two outputs. The inputs, presented in vector form, are the current input x_t, the previous hidden

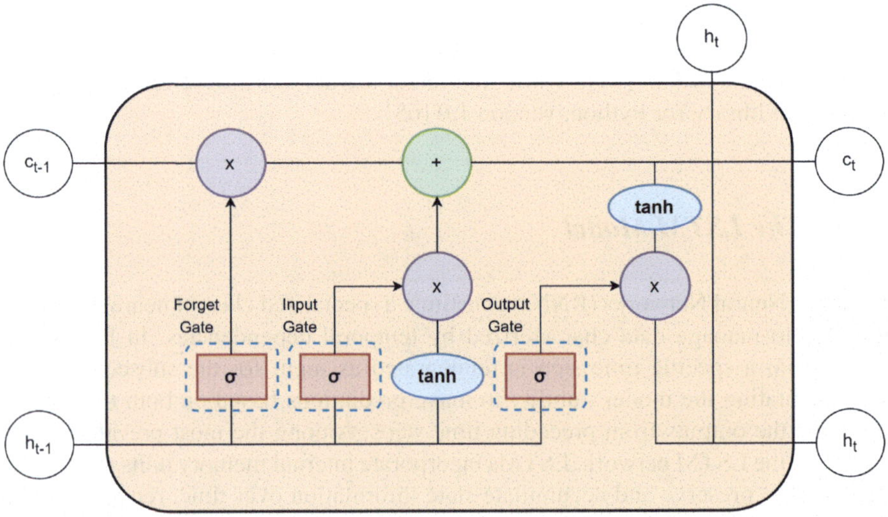

Fig. 4.2 LSTM overview

state h_{t-1}, and the previous cell state c_{t-1}. The fundamental operation of an LSTM is regulated by its three gates: the forget gate, the input gate, and the output gate.

The **forget gate** is the initial component in the LSTM architecture. Its primary function is to determine which parts of the information should be retained or discarded. It receives the previous hidden state h_{t-1} and the current input x_t as inputs, processing them through a sigmoid activation function σ_g. The sigmoid function generates outputs ranging from 0 (indicating complete forgetting) to 1 (indicating full retention), yielding the activation vector of the forget gate, f_t, as:

$$f_t = \sigma_g(W_f x_t + U_f h_{t-1} + b_f). \tag{4.1}$$

The input gate is responsible for updating the cell state. It operates in two stages. In the first stage, the previous hidden state and the current input are processed through a second sigmoid activation function σ_g. In the second stage, the same inputs are passed through the hyperbolic tangent function σ_c to adjust the network parameters. Ultimately, the cell state vector c_t is derived from the element-wise product of the cell's input activation vector $\tilde{c}_t$ and the activation vector of the update gate i_t. The activation functions of the input gate are represented by the following set of equations:

$$\tilde{c}_t = \sigma_c(W_c x_t + U_c h_{t-1} + b_c) \tag{4.2}$$

$$i_t = \sigma_g(W_i x_t + U_i h_{t-1} + b_i) \tag{4.3}$$

$$c_t = f_t \circ c_{t-1} + i_t \circ \tilde{c}_t \tag{4.4}$$

Finally, the output gate determines the subsequent hidden state h_t. The hidden state encapsulates information regarding prior inputs and is utilized for making predictions. The preceding hidden state h_{t-1} and the current input x_t are transmitted to the third sigmoid function σ_g. Subsequently, the modified cell state is processed through the hyperbolic tangent function σ_h. These outputs are multiplied element-wise, enabling the network to ascertain the information that the hidden state should retain.

$$o_t = \sigma_g(W_o x_t + U_o h_{t-1} + b_o) \tag{4.5}$$

$$h_t = o_t \circ \sigma_h(c_t) \tag{4.6}$$

Parameters $W \in \mathbb{R}^{h \times d}$, $U \in \mathbb{R}^{h \times h}$, and $displaystyleb \in \mathbb{R}^h$ in the preceding equations represent weight matrices and bias vector parameters, respectively, which are acquired during the training process. It is important to note that this description illustrates the functioning of a single LSTM unit.

4.3.3 Baseline Forecasting Models

Below, we present the neural network (NN) architectures of the four LSTM models examined for the development of the meta-learner. While the overall architecture of each model is initially general, the final architectures were selected for our experiments through a grid search optimization on key hyperparameters of the networks, including the number of layers, the number of LSTM cells per layer, and the number of filters (for CNNs).

Typical LSTM model architectures comprise a single fully connected layer of LSTM units, followed by a feedforward output layer. In this study, a stacked LSTM model (**StackLSTM**) is employed, which features multiple fully connected layers, each containing several LSTM cells. These layers are arranged sequentially, forming a deeper architecture capable of effectively handling more complex learning tasks [8]. Similar to feedforward neural networks (FFNNs), where stacking multiple layers enables a hierarchical representation of data features for model training, the same principle applies to deep LSTM models. Stacking LSTM layers allows the model to build upon representations learned by earlier layers and capture more intricate patterns, thereby adding a higher level of abstraction. As noted by Hermans and Schrauwen [30], each layer addresses a portion of the prediction problem and passes its intermediate output to the following layer. This architecture functions as a processing pipeline, with each layer contributing to the overall solution until the final layer produces the complete prediction. This structure equips the model to identify complex patterns and structures in time-series data efficiently.

In the StackLSTM model utilized in this study, each LSTM layer generates a sequence of vectors that serve as input for the subsequent LSTM layer. A single LSTM layer processes a three-dimensional input and yields a two-dimensional output, which consists of the hidden state from the final time step, thereby encoding the conclusion of the sequence. Nonetheless, the following LSTM layer also necessitates a three-dimensional input. To address this requirement, the output from the preceding LSTM layer, which includes the hidden states from all prior time steps, is transmitted to the next layer. This approach guarantees that all layers receive a three-dimensional input, thereby enabling the model to construct a hierarchical representation of the data across multiple layers [64].

The StackLSTM architecture employed in this study is illustrated in Fig. 4.3. Key parameters are defined as follows: *batchsize*, representing the number of training samples before updating the model's parameters; *numtimesteps*, denoting the number of prior time steps provided to the model; and *numfeatures*, indicating the length of the feature sequence (i.e., the number of model features). During the model fine-tuning process, two hyperparameters were evaluated: the number of layers, which could be 1, 2, or 3, and the size of each layer, with potential values of 24, 48, 96, or 192 cells. The final architecture comprises three LSTM layers containing 24, 48, and 96 cells, respectively.

In many forecasting tasks, it is advantageous to develop LSTM models that can learn temporal relationships in both directions: forward (from past to future) and

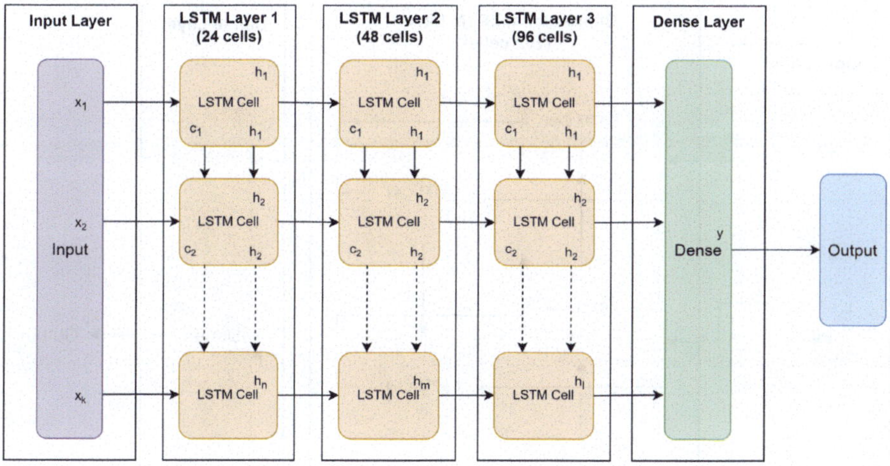

Fig. 4.3 The proposed architecture of the stacked LSTM (StackLSTM)

backward (from future to past). The primary benefit of this approach is that by combining both perspectives, predictions can leverage sequential information from data points preceding and following the target [25]. Such models are commonly referred to as bidirectional Recurrent Neural Networks (RNNs) [79], or bidirectional LSTMs **(BiLSTM)** when the layers comprise LSTM cells.

The key architectural distinction between BiLSTMs and standard LSTMs is that BiLSTMs incorporate two parallel layers that process data in both forward and backward directions [112]. While the forward and backward passes are performed similarly to those in regular neural networks, the dual layers in BiLSTMs enable the model to retain information from both temporal directions [100]. This unique feature provides BiLSTMs with a significant advantage, allowing them to overcome the limitation of conventional LSTMs, which can only utilize preceding context. Consequently, BiLSTMs have been effectively applied to various forecasting tasks, including speech recognition [73] and sentiment analysis [10], thereby making them highly suitable for time-series forecasting.

The configuration of the BiLSTM model was established based on the number of bidirectional LSTM layers and the number of cells per layer. Experiments were performed with 1, 2, and 3 layers, and the number of cells per layer was evaluated with values of 24, 48, 96, and 192. The ultimately selected architecture, as illustrated in Fig. 4.4, comprises a single bidirectional LSTM layer containing 192 cells.

A hybrid approach to time-series forecasting integrates convolutional neural networks (CNNs) with LSTM networks. CNNs, a specialized category of neural networks, are conventionally employed for tasks such as image analysis, facial recognition, image search, and augmented reality [44]. They can be regarded as regularized

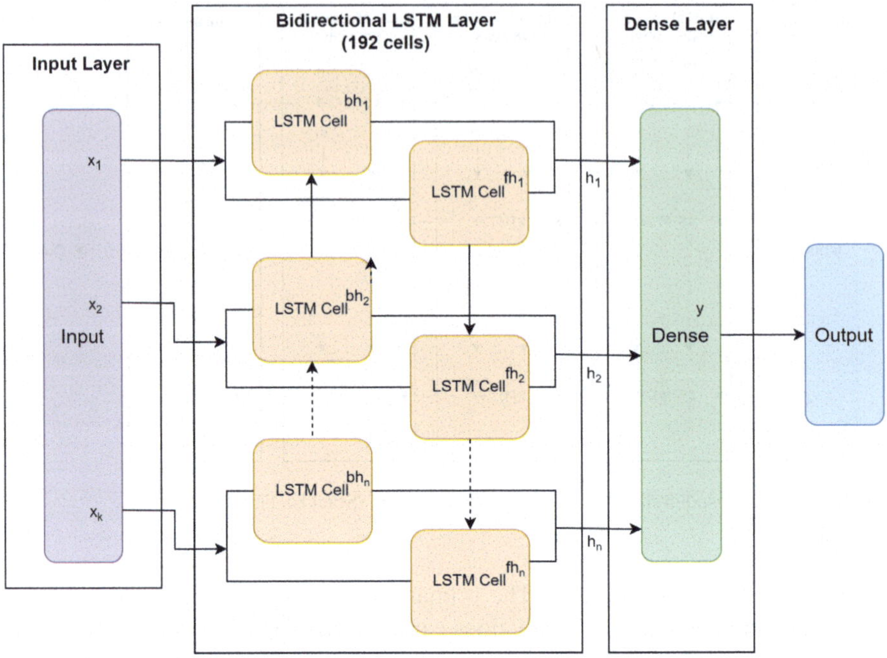

Fig. 4.4 The proposed architecture of the bidirectional LSTM (BiLSTM)

variants of multilayer perceptrons, specifically designed to capture hierarchical patterns in data by constructing progressively complex representations from simpler ones through the use of filters.

Convolutional layers offer the distinct advantage of learning internal representations and patterns within sequential data. One-dimensional convolutional neural networks (CNNs) have demonstrated considerable efficacy in extracting features from input sequences. The combination of CNNs with long short-term memory networks (LSTMs) capitalizes on the strengths of both architectural paradigms [46]. In this hybrid model, referred to as CNN-LSTM, convolutional layers extract patterns from the data, while LSTM layers utilize this information to generate predictions. Each convolutional layer operates as an encoder, learning the structure of smaller subsequences within the input data. These encoded subsequences are subsequently transmitted as time steps to the LSTM layer for further processing and prediction.

The CNN-LSTM model architecture is composed of two primary components. The first component includes convolutional and pooling layers that identify and extract features from the input data, serving as preprocessing layers to relay information to the fully connected network [68]. Typically, a convolutional layer is succeeded by a nonlinear activation function and a max pooling layer. The pooling layer employs a subsampling technique, extracting specific values from the convolved features to generate a lower-dimensional representation. Max pooling functions similarly to convolution, utilizing a sliding window to analyze sections of the convolved features and produce the maximum value from each section [68]. The second component consists of fully connected LSTM layers, which capture long-term dependencies within the extracted features. These layers feed into a dense layer that produces the final prediction.

From a technical perspective, the input sequences must be divided into two-dimensional subsequences for the CNN to process effectively. For time-series data with k time steps as input and one output, each input sample is split into $\frac{k}{2}$ subsequences, each containing two time steps. This division allows the CNN to process and interpret each subsequence individually before passing the results to the LSTM model for further analysis and prediction.

The model's finely tuned parameters included the number of filters (tested values: 32, 64, 128; selected value: 64), the number of LSTM layers (tested values: 1, 2, 3; selected value: 2), and the number of LSTM cells per layer (tested values: 25, 50, 100; selected values: 200 for the first layer and 100 for the second layer). The architecture of the finalized CNN-LSTM model is illustrated in Fig. 4.5.

In the CNN-LSTM architecture, the CNN captures spatial features through convolutional operations, while the LSTM learns sequential features using its memory units. Convolutional LSTM (ConvLSTM) combines the strengths of both architectures into a unified design by embedding convolutional operations directly within the LSTM cells, providing a more compact and efficient implementation. Unlike traditional CNN-LSTM models that rely on matrix multiplication, ConvLSTM applies convolutions to both hidden states and cell states [81], enabling the model to process input data convolutionally within each LSTM unit. While primarily designed for two-dimensional spatiotemporal data, the ConvLSTM architecture is also adaptable for multivariate time series forecasting.

As with the CNN-LSTM model, ConvLSTM requires preprocessing of the time series data before input. Each input sample of k time steps is divided into

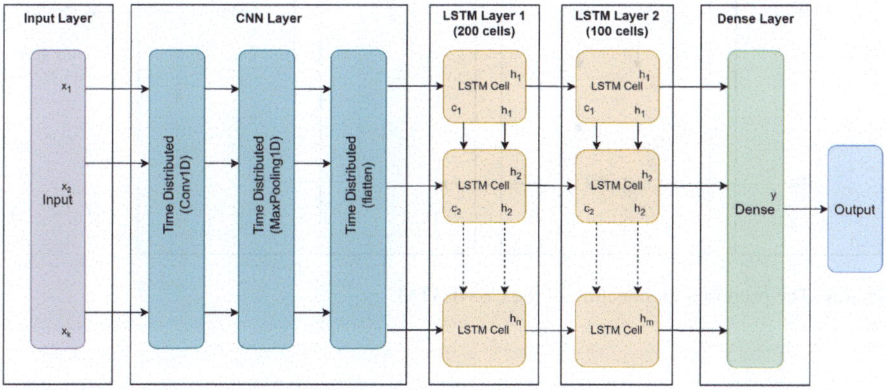

Fig. 4.5 The proposed architecture of the CNN-LSTM

$\frac{k}{2}$ subsequences, each containing two time steps. This preprocessing results in a two-dimensional input format with $\frac{k}{2}$ rows and two columns for each sample.

The only parameter that was finely tuned for the ConvLSTM model was the number of filters in the convolutional LSTM layer. Tested values included 16, 32, 64, and 128, with 64 selected as the optimal value. The finalized ConvLSTM model architecture is depicted in Fig. 4.6.

In summary, the four forecasting models discussed above have demonstrated strong predictive capabilities in PV energy production forecasting, each with distinct strengths and limitations. The Stacked-LSTM excels in processing sequential data, such as time series, effectively capturing long-term dependencies when deployed in deep architectures. The Bi-LSTM is adept at capturing relationships in both past and future data but may be less effective than the Stacked-LSTM for longer forecasting horizons. The CNN-LSTM shares the sequential data processing capabilities of the Stacked-LSTM while incorporating filters to extract and process more complex patterns. Lastly, the ConvLSTM is particularly suited for handling spatiotemporal data and can be adapted effectively for univariate time series forecasting. Given the inherent complexities of PV energy production forecasting, no single model is likely to consistently outperform others across all scenarios. Therefore, the next section introduces a meta-learner designed to combine the strengths of these four core models.

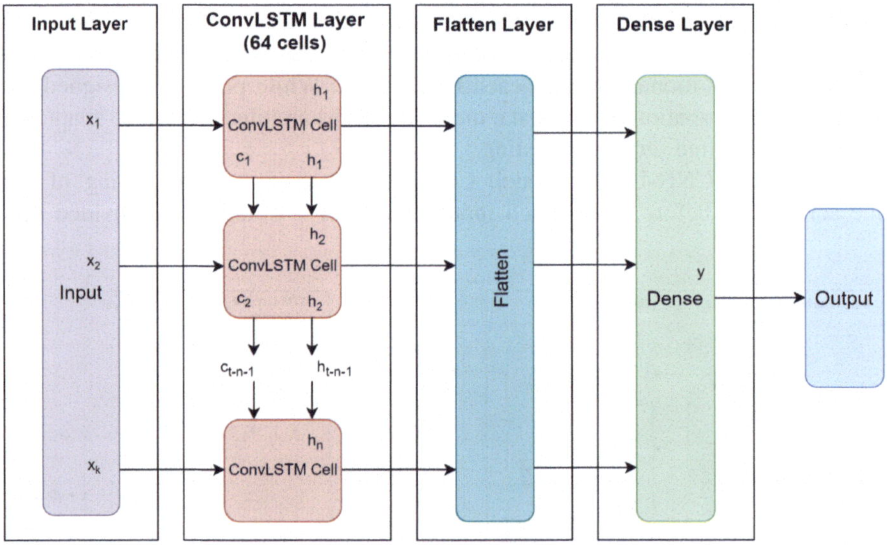

Fig. 4.6 The proposed architecture of the ConvLSTM

4.3.4 Meta-Learning Model

Meta-learning, also known as stacking, is the process of "learning to learn." It involves assigning a meta-model (MM) the task of enhancing the accuracy of a specific forecasting problem by using the predictions of other forecasting models as its base inputs (meta-data) [96]. This methodology is particularly beneficial in forecasting tasks where diverse predictive models, each with distinct characteristics, structures, and data assumptions, are likely to exhibit varying performance under different conditions [76]. For example, in PV energy production forecasting, certain models may excel during the summer when weather conditions are more stable but underperform in winter. Similarly, some models might be more accurate during peak production hours but less reliable during periods of low production.

While the final predictions (level-1 predictions) generated by the meta-learner may involve a combination of base predictions (level-0 predictions), whether linear or nonlinear, meta-learning extends beyond conventional prediction combination methods. Unlike traditional approaches that rely on the average performance of base models, meta-learning focuses on identifying the specific conditions under which an individual model or a combination of models performs optimally [94]. By learning these conditions, the meta-learner can combine predictions in a more efficient, data-driven manner, resulting in enhanced generalization [106]. As noted by Hospedales et al. [32], meta-learning also provides solutions to numerous challenges faced by traditional meta-modeling, including those related to data limitations and computational complexity.

The meta-learner is implemented using the Support Vector Regression (SVR) method [78]. SVR is a widely utilized meta-modeling algorithm that conducts regression in a manner analogous to traditional linear regression, wherein the predictive line is estimated as:

$$\hat{y} = \mathbf{W}^T \mathbf{x} + b, \tag{4.7}$$

where $\hat{y}$ is the target variable, $\mathbf{x}$ is the set of features utilized as predictor variables, $\mathbf{W}$ is a weight vector, and b is an intercept that accounts for bias. However, Support Vector Regression (SVR) employs a distinct method to compute the prediction line, referred to as a hyperplane. The data points located on either side of the hyperplane, which have the shortest distance to it, are termed support vectors. Traditional regression methods strive to minimize the error between actual values and predictions. In contrast, the SVR method seeks to fit the optimal line within a defined threshold value d, which delineates the distance between the hyperplane and the support vectors. Thus, the SVR method adheres to the condition $-d < y - \mathbf{W}^T \mathbf{x} + b < d$, utilizing points within this margin to generate predictions.

In this study, the Radial Basis Function (RBF) kernel is employed in Support Vector Regression (SVR) to map the sample data into a high-dimensional feature space, thereby facilitating nonlinear regression. SVR was selected as the meta-learner over other machine learning models due to its ability to combine the predictions of base models in a weighted manner while effectively capturing nonlinear relationships among features. Other commonly utilized options for meta-learning include linear regression, tree-based regression, and neural networks (NNs), each of these alternatives presents limitations when compared to SVR.

For instance, while linear regression is a straightforward and intuitive method for weighted prediction, its capability to model nonlinear relationships among the features is inherently limited. Tree-based models, such as gradient boosting trees and random forests, cannot directly integrate level-0 and level-1 predictions, since their final outputs are based on averages within nodes that adhere to specific decision rules. Neural networks, although highly effective for nonlinear regression tasks, typically necessitate substantially larger training datasets than Support Vector Regression (SVR) to operate effectively.

Furthermore, the proposed meta-learner employs a nonlinear regression method to generate final predictions, thereby enhancing the flexibility and generality of its mixing mechanism compared to standard meta-learners. Traditional meta-learners generally estimate "optimal" combination weights for base predictions and rely on a weighted average, whereas the SVR-based approach provides a more adaptable solution.

4.4 Case Study

This section delineates the case study conducted to evaluate the predictive performance of the proposed meta-learning method. It includes a comprehensive description of the dataset, the experimental design, the metrics utilized to assess prediction accuracy, and the forecasting methods selected as benchmarks. The results are presented in the concluding segment of this section.

4.4.1 Dataset

The dataset consists of three PV systems situated in Lisbon, Portugal, designated as "Joao", "Julia Moreira", and "Zarco". These systems are mounted on public buildings: a social solidarity institution, a dental unit, and an elementary school, with corresponding capacities of 40 kW, 27 kW, and 50 kW, respectively.

The dataset comprises hourly data covering a total of 21,958 h, equivalent to 2 years and six months, from August 1, 2018, to January 31, 2021. As anticipated, electricity production typically peaks between 12:00 and 15:00, coinciding with maximum solar radiation during daylight hours. Conversely, production is significantly lower during the early morning hours (09:00 to 11:00) and late afternoon (16:00 to 18:00). Production is absent at night due to the lack of solar radiation.

Nevertheless, PV production exhibits considerable variability across different days, contingent upon weather conditions. Furthermore, the distributions of the three

PV systems are observed to be analogous, as the installations are situated in proximity to one another and possess comparable orientations. Consequently, for the remainder of this subsection, the analysis will concentrate on the subset of data derived from the "Joao" PV system.

In addition to electricity production records measured in kilowatts (kW) for each park, the dataset comprises observed weather conditions, including solar radiation (measured in W/m^2), diffuse solar radiation (measured in W/m^2), and temperature (measured in °C). These meteorological variables are recorded on-site by weather stations installed at the PV parks and reflect the average conditions for each corresponding hour.

Figure 4.7 illustrates the distributions (histograms) of electricity production at the "Joao" PV park alongside three weather variables, as well as their pairwise correlations. The analysis reveals a strong correlation between PV production and both

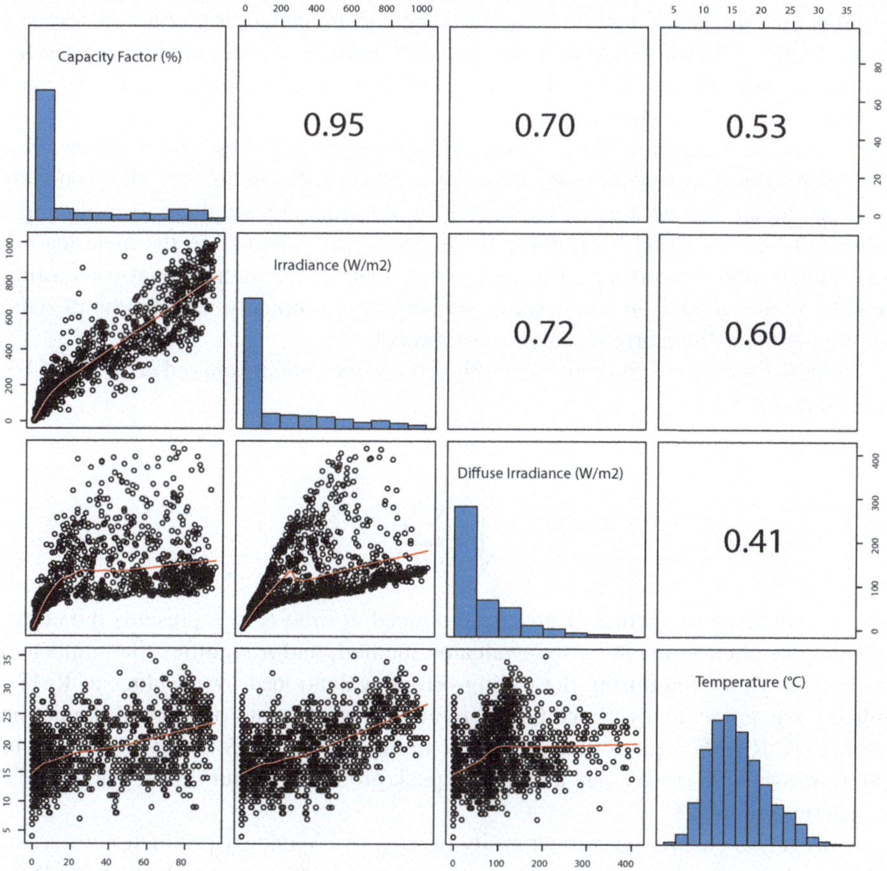

Fig. 4.7 Pairwise correlation among PV production and weather features for "Joao" PV park

solar radiation and diffuse solar radiation, with temperature also exhibiting a positive influence. Consequently, all meteorological variables are identified as essential inputs for forecasting PV production.

4.4.2 Experimental Modeling

To ensure that the results are representative and accurately reflect the dynamics of different time periods and varying weather conditions, we employed the rolling-origin evaluation approach [91]. This method, analogous to cross-validation for time-series data, initiates by utilizing a historical data period to train the forecasting method under evaluation. Subsequently, the method is applied to forecast PV production for the next hour (one-step-ahead prediction), with the accuracy of the forecast assessed against the observed production using a selected performance metric. The prediction origin is then advanced by one hour, allowing for the generation of a new forecast for evaluation. This iterative process continues until no additional data remain for evaluation, with the overall performance of the method calculated as the average accuracy across all evaluations.

In this case study, the initial year of data, comprising 8,760 observations, was utilized to train the foundational forecasting models. Subsequently, these models were employed to generate predictions for the subsequent year, thereby producing a sufficiently large dataset for training the meta-learner. Thereafter, the meta-learner was trained, and its performance, along with that of the foundational forecasting models, was assessed on a test set containing approximately six months of data, employing the rolling-origin evaluation approach.

For performance evaluation, we employed the root mean squared error (RMSE), defined as follows:

$$\text{RMSE} = \sqrt{\frac{1}{n}\sum_{t=1}^{n}(y_t - \hat{y}_t)^2},$$

where y_t denotes the actual electricity produced at time t, $\hat{y}_t$ represents the corresponding prediction made by the evaluated method, and n signifies the number of predictions generated during the rolling-origin evaluation. Lower values of RMSE indicate greater accuracy. Since the mean minimizes the sum of squared prediction errors [37], RMSE is particularly effective in identifying models that provide accurate forecasts both on average and during peak production hours, which are critical for energy producers.

However, to ensure the comparability of solar or PV energy production forecasting models across different locations and seasons [109], we supplement the Root Mean Square Error (RMSE) with the Forecast Skill Index (FSI) as an additional performance measure [108]. The FSI assesses the relative performance of a forecasting model by comparing it to a reference model, employing RMSE as the primary accuracy metric. For the reference model, we utilize the persistence model, which posits that future values (predictions) will mirror the current observed values (actuals) [74]. The persistence model is widely recognized as the standard baseline in the solar energy forecasting literature for validating model performance. The Forecast Skill Index is defined as follows:

$$\text{Forecast Skill Index} = 1 - \frac{\text{RMSE}_{proposed}}{\text{RMSE}_{reference}} \tag{4.8}$$

where $RMSE_{proposed}$ and $RMSE_{reference}$ represent the accuracy estimates of the proposed forecasting models and the persistence forecasting models, respectively, based on the RMSE metric.

We evaluate the forecasting accuracy of the four base models and the meta-learner discussed in Sect. 4.3. The successful implementation of the proposed methods significantly depends on the selection of appropriate features that yield meaningful input for the forecasting algorithms. Building on the insights from Sect. 4.4.1 and concentrating on the development of NWP-independent models, we employ the following set of features:

- The solar radiation intensity recorded by the weather stations over the previous six hours.
- Features representing seasonal and calendar variations in PV production, specifically the month of the year and the time of day. The month feature is encoded using one-hot encoding, while the time of day is transformed using sine and cosine functions. Like solar radiation, these seasonal and calendar features are derived for the previous six hours.
- Previous PV production observations (lags 1–6) to capture potential autocorrelations.
- A categorical variable representing the weather profile over the past six hours. The weather profile for each hour is determined by clustering dataset observations using solar radiation, diffuse solar radiation, and temperature as input variables for the k-means algorithm, which generates three clusters. The algorithm's output is then one-hot encoded, resulting in three binary features. Table 4.1 provides an

Table 4.1 Weather profiles created for the "Joao" PV plant. For each of the three profiles the average and the standard deviation (in parenthesis) of the variables used for creating the clusters are reported

Cluster	Irradiance	Diffuse irradiance	Temperature
1	709.95 (138.26)	168.75 (84.87)	23.34 (4.49)
2	316.18 (117.30)	129.78 (59.36)	17.76 (4.11)
3	13.13 (33.69)	8.01 (19.06)	14.94 (3.88)

overview of the information conveyed by the clusters to the forecasting models. As depicted, the first two clusters predominantly correspond to hours with high and medium solar radiation, respectively, while the third cluster represents hours with low radiation and temperature.

In total, the forecasting methods employ 19 features $\times$ 6 lags = 114 inputs, which are scaled to a range between zero and one to improve training efficiency and enhance generalization.

For the meta-learner to improve prediction accuracy, its forecasts must exceed those of the four base models contributing to the final predictions. Achieving this would indicate that no single base model is, on average, more accurate than the "optimal" predictions generated by the meta-learning method. Consequently, the four base models described in Sect. 4.3 serve as natural benchmarks for this study.

Moreover, as discussed in Sect. 4.1, forecast ensembles often produce more accurate results than their individual components. This is particularly true for simple combinations based on the mean or median of base predictions [50]. To assess whether the meta-learner provides additional value compared to such methods, a benchmark based on the equal-weighted combination of the four base models (Equal Weight - EW) is also included in the evaluation.

4.4.3 Results

Table 4.2 presents the overall accuracy of the meta-learner, the four base forecasting models, and the EW benchmark for each PV system, evaluated using RMSE and the Forecast Skill Index.

Table 4.2 Overall accuracy (RMSE and forecast skill index) per forecasting method and PV plant. The most accurate forecasting approach for each plant and accuracy measure is highlighted with bold

PV plant	Persistence	StackLSTM	BiLSTM	CNN-LSTM	ConvLSTM	EW	Meta
	RMSE						
Joao	4.555	2.393	2.569	2.435	2.415	2.377	**2.273**
Julia Moreira	2.764	1.587	1.645	1.719	1.634	1.601	**1.570**
Zarco	4.798	2.413	2.323	2.497	2.434	2.312	**2.298**
	Forecast skill index						
Joao	0	0.475	0.436	0.465	0.470	0.478	**0.500**
Julia Moreira	0	0.426	0.405	0.378	0.409	0.421	**0.432**
Zarco	0	0.497	0.516	0.479	0.493	0.518	**0.521**

Analyzing the performance of the base forecasting models, we observe that no single model consistently outperforms the others across all PV systems. The StackLSTM generally delivers more accurate results on average, followed by the ConvLSTM and BiLSTM. However, the differences in performance among the individual base models are neither consistent nor substantial. This highlights the importance of optimally combining predictions from multiple models to achieve more generalized and accurate forecasting solutions.

The meta-learner, evaluated using Root Mean Square Error (RMSE), consistently outperforms the base forecasting models across all PV systems, enhancing accuracy by up to 5% compared to the most accurate base model for each system. Similarly, the meta-learner surpasses the exponential weighting (EW) benchmark by 0.6% to 4.4%, dependent on the specific PV system analyzed. Conversely, while the EW benchmark is generally more accurate than individual base models, it occasionally does not exceed the performance of the best-performing base model. These findings indicate that the proposed meta-learning approach effectively integrates predictions from multiple models, outperforming standard yet competitive combination methods.

To assess whether the meta-learner provides significantly improved predictions, we utilize the multiple comparisons with the best (MCB) test [36]. This test computes the average ranks of forecasting methods based on root mean square error (RMSE) for hourly forecasts, specifically when actual PV production exceeds zero, and evaluates whether these ranks differ statistically. Figure 4.8 presents the results of this analysis for each PV system individually. Non-overlapping intervals between two methods signify statistically significant differences in performance. Methods whose intervals do not overlap with the gray interval in the figure are deemed significantly inferior to the best-performing method, and conversely.

The results of the MCB test are highly encouraging, indicating that the meta-learner provides significantly more accurate predictions than any of the examined benchmarks in two out of the three PV installations. For the third installation, "Julia

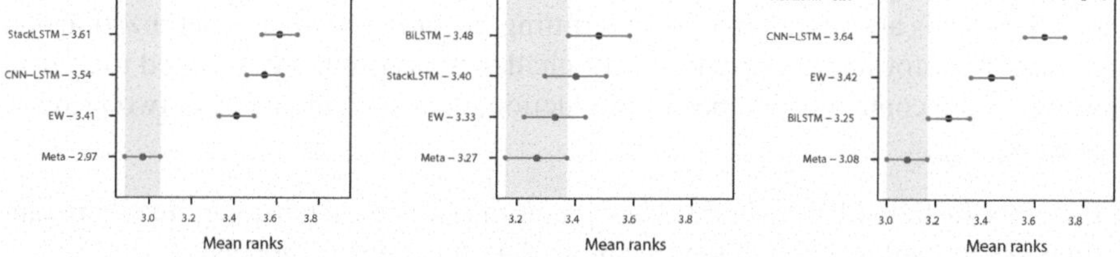

Fig. 4.8 Nemenyi test results at a 5% significance level for the RMSE. The tests involve records where the actual PV production is greater than zero

Moreira," the meta-learning approach also ranks first; however, the difference in performance between the meta-learner and the EW and StackLSTM methods is not statistically significant.

4.5 Transfer Learning Framework for PV Forecasting

This section introduces Transfer Learning (TL) techniques aimed at evaluating their effectiveness in forecasting PV production for installations with limited data availability. Three TL strategies, based on a stacked LSTM architecture, are developed and compared to a conventional non-TL approach. The proposed methods are tested on PV energy production systems located in diverse cities with varying nominal capacities. Experimental results indicate that all three TL strategies significantly exceed the non-TL approach in forecasting accuracy, as assessed by various error metrics. As discussed in previous sections, numerous approaches have been proposed to address the challenge of forecasting PV energy production. The simplest of these is the naive persistence method, which posits that the power output will remain unchanged from the previous observation. A more sophisticated variation is the smart persistence method, which asserts that the clear sky index [87, 88] of the current power output will remain constant into the future. These persistence methods are commonly employed as benchmark models to evaluate the efficacy of more advanced forecasting techniques.

In this study, a stacked LSTM architecture with multiple LSTM layers is chosen for two primary reasons. First, LSTM models excel at capturing the dynamic performance of systems, effectively handling sequential data with temporal relationships. These temporal, non-linear relationships are characteristic of the interactions between weather variables and PV power production, making LSTMs particularly suitable for this forecasting task. Second, as a weight-based model, LSTMs are well-suited for implementing TL approaches, unlike traditional statistical methods.

DL models are inherently data-intensive, typically requiring substantial datasets to achieve high predictive accuracy. Unlike traditional time series forecasting techniques and ML models, DL methods depend more heavily on data, a trade-off justified by their superior predictive capabilities. However, when trained on very limited data, DL models are susceptible to underfitting, resulting in poor generalization and high bias in performance estimates. This challenge is commonly referred to as data scarcity. In the context of PV power production, data scarcity arises in two primary scenarios:

- Recently Installed PV Systems: Newly commissioned systems require significant time to accumulate enough power output data for training models.
- Data Gaps: Missing values due to issues such as smart meter malfunctions can result in insufficient data for model training.

In both instances, the absence of adequate training data impedes the development of precise forecasting models. For PV production forecasting, a standard benchmark for adequate training data is one calendar year, which is essential for capturing seasonal patterns. To mitigate this limitation, a model developed for one location can be adapted for application at another site with minimal historical data. Humans naturally leverage knowledge gained from one task to solve similar problems-a concept reflected in DL through TL [95]. Traditional DL approaches involve training models from scratch using subject-specific data. In contrast, TL employs data from related applications to accelerate and enhance the learning process. TL has proven highly effective in addressing issues related to data insufficiency or incompleteness, which often arise from challenges such as inaccessible data, the high costs of data collection devices, or inadequate data storage systems. In the digital era, the availability and quality of data have significantly improved, allowing for the repurposing of extensive data repositories for similar tasks, thereby rendering TL a particularly suitable approach [104].

In the energy sector, TL has emerged as a prominent technique for addressing challenges associated with insufficient or low-quality data. For instance, the Hephaestus method utilizes TL for cross-building energy forecasting, taking into consideration seasonality and trends [70]. Additional examples include TL-based short-term energy forecasting for buildings [19] and energy consumption forecasting in the context of low-quality data [20]. However, few studies have investigated the application of TL for PV production forecasting [116], thus highlighting a significant opportunity for further research in this domain.

In this section, we examine three distinct TL strategies and evaluate their effectiveness in providing accurate forecasts of PV production. We compare the performance of traditional models with TL models across varying levels of data availability. The findings indicate that TL models significantly surpass conventional approaches, achieving an improvement of 24.8% in accuracy when trained on one year of data. The performance gap becomes even more pronounced when limited training data is available, with TL models exhibiting substantial advantages when trained on a dataset spanning three months. These results underscore the potential of TL to enhance forecasting for recently installed PV systems, establishing it as a reliable tool for prosumers in facilitating energy balancing and demand response management from the early stages of operation [93].

4.5.1 Proposed Transfer Learning Strategies

The core concept of TL is to address and resolve a problem by leveraging the steps and insights utilized to tackle a similar problem with comparable characteristics. Essentially, TL involves transferring knowledge from a model trained in a source domain to a target domain, relaxing the assumption that the data for both domains must be independent and identically distributed [89]. TL offers several benefits, including reduced training time [111], enhanced performance of neural networks, and, most critically, the capacity to achieve high accuracy even with limited data [34]. The overall framework of the TL process is illustrated in Fig. 4.9.

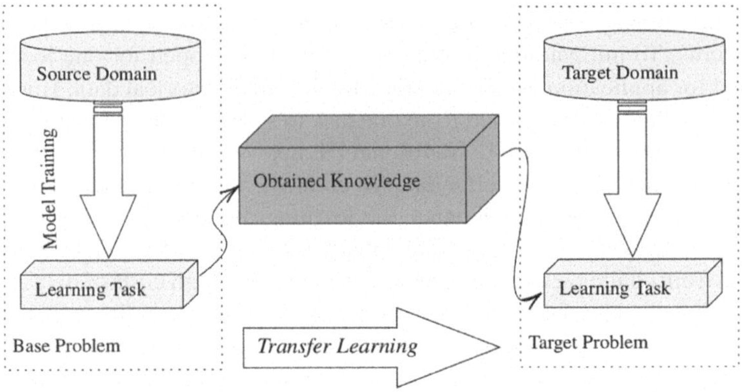

Fig. 4.9 Transfer Learning in a nutshell

According to the official definition of transfer learning (TL) proposed by Pan and Yang [63], "Given a source domain $\mathcal{D}_S$ and a learning task $\mathcal{T}_S$, a target domain $\mathcal{D}_T$ and a learning task $\mathcal{T}_T$, where $\mathcal{D}_S \neq \mathcal{D}_T$, or $\mathcal{T}_S \neq \mathcal{T}_T$, TL aims to enhance the learning of the target predictive function $f_T(\cdot)$ in $\mathcal{D}_T$ by leveraging knowledge from $\mathcal{D}_S$ and $\mathcal{T}_S$." This definition becomes clearer through the concepts of "*domain*" and "*task*. A domain $\mathcal{D}$ consists of a feature space $\mathcal{X}$ and a specific probability distribution $P(X)$, where $X = \{x_1, ..., x_n\} \in \mathcal{X}$.

The feature space is defined as a collection of features associated with the specific properties of the data provided as input to the model. In a specific domain, $\mathcal{D} = \{\mathcal{X}, P(X)\}$, a task comprises two components: a tag space $\mathcal{Y}$ and an objective prediction function $f : \mathcal{X} \rightarrow \mathcal{Y}$. The objective prediction function is designed to predict the tag $f(x)$ for each new instance x.

Finally, according to Lu et al. [47], there are three primary categories of TL methods. **Inductive** Transfer learning posits that the learning task in the target domain differs from that in the source domain. **Unsupervised** Additionally, TL also assumes a distinction between the learning tasks in the target and source domains; however, it specifically addresses unsupervised problems such as clustering and density estimation. Lastly, **transductive** transfer learning asserts that the learning tasks are identical in both domains, despite the fact that the source and target domains are distinct yet related. The proposed transfer learning approach for the problem of PV production forecasting falls within the realm of transductive transfer learning, as both the source and target tasks are the same (hourly PV forecasting), while the source and target domains vary in terms of location, nominal capacity, and weather conditions.

The original LSTM architecture consists of a single layer that processes input data and transfers the output to a single feed-forward layer. In this study, however, we propose an alternative architecture featuring multiple hidden layers, each composed of several LSTM units, followed by a feed-forward output layer. Instead of producing a single output value, each layer outputs a sequence of values that serve as input for the subsequent layer. This design, known as a stacked LSTM network, was first introduced by Graves et al. [24] in the context of speech recognition. In contrast to simple feed-forward networks, stacked LSTM networks yield deeper models that exhibit enhanced approximation accuracy. Furthermore, the architecture of LSTMs, which permits them to process sequential data by deriving their hidden states from all preceding hidden states, supports the development of frameworks that provide a more abstract representation of input data over time. This attribute allows the model to capture tasks across multiple time scales, thereby improving its ability to understand temporal patterns [4, 12].

TL involves the reuse of weights from a model that has been trained on data from a source domain to fine-tune a new model utilizing data from a target domain. The pre-trained model is referred to as the base model, while the adapted models in the target domain are known as TL models. The weights of each layer in the base model can be managed in various ways to optimize the performance of the TL model within the target domain. These approaches include: (a) fixing the layer weights, (b) fine-tuning the layer weights using data from the target domain, and (c) training the layer weights from scratch with data from the target domain. In this study, three TL strategies are developed and evaluated with respect to their forecasting accuracy for PV production prediction.

- **First TL Strategy:** In the first strategy, the weights of the initial layers are kept fixed, while only the weights of the final hidden layer are trainable. This approach, known as weight freezing, is commonly used to transfer features from the source domain to the target domain. It is particularly popular in image processing tasks, where the initial layers serve as feature extractors, and the final layers are adapted to new data.
- **Second TL Strategy:** In the second strategy, the base model serves as a weight initialization framework for the TL model. The weights of all layers in the TL model are initialized using data from the source domain and subsequently fine-tuned with data from the target domain. This method is widely applied in scenarios where the source domain has abundant data, while the target domain suffers from data scarcity. However, a high degree of similarity between the source and target domains is essential for the success of this approach.
- **Third TL Strategy:** In the third strategy, the initial layers of the TL model are frozen, while the final layer is trained from scratch. This is achieved by unfreezing the last layer of the base model and adding a new layer after the frozen layers. Although similar to the first strategy, this approach differs in that the weights of the final layer are not initialized with source domain data. As a result, the TL model functions as a feature extraction mechanism due to the frozen layers, while the randomly initialized final layer's weights allow for fine-tuning to the target domain.

4.5.2 Case Study

The results of this study are organized into three categories: the forecasting performance of the baseline stacked LSTM model, the performance comparison between the TL models and the conventional model in the target domain, and the outcomes of applying TL with different volumes of available data. In this context, the term conventional model refers to an LSTM model that does not incorporate TL and relies exclusively on training data from the target PV system.

The PV systems are located in four cities in Portugal, with four systems in Lisbon and one each in Setúbal, Faro, and Braga Table 4.3. The available data spans 14 to 30 months, depending on the PV system. The selected systems also vary in terms of nominal and maximum power. The base PV system has a nominal power of 23.52 kW, while the target domain PV systems range from 30 to 271.53 kW, exceeding the base system's power by more than tenfold. These specific PV systems were chosen to evaluate the model's performance on systems located in the same city as the base PV system (PV_2, PV_4, and PV_7) and in different cities (PV_3, PV_5, and PV_6). This allows for the assessment of potential differences in the forecasting accuracy of the TL models.

The selected features for the stacked LSTM model include temperature, humidity, solar radiation, PV production, one-hot encoding of the month of the year, and sine/cosine transformations of the time of day. The following preprocessing steps are applied to each dataset: The data are normalized to the range [0, 1]. Then, data are formatted into a "5-inputs - 1-output" structure for processing by the stacked LSTM model. Finally, the datasets are split into training and testing sets.

The base model is trained on the entire dataset from the source domain. TL models are trained using 12 months of data (equivalent to 8,760 hourly entries in the training set), with the remaining data reserved for testing. The testing period varies for each target PV system depending on the total number of available data points-PV_7 has the longest testing period, consisting of 13,172 hourly rows, while PV_3 has the shortest testing period, consisting of 910 hourly rows. To investigate the impact of TL under conditions of limited data, the same preprocessing routine is applied to all PV systems with training datasets of 3, 6, and 9 months, while maintaining a constant test set.

Table 4.3 Information of the PV plants that were used for the experimental application of the study

PV	Location	Latitude	Longitude	Nominal(kW)	Peak(kW)	Rows	From Date	To Date
$PV_1\,(base)$	Lisbon	38.728	−9.138	23.52	20.00	21956	01/08/2018	31/01/2021
PV_2	Lisbon	38.833	−9.191	46.00	40.00	21908	31/12/2018	01/01/2021
PV_3	Setubal	38.577	−8.872	271.53	216.00	9670	02/02/2020	10/03/2021
PV_4	Lisbon	38.725	−9.120	30.00	27.00	20588	01/08/2018	31/01/2021
PV_5	Faro	37.031	−7.893	60.48	50.00	15044	08/09/2019	30/04/2021
PV_6	Braga	41.493	−8.496	119.88	108.00	18045	10/01/2019	31/01/2021
PV_7	Lisbon	38.701	−9.236	55.65	50.00	21932	01/08/2019	31/01/2021

The experimental implementation is conducted using the Python programming language, leveraging open-source libraries such as NumPy and Pandas, along with the TensorFlow application programming interface. The source code is available on GitHub: Source Code. Based on existing literature, the ADAM optimizer is chosen with a learning rate of 0.001. The stacked LSTM model developed for this study consists of three fully connected intermediate layers: the first layer contains 24 neurons, the second layer has 48 neurons, and the third layer comprises 96 neurons, followed by an output layer. The number of epochs is set to 100, and the batch size is configured to 128, as determined through trial-and-error experimentation.

The stacked LSTM model incorporates the following lagging features: (a) measured power output, (b) air temperature, (c) global horizontal radiation, (d) humidity, (e) month of the year (one-hot encoded), and (f) time of day (sine/cosine transformation). These features are provided to the LSTM model in a "5-inputs - 1-output" hourly data format, where one data point for each feature from the preceding five hours is utilized as input, and the model predicts the photovoltaic power output for the subsequent hour, thereby facilitating a one-hour-ahead prediction.

Establishing an accurate baseline model is essential for achieving reliable predictions in the target domain. To evaluate the performance of the LSTM model for the baseline PV system, the following procedure is employed: the base PV dataset is divided into a training set and an evaluation set, utilizing an 80-20 split. The initial 80% of the data (17,563 observations) is designated for training, while the remaining 20% (4,391 observations) is reserved for evaluation. The LSTM model is trained on the training set, and its accuracy is assessed by calculating the Root Mean Square Error (RMSE), Mean Absolute Error (MAE), and the coefficient of determination (R^2) between the predictions and actual values over the evaluation period, as follows:

$$RMSE = \sqrt{\frac{1}{n}\sum_{t=1}^{n}(y_t - \hat{y}_t)^2} \tag{4.9}$$

$$MAE = \frac{1}{n}\sum_{t=1}^{n}|y_t - \hat{y}_t| \tag{4.10}$$

$$R^2 = 1 - \frac{\sum_{t=1}^{n}(y_t - \hat{y}_t)^2}{\sum_{t=1}^{n}(y_t - \bar{y})^2} \tag{4.11}$$

where y_t represents the actual value of the solar production time series at the hourly interval t during the evaluation period, $\hat{y}_t$ denotes the model's predicted value, and $\bar{y}$ signifies the average of the actual values. To provide a more comprehensive evaluation of the model, two additional metrics are computed: the Mean Bias Error (MBE) and the Normalized Root Mean Square Error (NRMSE). The MBE measures the systematic bias of the model, indicating whether it tends to underpredict or overpredict. Conversely, the NRMSE is particularly suitable for comparing models with different scales, as it normalizes the RMSE by the observed range of the variable. These metrics are calculated as follows:

$$MBE = \frac{1}{n}\sum_{t=1}^{n}(y_t - \hat{y}_t) \tag{4.12}$$

$$NRMSE = \frac{RMSE}{\bar{y}} \tag{4.13}$$

The model demonstrates high accuracy, successfully capturing the daily patterns of the key variables, as indicated by the evaluation metrics ($MAE = 0.467$, $RMSE = 0.992$, $MBE = -0.097$, $nRMSE = 0.301$, $R^2 = 96.254\%$). However, these five error metrics alone are insufficient to fully demonstrate the proposed model's capabilities in comparison to other models across different geographic locations. As highlighted by Yang et al. [110] and Yang [109], the accuracy of solar radiation forecasting models- which encompass both solar radiation and solar power forecasting, the latter being the focus of this study- should be comparable across various locations and time periods. This comparability can be achieved through the use of a standardized metric, the Forecasting Skill Score (FSS). The FSS assesses the performance of the proposed model by comparing it to a reference model using a specific error metric. In this study, the definition provided by Pedro and Coimbra is utilized, which estimates the expected power production under clear sky conditions [66]. The formula for the adopted smart persistence model is expressed by the following equation:

$$\hat{y}(t + \Delta t) = \begin{cases} y_{c-s}(t + \Delta t), & if\ y_{c-s}(t) = 0 \\ y_{c-s}(t + \Delta t)\frac{y(t)}{y_{c-s}(t)} & otherwise \end{cases} \tag{4.14}$$

where $y(t)$ represents the measured output power, and $y_{c-s}(t)$ indicates the expected output power under clear sky conditions. The purpose of this model is to decompose power production, illustrating that a portion of the power output relative to clear sky conditions remains constant over short time intervals. Moreover, during nighttime conditions, the smart persistence model forecasts that the power output is equivalent to the value observed under clear sky conditions. The approximate function of the clear sky model can be derived by averaging historical power output values based on the time of day (ranging from 0 to 23) and the day of the year (ranging from 0 to 255). The subsequent step involves constructing a smooth surface that encompasses this function [66].

4.5.3 Results

The performance of the smart persistence model is reflected in the following error metrics: $MAE = 0.582$, $RMSE = 1.274$, $MBE = 0.029$, $nRMSE = 0.387$, $R^2 = 93.811\%$. Although the smart persistence model demonstrates relatively strong performance in comparison to the naive persistence model ($RMSE_{Naive} = 1.985$, $MAE_{Naive} = 1.110$), it is evident that the LSTM model significantly outperforms the smart persistence model. This distinction is further emphasized by the forecasting skill score of the LSTM model, which is measured at 0.221. A positive forecasting skill score signifies that the proposed model surpasses the performance of the smart persistence model, whereas a negative score indicates the superiority of the smart persistence model. The results indicate that the model effectively captures seasonality, trends, and weather-related variations during both summer and winter, yielding forecasts that are substantially more accurate than those produced by the smart persistence model.

The TL models utilize the same features as the baseline model, leveraging the pre-trained baseline model to address the same problem with identical features and expected output, yet applied to a different PV system. Consequently, the features for the TL models are as follows: (a) measured power output, (b) air temperature, (c) global horizontal radiation, (d) humidity, (e) month of the year (one-hot encoded), and (f) time of day (sine/cosine transformation). The model output constitutes a one-hour-ahead prediction of PV power output.

The validation of the proposed TL strategies is conducted across six PV systems, which exhibit varying nominal and maximum power capacities, located in four cities in Portugal. Four architectures are compared: the proposed TL strategies and a conventional model devoid of TL components. For the TL models, pre-training is performed utilizing the entire dataset of the baseline PV system, which comprises 30 months of data. Subsequently, all four models are trained on one year of data, totaling 8,760 h, and are tested against the remaining dataset. The size of the test dataset varies for each PV system, contingent upon data availability, as demonstrated in Table 4.3. The models' accuracy is assessed based on their performance on the test data, utilizing the following metrics: $RMSE$, MBE, MAE, $NRMSE$, and R^2.

For each model, twenty training repetitions are conducted to mitigate randomness, a number that is frequently recommended in the literature. The forecasting performance of all models is summarized in Table 4.4, where the average values of $RMSE$, MBE, MAE, $NRMSE$, and R^2 are presented, thereby providing valuable insights.

Firstly, it is important to note that all LSTM models outperform the smart persistence model in terms of $RMSE$, indicating that the selected model and features are well-suited for this specific problem. The only instance in which the LSTM model underperforms compared to the smart persistence model is for the conventional LSTM of PV_2. However, even in this case, all three TL models exhibit lower error metrics than the smart persistence model. The forecasting skill score varies from -0.15 (indicating a negative value for PV_2) to 0.48 for the conventional model,

Table 4.4 Average forecasting performance (accuracy) of the smart persistence model and the stacked LSTM models with and without TL

PV	Metric	Without TL	Smart persistence	TL 1	TL 2	TL 3
PV_2	$RMSE$ (KWh)	3.77	3.27	2.34	2.37	2.36
	MBE (KWh)	−0.25	0.17	0.16	0.06	0.05
	MAE (KWh)	1.81	1.38	1.11	1.11	1.12
	$nRMSE$ (%)	53.14	46.18	33.06	33.38	33.28
	R^2 (%)	87.74	91.37	95.57	95.49	95.51
PV_3	$RMSE$	11.39	17.32	10.59	11.35	11.48
	MBE	0.83	0.02	0.79	0.81	0.80
	MAE	6.01	6.49	5.29	5.94	5.85
	$nRMSE$	40.78	62.15	37.94	40.63	41.13
	R^2	93.60	85.27	94.49	93.68	93.53
PV_4	$RMSE$	1.65	2.69	1.48	1.46	1.46
	MBE	−0.01	−0.82	0.11	0.06	0.07
	MAE	0.81	1.32	0.70	0.69	0.70
	$nRMSE$	35.57	57.84	31.94	31.42	31.51
	R^2	94.46	85.40	95.55	95.70	95.68
PV_5	$RMSE$	2.66	5.27	2.31	2.31	2.31
	MBE	−0.08	0.35	−0.04	−0.11	−0.16
	MAE	1.37	1.52	1.13	1.15	1.15
	$nRMSE$	31.68	62.75	27.56	27.46	27.54
	R^2	95.17	81.51	96.43	96.46	96.44
PV_6	$RMSE$	6.75	8.59	5.65	5.50	5.51
	MBE	1.35	−0.04	0.94	0.43	0.56
	MAE	3.54	4.25	2.74	2.79	2.77
	$nRMSE$	40.92	52.11	34.23	33.37	33.40
	R^2	92.72	88.58	95.07	95.33	95.31
PV_7	$RMSE$	3.64	4.59	2.47	2.40	2.41
	MBE	−0.16	0.15	0.24	−0.02	0.03
	MAE	1.75	1.64	1.15	1.11	1.12
	$nRMSE$	40.87	51.49	27.78	26.97	27.08
	R^2	92.23	88.19	96.56	96.76	96.74

while the TL strategies range from 0.28 to 0.56. On average, the forecasting skill score improves by 16.3% when utilizing the TL models compared to the conventional LSTM. Lastly, the MBE metric indicates that none of the developed models demonstrate any signs of bias.

When comparing the conventional LSTM model to the three TL models, the influence of TL is evident, as the TL strategies consistently exhibit superior accuracy across all six PV systems. In contrast, the models trained using the three TL strategies demonstrate minimal variability, providing more accurate and unbiased predictions. A notable exception is PV_3, where the evaluation period is restricted to merely 38 days, encompassing 910 hourly predictions. In this instance, the three TL models do not surpass the conventional model to the same degree as observed in other PV systems. This discrepancy can be attributed to the evaluation being conducted exclusively in March, a winter month, which poses additional challenges due to frequently disrupted weather phenomena. Another indication of the forecasting difficulty for PV_3 is evidenced by the smart persistence model's inability to yield more accurate predictions.

As previously discussed, a minimum of one calendar year of data is necessary to adequately train a model to capture all seasonal and weather patterns associated with the problem. The results presented herein indicate that TL models can outperform conventional models when one year of training data is available, whereas conventional models still exceed the performance of baseline smart persistence models. Nonetheless, TL has the potential to develop reliable and accurate prediction models even when the available training data for the target domain is less than one year. In this context, the proposed architectures are compared for target PV systems utilizing varying training periods-specifically, 3 months, 6 months, and 9 months of data. It is crucial to note that while the training period varies, the testing period remains constant to ensure a fair comparison across different scenarios.

Figure 4.10 displays the RMSE metric for the four models across different training periods. The results indicate that TL models exhibit greater robustness with varying amounts of training data, with their performance gradually improving as more data become available. This enhancement can be attributed to their prior training on the baseline PV system utilizing three years of hourly data. In contrast, the impact of data scarcity is evident in the conventional LSTM model, which demonstrates significant improvement as the training period extends, thereby enabling it to more effectively identify seasonal and weather patterns. Notably, none of the conventional models trained on three months of data surpass the performance of the smart persistence model, and only three of the conventional models trained on six months of data achieve greater accuracy than the smart persistence model. Conversely, the TL models trained on three months of data exhibit a lower RMSE than both the conventional LSTM and the smart persistence models.

Finally, the Root Mean Square Error (RMSE) gap between the conventional model and the optimal TL model decreases as the volume of training data increases. This trend is consistent across all six target PV systems. For instance, the RMSE difference for PV_5 diminishes from 150.5% (with 3-month-trained models) to 15.1%, nearly tenfold reduction. Similar reductions are observed across the other five PV systems, highlighting the effectiveness of TL, particularly when less than a full calendar year of data is available.

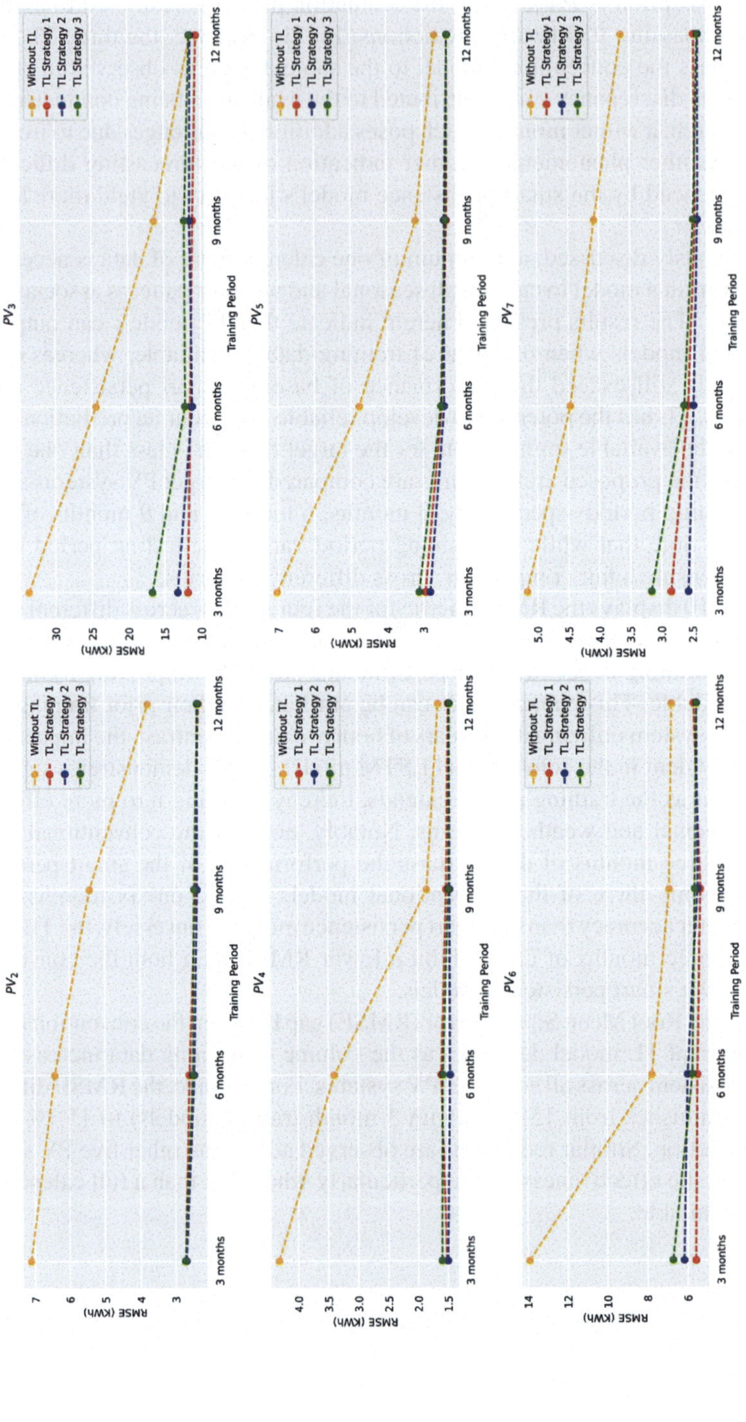

Fig. 4.10 Comparative performance of the four models based on RMSE index for 3 month, 6 month, 9 month and 12 month training period for the six target domain PVs

4.6 Conclusions

This chapter presents a comprehensive approach to short-term PV production forecasting. The developed method introduces four LSTM models with distinct deep architectures capable of accurately predicting PV production for the subsequent hour. The forecasts generated by these models serve as the foundation for a meta-learning framework that optimally integrates these predictions while incorporating additional external features, such as meteorological and seasonal variables. The meta-learner amalgamates base predictions by leveraging the strengths of each individual model, enhancing the overall accuracy and generalizability of the solution. Furthermore, the proposed method is independent of numerical weather prediction (NWP), as the weather features utilized are derived from preceding time steps, thus rendering the model more reliable and expedient to deploy.

Given the lengthy process of collecting sufficient data from newly installed solar stations, the idea of leveraging data from established solar systems to provide accurate predictions for newly installed systems becomes increasingly relevant. This study investigates whether TL can effectively be utilized for forecasting PV production in installations with limited data. Three TL strategies based on a stacked LSTM architecture are developed and compared to a non-TL approach. The methodology is tested on PV power production systems located in different cities with varying nominal capacities. The experimental results demonstrate that all three TL strategies significantly outperform the non-TL model in terms of forecasting accuracy, as evaluated using various error metrics.

The models are also compared to a smart persistence model based on power production under clear sky conditions. The TL models significantly outperform the baseline model, achieving forecasting skill scores between 0.28 and 0.56, which are regarded as satisfactory according to existing literature. In contrast, the non-TL LSTM model exhibits limited forecast accuracy, with an average decrease in forecasting skill of 16.3%. These results are based on the scenario in which one calendar year of data is available. Additional experiments involving varying volumes of training data indicate that the performance gap between the TL models and the non-TL approach widens as data scarcity increases. Specifically, when only three months of training data are available, the conventional LSTM model fails to exceed that of the baseline model, whereas the TL models continue to demonstrate superior accuracy.

One critical aspect of TL is the reproducibility of the strategies presented. To evaluate this, an experimental application was conducted on three PV systems located in the same city as the base PV system, as well as three PV systems located in different cities. This setup enabled a comparison of forecasting accuracy between the two groups. The geographical differences and varying nominal capacities of these PV systems necessitated the use of the forecasting skill score for comparison, in accordance with the recommended verification guidelines for deterministic solar forecasts.

Future research could explore the benefits of meta-learning for multi-step PV production forecasting models. It is well established that the performance of NWP-independent models deteriorates over time, rendering them less reliable than NWP-based models for tasks such as day-ahead forecasting. However, meta-learning has the potential to enhance the performance of base models used for such forecasts. Additionally, further studies could investigate various types of models as meta-learners and compare their performance with the proposed SVR-based approach. Although the SVR model has demonstrated effectiveness, exploring different features and base predictions as inputs to the meta-learner may influence the selection of the meta-learner. Moreover, an increased emphasis on the selection and development of base models, including the examination of diverse machine learning architectures, feature selection, and data preprocessing techniques, could represent valuable areas for future research.

Moreover, future studies should take into account geographic factors such as altitude and the orientation of solar installations when comparing the base and target domains. Lastly, the potential of utilizing transfer learning for solar power forecasting presents an opportunity for future research focused on applying knowledge transfer to related challenges, including forecasting energy consumption across interconnected buildings, wind power forecasting, and predicting the output of hydroelectric installations.

References

1. M. Abdel-Basset, H. Hawash, R.K. Chakrabortty, M. Ryan, Pv-net: an innovative deep learning approach for efficient forecasting of short-term photovoltaic energy production. J. Clean. Prod. **303**, 127037 (2021)
2. R. Ahmed, V. Sreeram, Y. Mishra, M. Arif, A review and evaluation of the state-of-the-art in pv solar power forecasting: techniques and optimization. Renew. Sustain. Energy Rev. **124**, 109792 (2020)
3. M.N. Akhter, S. Mekhilef, H. Mokhlis, N. Mohamed Shah, Review on forecasting of photovoltaic power generation based on machine learning and metaheuristic techniques. IET Renew. Power Gener. **13**(7), 1009–1023 (2019)
4. M. Amirul Islam, M. Rochan, N.D. Bruce, Y. Wang, Gated feedback refinement network for dense image labeling, in *Proceedings of the IEEE Conference on Computer Vision and Pattern Recognition*, pp. 3751–3759 (2017)
5. J. Antonanzas, N. Osorio, R. Escobar, R. Urraca, F.J. Martinez-de Pison, F. Antonanzas-Torres, Review of photovoltaic power forecasting. Sol. Energy **136**, 78–111 (2016)
6. S. Aslam, H. Herodotou, S.M. Mohsin, N. Javaid, N. Ashraf, S. Aslam, A survey on deep learning methods for power load and renewable energy forecasting in smart microgrids. Renew. Sustain. Energy Rev. **144**, 110992 (2021)
7. P. Bacher, H. Madsen, H.A. Nielsen, Online short-term solar power forecasting. Sol. Energy **83**(10), 1772–1783 (2009)
8. Y. Bengio et al., Learning deep architectures for ai. Found. Trends® Mach. Learn. **2**(1), 1–127 (2009)
9. D. Chaturvedi, I. Isha, Solar power forecasting: a review. Int. J. Comput. Appl. **145**(6), 28–50 (2016)

10. T. Chen, R. Xu, Y. He, X. Wang, Improving sentiment analysis via sentence type classification using bilstm-crf and cnn. Expert Syst. Appl. **72**, 221–230 (2017)
11. F. Chollet et al., Keras (2015). https://keras.io
12. J. Chung, C. Gulcehre, K. Cho, Y. Bengio, Gated feedback recurrent neural networks, in *International Conference on Machine Learning* (PMLR, 2015), pp. 2067–2075
13. U.K. Das, K.S. Tey, M. Seyedmahmoudian, S. Mekhilef, M.Y.I. Idris, W. Van Deventer, B. Horan, A. Stojcevski, Forecasting of photovoltaic power generation and model optimization: a review. Renew. Sustain. Energy Rev. **81**, 912–928 (2018)
14. M. De Benedetti, F. Leonardi, F. Messina, C. Santoro, A. Vasilakos, Anomaly detection and predictive maintenance for photovoltaic systems. Neurocomputing **310**, 59–68 (2018)
15. I.C. Dimitriou, A. Arsenopoulos, G.P. Trachanas, E. Sarmas, V. Marinakis, Assessing the adoption level of solar pv installations, on district scale of urban environment. Renew. Energy **237**, 121676 (2024)
16. A. Dolara, F. Grimaccia, S. Leva, M. Mussetta, E. Ogliari, A physical hybrid artificial neural network for short term forecasting of pv plant power output. Energies **8**(2), 1138–1153 (2015)
17. J. Donahue, L. Anne Hendricks, S. Guadarrama, M. Rohrbach, S. Venugopalan, K. Saenko, T. Darrell, Long-term recurrent convolutional networks for visual recognition and description, in *Proceedings of the IEEE Conference on Computer Vision and Pattern Recognition* (2015), pp. 2625–2634
18. H. Eom, Y. Son, S. Choi, Feature-selective ensemble learning-based long-term regional pv generation forecasting. IEEE Access **8**, 54620–54630 (2020)
19. C. Fan, Y. Sun, F. Xiao, J. Ma, D. Lee, J. Wang, Y.C. Tseng, Statistical investigations of transfer learning-based methodology for short-term building energy predictions. Appl. Energy **262**, 114499 (2020)
20. Y. Gao, Y. Ruan, C. Fang, S. Yin, Deep learning and transfer learning models of energy consumption forecasting for a building with poor information data. Energy Build. **223**, 110156 (2020)
21. J.Á. González Ordiano, S. Waczowicz, M. Reischl, R. Mikut, V. Hagenmeyer, Photovoltaic power forecasting using simple data-driven models without weather data. Comput. Sci.-Res. Dev. **32**(1), 237–246 (2017)
22. A. Graves, Generating sequences with recurrent neural networks (2013). arXiv:1308.0850
23. A. Graves, M. Liwicki, S. Fernández, R. Bertolami, H. Bunke, J. Schmidhuber, A novel connectionist system for unconstrained handwriting recognition. IEEE Trans. Pattern Anal. Mach. Intell. **31**(5), 855–868 (2008)
24. A. Graves, A.-R. Mohamed, G. Hinton, Speech recognition with deep recurrent neural networks, in *2013 IEEE International Conference on Acoustics, Speech and Signal Processing* (IEEE, 2013), pp. 6645–6649
25. A. Graves, J. Schmidhuber, Framewise phoneme classification with bidirectional lstm and other neural network architectures. Neural Netw. **18**(5–6), 602–610 (2005)
26. P. Gupta, R. Singh, Pv power forecasting based on data-driven models: a review. Int. J. Sustain. Eng. **14**(6), 1733–1755 (2021)
27. M. Gürtler, T. Paulsen, The effect of wind and solar power forecasts on day-ahead and intraday electricity prices in Germany. Energy Econ. **75**, 150–162 (2018)
28. S.A. Haider, M. Sajid, H. Sajid, E. Uddin, Y. Ayaz, Deep learning and statistical methods for short-and long-term solar irradiance forecasting for Islamabad. Renew. Energy **198**, 51–60 (2022)
29. K. He, X. Zhang, S. Ren, J. Sun, Deep residual learning for image recognition, in *Proceedings of the IEEE Conference on Computer Vision and Pattern Recognition* (2016), pp. 770–778
30. M. Hermans, B. Schrauwen, Training and analysing deep recurrent neural networks. *Advances in Neural Information Processing Systems*, 26 (2013)
31. S. Hochreiter, The vanishing gradient problem during learning recurrent neural nets and problem solutions. Int. J. Uncertain. Fuzziness Knowl.-Based Syst. **6**(02), 107–116 (1998)
32. T. Hospedales, A. Antoniou, P. Micaelli, A. Storkey, Meta-learning in neural networks: a survey (2020). arXiv:2004.05439

33. G. Huang, Z. Liu, L. Van Der Maaten, K.Q. Weinberger, Densely connected convolutional networks, in *Proceedings of the IEEE Conference on Computer Vision and Pattern Recognition* (2017), pp. 4700–4708
34. Z. Huang, Z. Pan, B. Lei, Transfer learning with deep convolutional neural network for sar target classification with limited labeled data. Remote Sensing **9**(9), 907 (2017)
35. S. Jain, T. Sharma, A.K. Gupta, End-of-life management of solar pv waste in India: situation analysis and proposed policy framework. Renew. Sustain. Energy Rev. **153**, 111774 (2022)
36. A.J. Koning, P.H. Franses, M. Hibon, H. Stekler, The M3 competition: statistical tests of the results. Int. J. Forecast. **21**(3), 397–409 (2005)
37. D. Koutsandreas, E. Spiliotis, F. Petropoulos, V. Assimakopoulos, On the selection of forecasting accuracy measures. J. Oper. Res. Soc. **73**(5), 937–954 (2022)
38. A.A. Lateko, H.-T. Yang, C.-M. Huang, H. Aprillia, C.-Y. Hsu, J.-L. Zhong, N.H. Phuong, Stacking ensemble method with the rnn meta-learner for short-term pv power forecasting. Energies **14**(16), 4733 (2021)
39. Y. LeCun, Y. Bengio, G. Hinton, Deep learning. Nature **521**(7553), 436–444 (2015)
40. D. Lee, K. Kim, Pv power prediction in a peak zone using recurrent neural networks in the absence of future meteorological information. Renew. Energy **173**, 1098–1110 (2021)
41. J. Lee, M.M. Shepley, Benefits of solar photovoltaic systems for low-income families in social housing of Korea: Renewable energy applications as solutions to energy poverty. J. Build. Eng. **28**, 101016 (2020)
42. W. Lee, K. Kim, J. Park, J. Kim, Y. Kim, Forecasting solar power using long-short term memory and convolutional neural networks. IEEE Access **6**, 73068–73080 (2018)
43. M. Lei, L. Shiyan, J. Chuanwen, L. Hongling, Z. Yan, A review on the forecasting of wind speed and generated power. Renew. Sustain. Energy Rev. **13**(4), 915–920 (2009)
44. Z. Li, F. Liu, W. Yang, S. Peng, J. Zhou, A survey of convolutional neural networks: analysis, applications, and prospects. IEEE Trans. Neural Netw. Learn. Syst. (2021)
45. L. Liu, M. Zhan, Y. Bai, A recursive ensemble model for forecasting the power output of photovoltaic systems. Sol. Energy **189**, 291–298 (2019)
46. I.E. Livieris, E. Pintelas, P. Pintelas, A cnn-lstm model for gold price time-series forecasting. Neural Comput. Appl. **32**(23), 17351–17360 (2020)
47. J. Lu, V. Behbood, P. Hao, H. Zuo, S. Xue, G. Zhang, Transfer learning using computational intelligence: a survey. Knowl.-Based Syst. **80**, 14–23 (2015)
48. X. Luo, D. Zhang, X. Zhu, Combining transfer learning and constrained long short-term memory for power generation forecasting of newly-constructed photovoltaic plants. Renew. Energy **185**, 1062–1077 (2022)
49. I. Maity, S. Rao, Simulation and pricing mechanism analysis of a solar-powered electrical microgrid. IEEE Syst. J. **4**(3), 275–284 (2010)
50. S. Makridakis, E. Spiliotis, V. Assimakopoulos, The M4 Competition: 100,000 time series and 61 forecasting methods. Int. J. Forecast. **36**(1), 54–74 (2020)
51. D. Markovics, M.J. Mayer, Comparison of machine learning methods for photovoltaic power forecasting based on numerical weather prediction. Renew. Sustain. Energy Rev. **161**, 112364 (2022)
52. P. Mathiesen, J. Kleissl, Evaluation of numerical weather prediction for intra-day solar forecasting in the continental united states. Sol. Energy **85**(5), 967–977 (2011)
53. M.J. Mayer, Benefits of physical and machine learning hybridization for photovoltaic power forecasting. Renew. Sustain. Energy Rev. **168**, 112772 (2022)
54. M.J. Mayer, G. Gróf, Extensive comparison of physical models for photovoltaic power forecasting. Appl. Energy **283**, 116239 (2021)
55. A. Mellit, A.M. Pavan, A 24-h forecast of solar irradiance using artificial neural network: application for performance prediction of a grid-connected pv plant at trieste, italy. Sol. Energy **84**(5), 807–821 (2010)
56. A. Mellit, A.M. Pavan, V. Lughi, Short-term forecasting of power production in a large-scale photovoltaic plant. Sol. Energy **105**, 401–413 (2014)

57. A. Mellit, A.M. Pavan, V. Lughi, Deep learning neural networks for short-term photovoltaic power forecasting. Renew. Energy **172**, 276–288 (2021)
58. V. Michalakopoulos, E. Sarantinopoulos, E. Sarmas, V. Marinakis, Empowering federated learning techniques for privacy-preserving pv forecasting. Energy Rep. **12**, 2244–2256 (2024)
59. P. Montero-Manso, G. Athanasopoulos, R.J. Hyndman, T.S. Talagala, FFORMA: feature-based forecast model averaging. Int. J. Forecast. **36**(1), 86–92 (2020)
60. E. Ogliari, A. Dolara, G. Manzolini, S. Leva, Physical and hybrid methods comparison for the day ahead pv output power forecast. Renew. Energy **113**, 11–21 (2017)
61. A. Olabi, M.A. Abdelkareem, Renewable energy and climate change. Renew. Sustain. Energy Rev. **158**, 112111 (2022)
62. C. Olah, Understanding lstm networks (2015)
63. S.J. Pan, Q. Yang, A survey on transfer learning. IEEE Trans. Knowl. Data Eng. **22**(10), 1345–1359 (2009)
64. R. Pascanu, C. Gulcehre, K. Cho, Y. Bengio, How to construct deep recurrent neural networks (2013). arXiv:1312.6026
65. F. Pedregosa, G. Varoquaux, A. Gramfort, V. Michel, B. Thirion, O. Grisel, M. Blondel, P. Prettenhofer, R. Weiss, V. Dubourg, J. Vanderplas, A. Passos, D. Cournapeau, M. Brucher, M. Perrot, E. Duchesnay, Scikit-learn: machine learning in Python. J. Mach. Learn. Res. **12**, 2825–2830 (2011)
66. H.T. Pedro, C.F. Coimbra, Assessment of forecasting techniques for solar power production with no exogenous inputs. Sol. Energy **86**(7), 2017–2028 (2012)
67. A. Rai, A. Shrivastava, K.C. Jana, A robust auto encoder-gated recurrent unit (ae-gru) based deep learning approach for short term solar power forecasting. Optik **252**, 168515 (2022)
68. W. Rawat, Z. Wang, Deep convolutional neural networks for image classification: a comprehensive review. Neural Comput. **29**(9), 2352–2449 (2017)
69. K. Reindl, J. Palm, Installing pv: Barriers and enablers experienced by non-residential property owners. Renew. Sustain. Energy Rev. **141**, 110829 (2021)
70. M. Ribeiro, K. Grolinger, H.F. ElYamany, W.A. Higashino, M.A. Capretz, Transfer learning with seasonal and trend adjustment for cross-building energy forecasting. Energy Build. **165**, 352–363 (2018)
71. J. Romaní, M. Belusko, A. Alemu, L.F. Cabeza, A. de Gracia, F. Bruno, Control concepts of a radiant wall working as thermal energy storage for peak load shifting of a heat pump coupled to a PV array. Renew. Energy **118**, 489–501 (2018)
72. A. Sagheer, M. Kotb, Time series forecasting of petroleum production using deep LSTM recurrent networks. Neurocomputing **323**, 203–213 (2019)
73. M. Sajjad, S. Kwon et al., Clustering-based speech emotion recognition by incorporating learned features and deep bilstm. IEEE Access **8**, 79861–79875 (2020)
74. E. Sarmas, N. Dimitropoulos, V. Marinakis, Z. Mylona, H. Doukas, Transfer learning strategies for solar power forecasting under data scarcity. Sci. Rep. **12**(1), 1–13 (2022)
75. E. Sarmas, E. Spiliotis, V. Marinakis, M.A. Bucarelli, F. Santori, H. Doukas, Revving up energy autonomy: a forecast-driven framework for reducing reverse power flow in microgrids. Sustain. Energy Grids Netw. **38**, 101376 (2024)
76. E. Sarmas, E. Spiliotis, V. Marinakis, T. Koutselis, H. Doukas, A meta-learning classification model for supporting decisions on energy efficiency investments. Energy Build. **258**, 111836 (2022)
77. E. Sarmas, E. Spiliotis, V. Marinakis, G. Tzanes, J.K. Kaldellis, H. Doukas, Ml-based energy management of water pumping systems for the application of peak shaving in small-scale islands. Sustain. Cities Soc. 103873 (2022)
78. B. Schölkopf, A.J. Smola, *Learning with kernel: Support Vector Machines, Regularization, Optimization and Beyond* (The MIT Press, 2001)
79. M. Schuster, K.K. Paliwal, Bidirectional recurrent neural networks. IEEE Trans. Signal Process. **45**(11), 2673–2681 (1997)
80. H. Sharadga, S. Hajimirza, R.S. Balog, Time series forecasting of solar power generation for large-scale photovoltaic plants. Renew. Energy **150**, 797–807 (2020)

81. X. Shi, Z. Chen, H. Wang, D.-Y. Yeung, W.-K. Wong, W.-C. Woo, Convolutional lstm network: a machine learning approach for precipitation nowcasting, in *Advances in Neural Information Processing Systems*, 28 (2015)
82. P. Skaloumpakas, E. Sarmas, M. Rachmanidis, V. Marinakis, Reshaping the energy landscape of crete through renewable energy valleys. Sci. Rep. **14**(1), 8038 (2024)
83. S. Sobri, S. Koohi-Kamali, N.A. Rahim, Solar photovoltaic generation forecasting methods: a review. Energy Convers. Manag. **156**, 459–497 (2018)
84. H. Song, N. Al Khafaf, A. Kamoona, S.S. Sajjadi, A.M. Amani, M. Jalili, X. Yu, P. McTaggart, Multitasking recurrent neural network for photovoltaic power generation prediction. Energy Rep. **9**, 369–376 (2023)
85. E. Spiliotis, N.Z. Legaki, V. Assimakopoulos, H. Doukas, M.S. El Moursi, Tracking the performance of photovoltaic systems: a tool for minimising the risk of malfunctions and deterioration. IET Renew. Power Gener. **12**(7), 815–822 (2018)
86. D. Stoian, E. Spiliotis, E. Stamatopoulos, E. Sarmas, P. Rekomaa, V. Marinakis, A neural network-based approach for hierarchical electricity consumption forecasting in buildings, in *2024 IEEE International Workshop on Metrology for Living Environment (MetroLivEnv)* (IEEE, 2024), pp. 274–279
87. X. Sun, J.M. Bright, C.A. Gueymard, B. Acord, P. Wang, N.A. Engerer, Worldwide performance assessment of 75 global clear-sky irradiance models using principal component analysis. Renew. Sustain. Energy Rev. **111**, 550–570 (2019)
88. X. Sun, J.M. Bright, C.A. Gueymard, X. Bai, B. Acord, P. Wang, Worldwide performance assessment of 95 direct and diffuse clear-sky irradiance models using principal component analysis. Renew. Sustain. Energy Rev. **135**, 110087 (2021)
89. C. Tan, F. Sun, T. Kong, W. Zhang, C. Yang, C. Liu, A survey on deep transfer learning, in *International Conference on Artificial Neural Networks* (Springer, 2018), pp. 270–279
90. A. Tascikaraoglu, O. Erdinc, M. Uzunoglu, A. Karakas, An adaptive load dispatching and forecasting strategy for a virtual power plant including renewable energy conversion units. Appl. Energy **119**, 445–453 (2014)
91. L.J. Tashman, Out-of-sample tests of forecasting accuracy: an analysis and review. Int. J. Forecast. **16**(4), 437–450 (2000)
92. R. Tassenoy, K. Couvreur, W. Beyne, M. De Paepe, S. Lecompte, Techno-economic assessment of Carnot batteries for load-shifting of solar PV production of an office building. Renew. Energy **199**, 1133–1144 (2022)
93. L. Toderean, T. Cioara, I. Anghel, E. Sarmas, V. Michalakopoulos, V. Marinakis, Demand response optimization for smart grid integrated buildings: review of technology enablers landscape and innovation challenges. Energy Build. 115067 (2024)
94. L. Todorovski, S. Džeroski, Combining multiple models with meta decision trees, in *European Conference on Principles of Data Mining and Knowledge Discovery* (Springer, 2000), pp. 54–64
95. L. Torrey, J. Shavlik, Transfer learning, in *Handbook of Research on Machine Learning Applications and Trends: Algorithms, Methods, and Techniques* (IGI Global, 2010), pp. 242–264
96. J. Vanschoren, Meta-learning, in *Automated Machine Learning*. (Springer, Cham, 2019), pp.35–61
97. L. Visser, T. AlSkaif, W. van Sark, Operational day-ahead solar power forecasting for aggregated PV systems with a varying spatial distribution. Renew. Energy **183**, 267–282 (2022)
98. B.H. Vu, I.-Y. Chung, Optimal generation scheduling and operating reserve management for PV generation using RNN-based forecasting models for stand-alone microgrids. Renew. Energy **195**, 1137–1154 (2022)
99. C. Wan, J. Zhao, Y. Song, Z. Xu, J. Lin, Z. Hu, Photovoltaic and solar power forecasting for smart grid energy management. CSEE J. Power Energy Syst. **1**(4), 38–46 (2015)

100. D. Wang, E. Nyberg, A long short-term memory model for answer sentence selection in question answering, in *Proceedings of the 53rd Annual Meeting of the Association for Computational Linguistics and the 7th International Joint Conference on Natural Language Processing (Volume 2: Short Papers)* (2015), pp. 707–712
101. J. Wang, P. Li, R. Ran, Y. Che, Y. Zhou, A short-term photovoltaic power prediction model based on the gradient boost decision tree. Appl. Sci. **8**(5), 689 (2018)
102. K. Wang, X. Qi, H. Liu, A comparison of day-ahead photovoltaic power forecasting models based on deep learning neural network. Appl. Energy **251**, 113315 (2019)
103. Y. Wang, W. Liao, Y. Chang, Gated recurrent unit network-based short-term photovoltaic forecasting. Energies **11**(8), 2163 (2018)
104. K. Weiss, T.M. Khoshgoftaar, D. Wang, A survey of transfer learning. J. Big data **3**(1), 1–40 (2016)
105. L. Wen, K. Zhou, S. Yang, X. Lu, Optimal load dispatch of community microgrid with deep learning based solar power and load forecasting. Energy **171**, 1053–1065 (2019)
106. D.H. Wolpert, Stacked generalization. Neural Netw. **5**(2), 241–259 (1992)
107. Y. Wu, Z. Liu, B. Li, J. Liu, L. Zhang, Energy management strategy and optimal battery capacity for flexible PV-battery system under time-of-use tariff. Renew. Energy **200**, 558–570 (2022)
108. D. Yang, A guideline to solar forecasting research practice: reproducible, operational, probabilistic or physically-based, ensemble, and skill (ropes). J. Renew. Sustain. Energy **11**(2), 022701 (2019)
109. D. Yang, Making reference solar forecasts with climatology, persistence, and their optimal convex combination. Sol. Energy **193**, 981–985 (2019)
110. D. Yang, S. Alessandrini, J. Antonanzas, F. Antonanzas-Torres, V. Badescu, H.G. Beyer, R. Blaga, J. Boland, J.M. Bright, C.F. Coimbra et al., Verification of deterministic solar forecasts. Sol. Energy **210**, 20–37 (2020)
111. Y. Yao, G. Doretto, Boosting for transfer learning with multiple sources, in *2010 IEEE Computer Society Conference on Computer Vision and Pattern Recognition* (IEEE, 2010), pp. 1855–1862
112. Y. Yao, Z. Huang, Bi-directional lstm recurrent neural network for Chinese word segmentation, in *International Conference on Neural Information Processing* (Springer, 2016), pp. 345–353
113. A. Zameer, J. Arshad, A. Khan, M.A.Z. Raja, Intelligent and robust prediction of short term wind power using genetic programming based ensemble of neural networks. Energy Convers. Manag. **134**, 361–372 (2017)
114. H. Zang, L. Cheng, T. Ding, K.W. Cheung, Z. Wei, G. Sun, Day-ahead photovoltaic power forecasting approach based on deep convolutional neural networks and meta learning. Int. J. Electr. Power & Energy Syst. **118**, 105790 (2020)
115. J. Zhao, Z.-H. Guo, Z.-Y. Su, Z.-Y. Zhao, X. Xiao, F. Liu, An improved multi-step forecasting model based on WRF ensembles and creative fuzzy systems for wind speed. Appl. Energy **162**, 808–826 (2016)
116. S. Zhou, L. Zhou, M. Mao, X. Xi, Transfer learning for photovoltaic power forecasting with long short-term memory neural network, in *2020 IEEE International Conference on Big Data and Smart Computing (BigComp)* (IEEE, 2020), pp. 125–132
117. R. Zhu, W. Guo, X. Gong, Short-term photovoltaic power output prediction based on k-fold cross-validation and an ensemble model. Energies **12**(7), 1220 (2019)

Chapter 5
Machine Learning-Driven Energy Consumption Forecasting for Building Profiling

Abstract This chapter investigates the application of Artificial Intelligence (AI) models to forecast energy consumption in buildings, particularly focusing on estimating potential energy savings resulting from renovation actions. Accurate estimation of energy savings is essential for the effective implementation of energy efficiency measures. To address this need, an ensemble model is proposed for precise baseline energy consumption estimation, utilizing tree-based algorithms such as Random Forest, XGBoost, and LightGBM. A key feature of the model is its emphasis on explainability, which provides transparency and insights into the primary factors influencing baseline energy consumption. Experimental evaluations conducted on a cluster of buildings in Latvia validate the accuracy of the model. The results demonstrate the superiority of the ensemble model over individual models designed for energy consumption estimation, thereby offering a reliable and interpretable approach for assessing energy savings in the building sector.

Keywords Measurement and verification · Baseline energy modeling · Energy efficiency · Explainable artificial intelligence · Building renovation

5.1 Introduction

Despite the significant contribution of the building sector to global energy consumption (35%) and energy-related CO2 emissions (38%) in 2019, energy use and emissions levels in this sector have remained stable since 2018. However, global operational energy consumption in buildings continues to approximate 130 EJ, which accounts for approximately 30% of total final energy consumption, while electricity usage in buildings constitutes around 55% of global electricity consumption [16]. These statistics underscore the critical need to reduce energy use and emissions in the building sector to combat climate change. Building renovation emerges as a promising solution, with energy-efficient upgrades, such as enhanced insulation and lighting, providing substantial energy savings and emissions reductions [39, 45]. Despite these advantages, progress in renovations remains slow, leaving a significant portion of the building stock energy inefficient. The COVID-19 pandemic has further

E. Sarmas et al., *Artificial Intelligence for Energy Systems*, Learning and Analytics in Intelligent Systems 46, https://doi.org/10.1007/978-3-031-85209-1_5

emphasized the importance of healthy and sustainable buildings, accentuating the urgency for action. Governments and policymakers play a crucial role in accelerating building renovation through regulations, financing, and policy tools, thereby unlocking its potential as a key strategy for reducing energy use, mitigating climate change, and fostering a healthier built environment [4, 15].

Energy Conservation Measures (ECMs) are essential for reducing energy consumption and greenhouse gas emissions across various sectors, including buildings, industry, and transportation [24]. These measures encompass equipment upgrades, improved insulation, energy-efficient lighting, and optimized operational practices. However, to ensure the effectiveness of ECMs, accurate measurement and verification (M&V) of energy savings is imperative. M&V has become increasingly important in light of the emergence of global energy policies, such as the EU Energy Efficiency Directive, which requires reliable estimates of energy savings for effective implementation [3]. Accurate estimation of savings entails comparing energy consumption during a reference period (post-ECM) with a hypothetical baseline scenario (pre-ECM conditions), necessitating the normalization of consumption [29]. This process often employs engineering or statistical models, which must incorporate a measure of uncertainty to ensure reliability [13].

Quantifying energy savings accurately is a complex task that involves multiple variables and inherent uncertainties. M&V has garnered considerable attention as efforts to enhance the accuracy of savings estimates intensify. Key challenges include the development of robust models that account for factors such as climate, usage patterns, and energy system performance. Advancements in digitization and the proliferation of Internet of Things (IoT) devices present opportunities for sophisticated machine learning (ML) models, which are capable of integrating diverse datasets to produce more accurate energy savings estimates [14, 43]. The adoption of ML models can enhance the effectiveness of ECMs and support ambitious energy policy goals [2, 33], successful implementation relies on access to high-quality data and expertise in data analysis and modeling.

This chapter examines the challenges associated with accurately estimating energy savings through the analysis of various ML models for baseline energy modeling. It investigates the implementation of ensemble methods that utilize high-quality big data to improve the accuracy of energy savings estimations. Comprehensive testing on real-world buildings in Latvia offers insights into the practical applicability of the proposed models. By emphasizing ML and deep learning (DL) models, this research contributes to the enhancement of ECM effectiveness and the attainment of ambitious energy policy objectives.

Key contributions of this chapter include:

- The study explores and evaluates various ML models for estimating baseline energy consumption in buildings, utilizing high-quality big data generated by IoT devices.
- The study investigates the potential of ML model ensembles to improve the accuracy of energy consumption estimation.

- The proposed models are evaluated using a large dataset from real-world buildings in Latvia, offering valuable insights into their effectiveness and generalizability in practical applications.

The remainder of the chapter is organized as follows. Section 5.2 presents the problem framework and the literature review. Section 5.3 outlines the developed methodology. Section 5.4 introduces an experimental application derived from a real-world case study in Latvia. Finally, Sect. 5.5 offers concluding remarks.

5.2 Problem Formulation and Related Work

5.2.1 Baseline Consumption Modeling

Legislative initiatives have primarily focused on the implementation of ECMs to decrease energy consumption. The effectiveness of these initiatives significantly depends on the precision of M&V processes for each ECM, which ensures an accurate estimation of energy savings. Nonetheless, a considerable challenge is the risk of overestimating savings at the project level, which could hinder efforts to address climate change. Consequently, there exists an urgent need for methodologies that can overcome barriers to precise M&V in building facilities, including issues related to cost, resource constraints, and time limitations. Previous studies have recognized ML as a promising solution to these challenges, underlining its potential in this context [12, 36, 44].

M&V projects typically involve three key phases: the baseline period, the implementation period, and the reporting period. These phases occur sequentially, with durations varying based on project-specific parameters. The baseline period, which precedes the implementation of ECMs, establishes a reference point for comparison, while the reporting period follows the application of ECMs. A critical component of M&V involves estimating the adjusted baseline during the reporting period by normalizing energy consumption to the baseline conditions. This task often necessitates the use of engineering or statistical models to develop a baseline model capable of achieving accurate normalization. Given the inherent imprecision of M&V processes, ensuring accuracy throughout is essential for reliable outcomes.

Uncertainty in M&V arises primarily from sampling, measurement, and modeling. To mitigate this uncertainty, the International Performance Measurement and Verification Protocol (IPMVP) provides a framework for quantifying project-related uncertainty. The protocol specifies a minimum acceptable level of uncertainty, defined as the threshold at which savings exceed twice the standard error of the baseline value (Fig. 5.1).

The IPMVP serves as the predominant and globally adopted methodology for M&V, comprising four distinct approaches that address a wide range of retrofit actions. Options A and B establish a project boundary that isolates the retrofit and

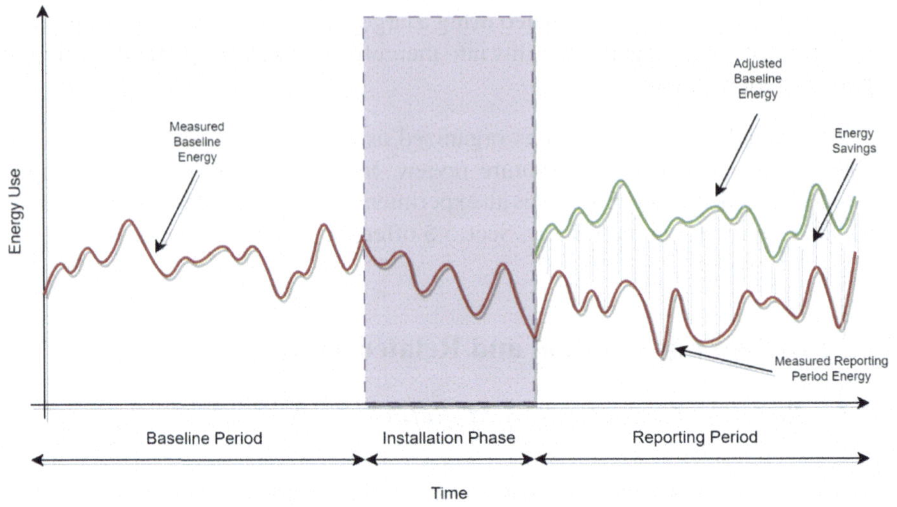

Fig. 5.1 Energy savings calculation according to the IPMVP framework. Savings can be estimated with measurements before and after the ECM installation, but a normalisation of the baseline energy is also required to incorporate weather and other effects

includes the affected equipment. Option C adopts a whole-building approach, suitable for scenarios where savings exceed 10% of the facility's total energy consumption. Option D involves a calibrated simulation of energy systems, proving beneficial in cases where baseline data is unavailable. Complementing the IPMVP, ASHRAE Guideline 14 and ISO 50015 provide methodologies rooted in its core principles, offering robust frameworks. However, a notable drawback of these generalized methodologies is the recognized lack of guidance on the calculation process.

The determination of savings can encompass either the entire facility or a specific portion of it, depending on the characteristics of the Energy Efficiency Measures (EEMs) and the intended purpose of the reporting. To isolate equipment and the associated energy use impacted by the EEMs from unaffected equipment, a measurement boundary is defined. Within this boundary, all energy consumed or produced must be measured or estimated using meters. It is crucial to account for all energy flows across the measurement boundary, including instances where energy may flow in reverse, such as in the case of on-site solar generation. The selection of a measurement boundary aligns with one or more of the four International Performance Measurement and Verification Protocol (IPMVP) options, which influence the level of detail in the reported savings and the required measurements. The choice of options should consider the purpose of the M&V reporting.

If the objective of the reporting is to verify the savings resulting from the impact of Energy Efficiency Measures (EEMs) on specific equipment, the measurement boundary should encompass the particular equipment, thereby facilitating the determination of measurement requirements within that boundary. This methodology aligns with the retrofit isolation option (Option A or B). In instances where the reporting objective involves verifying and/or managing the overall energy performance of the entire facility or validating savings derived from multiple EEMs with interactive effects, it is feasible to employ meters that measure the energy supply to the entire facility to evaluate performance and savings. In this scenario, the measurement boundary includes the entire facility, and the approach adopted is Option C: Whole Facility. When reliable or available data for the baseline or reporting period is lacking (e.g., in new construction projects), calibrated simulation models can furnish energy data for either a portion or the entirety of the facility. The measurement boundary can be defined accordingly, and the approach utilized is Option D: Calibrated Simulation.

5.2.2 Related Literature

Numerous approaches have been explored in the literature for modeling building energy consumption during the baseline period. ML-based energy modeling methods have consistently demonstrated superior performance compared to traditional linear regression techniques [26, 49], with their effectiveness validated across a wide array of studies in commercial [11, 13, 20, 23, 28, 35, 40, 48], residential [1, 10, 17], and industrial applications [12, 22, 32]. To identify the most effective ML approach and account for factors influencing model performance-such as input variables, training data quality and quantity, data granularity, and prediction requirements-it is essential to compare various modeling techniques under consistent conditions tailored to the project context.

In this vein, several studies have evaluated different ML techniques within the same application settings. For instance, Gallagher et al. [12] applied six ML methods-two- and multi-variable least squares regression, decision trees, k-nearest neighbors, artificial neural networks, and support vector machines-to M&V for energy savings in industrial buildings (specifically, a biomedical manufacturing facility). A single-layer feed-forward neural network trained on monthly data emerged as the most accurate model, reducing error by 51% compared to traditional methods. Similarly, Touzani et al. [42] introduced a gradient boosting machine and tested its performance on a large dataset of commercial buildings, comparing it to piecewise linear regression and a random forest (RF) algorithm. The gradient boosting machine demonstrated superior prediction accuracy in over 80% of test cases.

In a separate study, Li et al. [23] evaluated several modeling techniques, including backpropagation neural networks (BPNN), radial basis function neural networks (RBFNN), generalized regression neural networks (GRNN), and support vector machines (SVM), to predict the cooling load of an office building in China. The results indicated that SVM and GRNN achieved higher accuracy compared to BPNN and RBFNN. Furthermore, Agenis-Nevers et al. [1] examined eleven linear and nonlinear algorithms, incorporating four ML methods: k-nearest neighbors, support vector regression, locally weighted smoothing splines, and cubist tree-based algorithms, along with RFs. The analysis revealed that model performance varied across different building types, with no single algorithm demonstrating consistent superiority in all scenarios.

The accuracy of energy savings estimates heavily relies on the selection of input variables for the model. Key factors influencing building energy consumption must be meticulously selected and incorporated into baseline models utilizing the most pertinent measurements. These factors may include climate and weather conditions such as humidity, cooling or heating degree days, wind speed, and solar radiation [9, 19]; building usage characteristics like occupancy levels, building type, the number of working days per month, and operational schedules [19, 38]; temporal aspects such as season, month, day of the week, or time of day; and building-specific features like total heat loss coefficient, orientation, and insulation thickness [11]. Additionally, these variables can interact or combine, leading to the creation of new features for the model.

For instance, outdoor air humidity can be represented as dew point temperature, wet-bulb temperature, or enthalpy [31]. These variables can be further combined to derive new climatic metrics, such as "moist air enthalpy" (which integrates temperature and humidity) and "potential evapotranspiration," a composite variable proposed by Agenis-Nevers et al. [1] that incorporates air pressure, humidity, temperature, wind speed, and solar radiation. To identify the most effective variable sets, Agenis-Nevers et al. [1] proposed an automated selection method tailored to the characteristics of the analyzed buildings. Similarly, Catalina et al. [7] introduced novel model features, including the building's global heat loss coefficient, southern equivalent area, and the temperature differential between indoor set-points and solar temperature, to predict heating energy demand.

Other factors impacting prediction performance include the model's training period and the frequency of measurements and predictions. Gallagher et al. [12] investigated how training period characteristics affect model accuracy, finding that higher measurement frequencies reduced error variance in tested models. However, they also noted that more detailed data did not always lead to better performance. For example, in Jain et al. [17]'s study of multi-family residential buildings, hourly granularity produced the most accurate predictions compared to 10-minute or daily measurements. Agenis-Nevers et al. [1] also compared models with different measurement frequencies (monthly and daily consumption), though the comparisons were conducted on different buildings, limiting direct comparability.

In practical scenarios, selecting parameters and configuring models must account for the variability in dataset quality, sampling frequency, and resolution, as these factors can differ significantly in real-world applications.

5.3 Proposed Methodology

In this section, we outline the methodology developed for baseline energy modeling. First, we introduce the selected ML models-RF, XGBoost, and LightGBM-used for baseline energy prediction. Next, we describe the overarching methodological framework, which emphasizes the benefits of aggregating multiple ML models to improve prediction accuracy.

5.3.1 Model Architecture

In forecasting, it is well established that no single model consistently outperforms others across all predictive scenarios. This principle is equally applicable to M&V and baseline energy modeling. Consequently, integrating predictions from multiple models-each based on distinct assumptions and configurations-has emerged as a prevalent strategy to enhance prediction accuracy. The key rationale behind this approach is that errors from individual models can offset one another, leading to improved overall performance. Various forecast combination methods have been proposed, ranging from simple averaging techniques (e.g., mean, median, or mode) to more advanced approaches involving in-sample and out-of-sample error analyses, information criteria, Bayesian inference, and regression-based strategies.

In our study, which focuses on baseline energy prediction following ECM implementation, we propose a stacking ensemble model. This ensemble integrates three ML models-RF, XGBoost, and LightGBM-each offering unique strengths in capturing different aspects of energy consumption patterns. By combining the predictive capabilities of these models, we aim to maximize the accuracy of baseline energy estimates.

The stacking ensemble model aggregates predictions from the individual ML models. To further refine this process, we incorporate a filtering step using an averaging-based meta-model, as illustrated in Fig. 5.2. This additional layer optimizes the merging of predictions from the base models, leveraging their complementary strengths. Through this iterative process, the ensemble model achieves enhanced accuracy and robustness in its predictions.

The adoption of this stacking ensemble approach represents a significant advancement in the field of M&V and baseline energy modeling. By effectively synthesizing the strengths of diverse ML models, the proposed methodology addresses the limitations of individual models and enhances the accuracy of energy savings estimation.

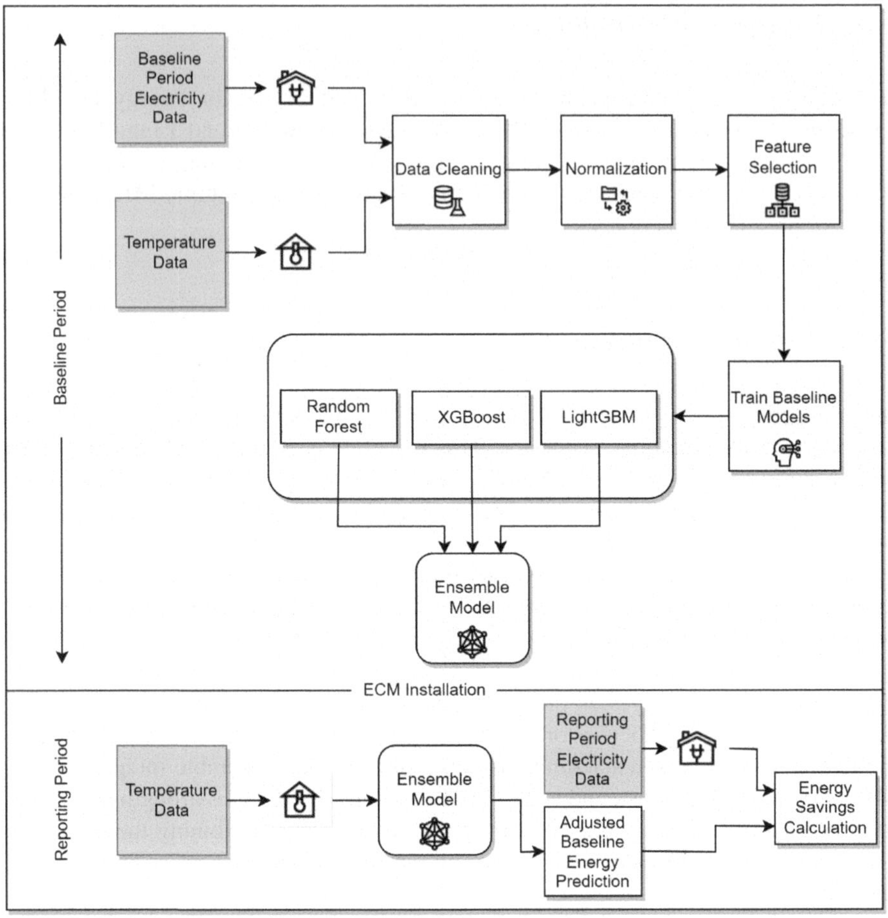

Fig. 5.2 The proposed methodological framework for estimating energy savings based on ML models and ensembling

5.3.2 *Baseline Regression Models*

Random Forest

The RF algorithm is a widely recognized and frequently employed technique in data science [6]. As a supervised machine learning algorithm, it is extensively utilized for addressing classification and regression problems. By constructing multiple decision trees based on discrete samples and integrating their outputs through majority voting or averaging, RF attains notable accuracy.

One of the key strengths of RF is its ability to process datasets containing both continuous and categorical variables, offering exceptional versatility for a broad array of classification and regression tasks. In the upcoming seminar, we will explore the detailed mechanics of the RF algorithm and showcase its application to an engaging classification challenge.

At the heart of the RF algorithm lies ensemble learning, a fundamental concept that enhances its performance. Ensemble learning combines multiple models to avoid dependence on a single one. In the case of RF, this approach leverages the principles of bagging [5], creating multiple training subsets from the original dataset using sampling with replacement. Each subset is used to train an independent decision tree, and the final prediction is derived by aggregating the outputs of all the trees [30].

RF offers several advantages for data practitioners. It excels at handling nonlinear relationships and noisy datasets [41], while requiring minimal data preprocessing compared to some other algorithms. However, it is important to note that RF prioritizes predictive accuracy over the interpretability of individual trees. Furthermore, the algorithm's complexity necessitates extensive parameter tuning to optimize performance.

XGBoost

Gradient Boosting (GB) is a flexible ensemble learning algorithm designed for both regression and classification tasks [27]. Its primary objective is to build a "strong learner" by iteratively combining multiple "weak learner" models, with each successive model aiming to correct the errors made by its predecessor [37]. Like other boosting methods, GB employs decision trees (either regressors or classifiers) trained on the negative gradient of the loss function at each iteration. This approach allows new models to specifically address the weaknesses of earlier ones, leading to a progressive improvement in overall performance. Once all the individual trees are incorporated, the final ensemble model predicts new samples.

XGBoost, an advanced variant of GB, stands out within the GB family due to its integration of sophisticated regularization techniques, such as $L1$ and $L2$ regularization, which enhance the model's ability to generalize. As an optimized implementation of the GB algorithm, XGBoost provides significant computational advantages over traditional GB methods, including reduced training times via parallelization across clusters [8]. Like RF, XGBoost aggregates ensembles of decision trees, but the key distinction lies in its sequential addition of trees to minimize prediction errors from previous iterations-highlighting the fundamental difference between bagging and boosting. The term "gradient boosting" reflects its use of the gradient descent optimization algorithm to minimize errors.

XGBoost offers several notable advantages that have established it as a leading GB implementation. For example, it employs a customized pruning strategy for decision trees, which accelerates training and enhances its capacity to handle large datasets [46]. XGBoost also features an approximate weighted quantile algorithm for node splitting, allowing it to bypass exhaustive evaluations of all potential splits. Furthermore, it partitions data into smaller segments for parallel processing and utilizes cache memory to expedite computations [47]. These optimizations have made XGBoost a preferred choice for a variety of classification and regression tasks in recent years.

As with RF, achieving high-accuracy predictions with XGBoost depends on tuning a critical set of hyperparameters. These include parameters governing the GB process, such as the number of trees, as well as those related to the structure of individual trees, like tree depth and the minimum loss reduction required for further node splits. For this study, the original XGBoost implementation library was employed.

LightGBM

LightGBM, introduced by Ke et al. [21], is a prominent gradient boosting (GB) framework similar to XGBoost. While both leverage the GB algorithm and decision trees, LightGBM emphasizes performance, scalability, and memory efficiency. It differs from XGBoost in its decision tree construction by employing a leaf-wise growth strategy, which selects the leaf with the highest potential loss reduction. Additionally, LightGBM uses a histogram-based learning algorithm to identify optimal split points, improving computational efficiency and reducing memory usage. Developed by researchers at Microsoft, LightGBM constructs decision trees sequentially, with each new tree refining the model by addressing residual errors from the previous tree. Its superior accuracy has been showcased in numerous machine learning competitions [25].

LightGBM's development was motivated by the need to optimize the decision tree construction process, which splits observations based on feature values to maximize information gain and reduce randomness. Identifying the best splits, a computationally intensive step, is central to tree-based learning [18]. Traditional GB implementations use either pre-sorted or histogram-based algorithms for split discovery. The pre-sorted method sorts feature values and evaluates all potential split points, while the histogram-based method discretizes continuous features into bins to create feature histograms. LightGBM adopts the histogram-based approach to address the efficiency challenges posed by large datasets.

To further enhance decision tree construction efficiency, LightGBM incorporates two innovative techniques: Gradient One-Side Sampling (GOSS) and Exclusive Feature Bundling (EFB). GOSS focuses on instances with larger gradient magnitudes to accelerate information gain, while EFB reduces the feature space by bundling mutually exclusive features. These optimizations make LightGBM highly effective for a wide range of machine learning tasks, including ranking, classification, and regression.

5.4 Case Study

This section presents a comprehensive experimental application of the proposed methodology utilizing data derived from actual renovation projects. It provides a detailed account of the training process for the primary classification methods, including the optimization of hyperparameters, followed by an evaluation of the meta-learning model. Finally, the application results are presented and analyzed in depth.

5.4.1 Dataset

The evaluation comprised data from eight actual buildings, and the subsequent subsection will focus on analyzing the accuracy metrics of the models. In this section, we concentrate on a specific building: a public library located in Gulbene, a city in northeastern Latvia. Similar renovations were conducted on the remaining seven buildings, including external wall insulation, roof insulation, modernization of the drainage system, replacement of doors and windows, and restoration of the insulation for the heating system's pipes.

The dataset consists of 1.5 years of hourly electricity consumption data collected through smart meters installed in the library. Additionally, hourly meteorological data were sourced from a weather station in Aluksne, a nearby city. To ensure data quality and reliability, preprocessing was conducted, included outlier removal and the handling of missing values. Missing data were addressed using linear interpolation, replacing gaps with values derived from the same hour on adjacent days. This dataset provides a robust foundation for analyzing energy consumption patterns and exploring their relationship with temperature variations. Figure 5.3 visually illustrates the temporal changes in temperature and electricity consumption. The upper section of the figure presents temperature data, while the lower section depicts electricity usage. This visualization facilitates the identification of significant patterns, trends, or correlations between the two variables.

The electricity consumption time series exhibits significant autocorrelation, as highlighted in Fig. 5.4, which illustrates the seasonal decomposition of the time series data. This decomposition categorizes electricity consumption into distinct components: trend, seasonality, and residuals. The top subplot presents the observed electricity values, followed by the trend component in the second subplot. The lower subplots depict the seasonality and residual components. This analysis offers valuable insights into the underlying patterns and variations within the electricity consumption data, confirming the presence of strong daily and weekly autocorrelation.

5.4.2 Experimental Design

This subsection outlines the training process for the electricity consumption prediction model and provides an overview of its features. The dataset used for training and testing spans 1.5 years of electricity consumption data, with one year dedicated to training the model and the remaining data reserved for testing.

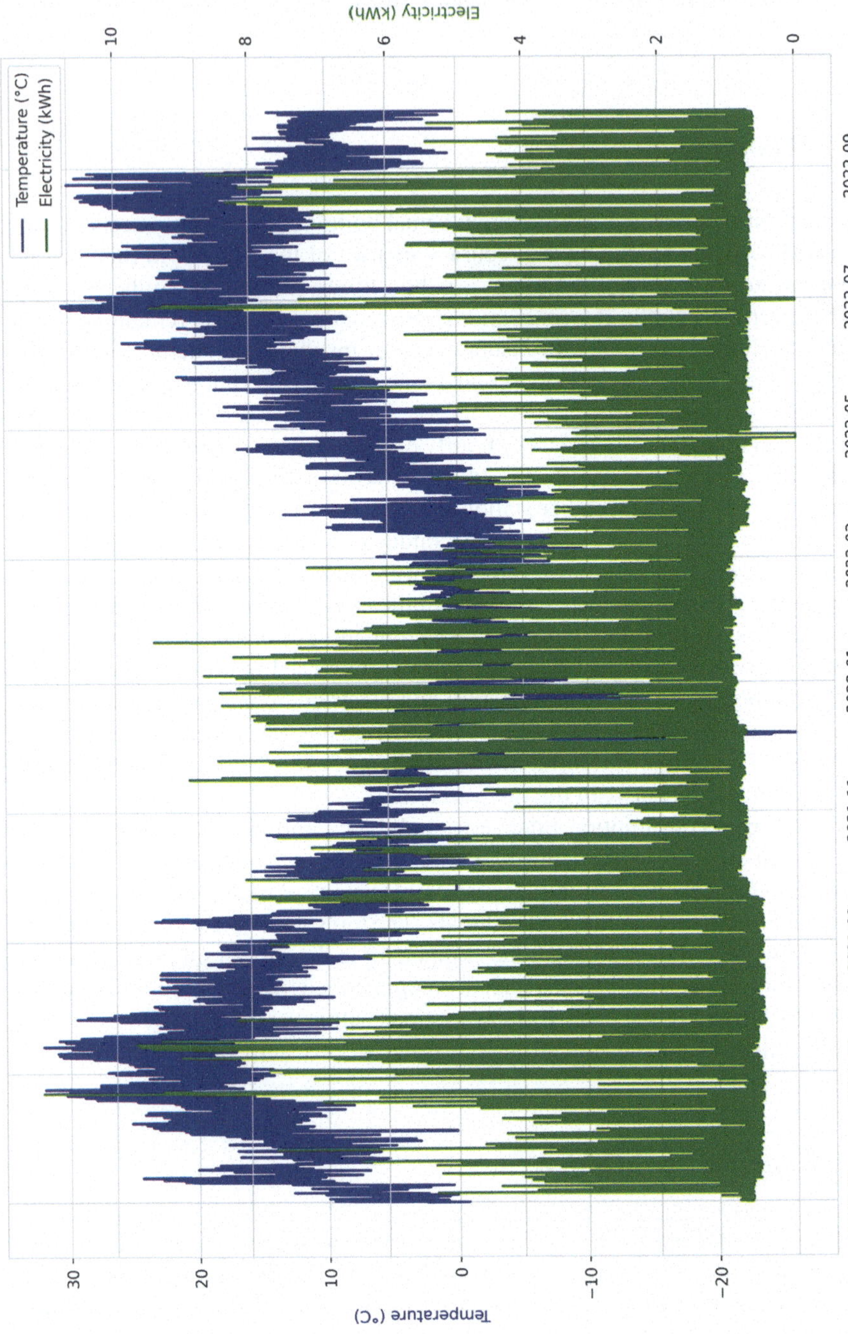

Fig. 5.3 Time-series of temperature (blue color) and electricity consumption (green color) over time

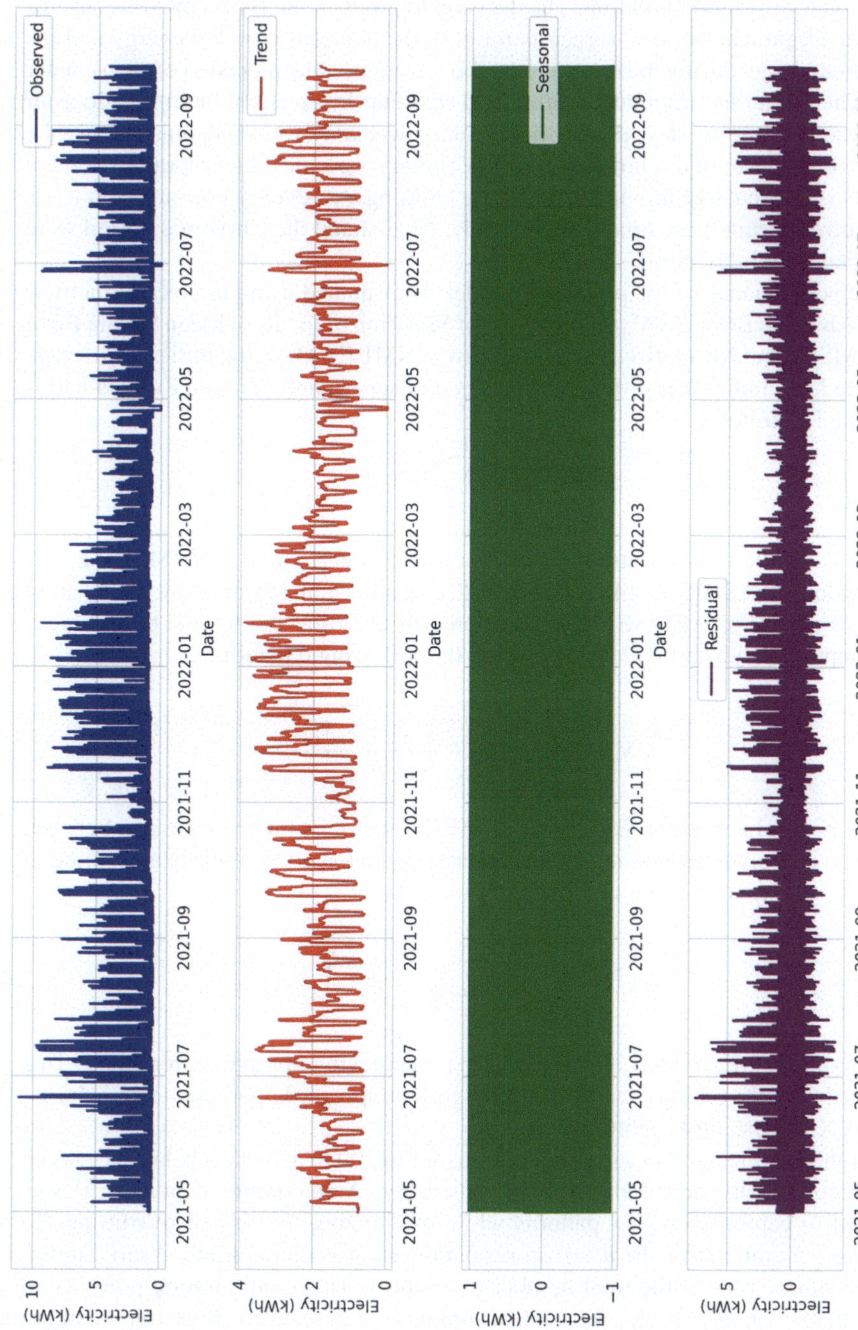

Fig. 5.4 Seasonal Decomposition of the Electricity consumption time-series

The prediction model is designed to generate hourly forecasts of energy consumption for the target building. The decision to produce hourly predictions ensures accurate alignment of point forecasts for both the adjusted baseline energy and the measured energy during the reporting period, leveraging the precision of smart meter data. The model incorporates weather and seasonal features, including outdoor air temperature, time of day, day of the week, a weekday/weekend binary variable, month of the year, and a holiday indicator (binary variable). Incorporating holiday periods is particularly important for office buildings, as energy consumption tends to be lower during these times. In this study, August and the Christmas period were identified as holiday periods [34].

The performance of the prediction models is evaluated using three error metrics: Mean Absolute Error (MAE), Coefficient of Variation of the Root Mean Square Error (CVRMSE), and Normalized Mean Bias Error (NMBE). MAE quantifies the average magnitude of individual differences between observed and predicted values and is computed as follows:

$$\mathrm{MAE} = \frac{1}{n}\sum_{i=1}^{n}\left|y_i - \hat{y}_i\right|$$

The Coefficient of Variation of the Root Mean Square Error (CVRMSE), as recommended by ASHRAE 14 (Ashrae, 2020), is a normalized measure of the Root Mean Square Error (RMSE). It is calculated by dividing the RMSE by the mean consumption value and is expressed using the following equation:

$$\mathrm{CVRMSE} = \frac{1}{\bar{y}}\sqrt{\frac{\sum_{i=1}^{n}(y_i - \hat{y}_i)^2}{n}}$$

The Normalized Mean Bias Error (NMBE) quantifies the average bias between observed and predicted values, expressed as a percentage of the mean consumption value. It is computed using the following formula:

$$\mathrm{NMBE} = \frac{1}{\bar{y}}\sum_{i=1}^{n}\left(\frac{y_i - \hat{y}_i}{y_i}\right)$$

Fine-tuning ML models is a critical step for achieving optimal performance. This section details the tuning process and highlights the specific hyperparameters selected for RF, XGBoost, and LightGBM.

For the RF, the number of estimators was set to 200 to achieve a balance between model complexity and computational efficiency. A maximum depth of 10 was selected to capture complex patterns while minimizing the risk of overfitting. To enhance generalization, the *maxfeatures* parameter was configured to "sqrt", utilizing the square root of the total number of features. This configuration prevents an over-reliance on any single feature. Bootstrapping was enabled (bootstrap=True) to introduce diversity into the training data and improve the robustness of the model.

For XGBoost, 500 estimators were employed to provide the model with adequate opportunities to capture complex patterns. A maximum depth of 6 was selected to balance the model's capability to capture intricate relationships while mitigating the risk of overfitting. The learning rate was set to 0.1, facilitating gradual weight updates to enhance convergence and improve generalization. For LightGBM, 300 estimators were used to allow the model to capture complex patterns while maintaining manageable training times. A maximum depth of 8 was selected to mitigate overfitting, consistent with best practices observed in RF and XGBoost. The number of leaves was set to 64, aligning with $2^{\text{Maximum Depth}}$, to balance model complexity and generalization. A learning rate of 0.05 was chosen to facilitate gradual and precise weight updates, promoting convergence and reducing the risk of instability.

5.4.3 Results

The performance evaluation focused on assessing five different models for predicting electricity consumption across eight buildings in Latvia. These models included a deep learning-based Artificial Neural Network (ANN), along with RF, XGBoost, LightGBM, and an ensemble model combining all the individual tree-based models.

The experimental results, summarized in Table 5.1, reveal that the ensemble model consistently outperformed RF, XGBoost, and LightGBM for all buildings. This superior performance is evidenced by the ensemble model's consistently lower CV(RMSE) values, highlighting the benefits of combining multiple models to improve the accuracy of electricity consumption predictions.

Interestingly, the performance of the deep learning model exhibited variability depending on the specific buildings analyzed. While the ANN outperformed RF, XGBoost, and LightGBM in some cases, its performance was comparable or slightly lower in others. This suggests that the effectiveness of deep learning architectures like ANN for electricity consumption prediction may be influenced by the unique characteristics of each building and the associated dataset.

These findings underscore the limitations of relying exclusively on deep learning models for electricity consumption forecasting. A more effective approach involves leveraging ensemble models that integrate non-deep learning methods. By combining diverse models, ensembles provide a robust and reliable framework for achieving accurate predictions. This highlights the importance of utilizing a variety of models and customizing the ensemble configuration to suit specific building scenarios.

Although deep learning models like ANN have demonstrated potential in certain cases, the ensemble model consistently outperforms both deep learning and other standalone models. This emphasizes the substantial advantages of ensemble methods in delivering precise and efficient energy consumption forecasts. The outcomes of this study offer valuable guidance on model selection and configuration for electricity consumption prediction, supporting enhanced energy management strategies and promoting greater energy efficiency in buildings.

Table 5.1 Building performance comparison using different forecasting models

Building	Error metric	ML models				
		ANN	RF	XGBoost	LightGBM	Ensemble
Building 1	CV(RMSE)	**0.10**	0.12	0.13	0.11	**0.10**
	NMBE	−0.02	**−0.01**	−0.03	−0.02	**−0.01**
	MAE	5.25	5.28	**5.23**	5.27	**5.23**
Building 2	CV(RMSE)	0.11	0.12	0.11	**0.10**	**0.10**
	NMBE	−0.01	−0.02	**−0.01**	0.00	0.00
	MAE	4.92	4.95	**4.88**	4.91	4.91
Building 3	CV(RMSE)	**0.09**	0.11	0.12	0.10	**0.09**
	NMBE	−0.01	−0.03	−0.02	−0.01	−0.01
	MAE	6.03	6.08	**5.97**	6.02	6.02
Building 4	CV(RMSE)	0.14	0.15	0.16	0.14	**0.13**
	NMBE	0.01	0.02	0.03	0.01	**0.00**
	MAE	5.70	5.73	**5.65**	5.69	5.70
Building 5	CV(RMSE)	**0.12**	0.13	0.14	0.12	**0.12**
	NMBE	−0.02	**−0.01**	−0.03	−0.02	−0.02
	MAE	6.18	6.22	**6.15**	6.19	6.19
Building 6	CV(RMSE)	0.13	0.15	0.15	0.14	**0.12**
	NMBE	−0.02	−0.01	−0.03	−0.02	**−0.01**
	MAE	5.42	5.45	**5.38**	5.41	5.39
Building 7	CV(RMSE)	0.10	0.12	0.11	**0.09**	**0.09**
	NMBE	0.00	0.01	−0.01	0.00	**0.00**
	MAE	4.97	5.00	4.96	**4.95**	4.95
Building 8	CV(RMSE)	**0.08**	0.10	0.11	0.09	**0.08**
	NMBE	−0.01	−0.02	−0.03	−0.02	**−0.01**
	MAE	5.88	5.92	**5.84**	5.89	5.85

5.5 Conclusions

Estimating energy savings from building upgrades is essential for multiple reasons. Leveraging ML models facilitates the optimization of energy consumption and enhances overall building energy efficiency, yielding significant economic and environmental benefits. This study demonstrates that ensemble models outperform individual tree-based models, showcasing their ability to combine multiple algorithms for increased accuracy and robustness in energy savings estimation. By harnessing the predictive strengths of different algorithms, ensemble techniques significantly improve overall performance. The success of the ensemble model lies in the complementary nature of its constituent models, which address individual weaknesses and enhance predictive capabilities. Integrating the diverse perspectives and learning strategies of XGBoost, RF, and LightGBM enables the ensemble model to effectively capture complex patterns and relationships within energy data, resulting in more precise estimations.

To advance energy savings estimation in buildings, several critical steps are required. First, standardizing data collection protocols is essential to ensure consistency and comparability across studies. This includes harmonizing data formats, defining common variables, and establishing standardized M&V procedures. Standardized practices promote data sharing, collaborative methodologies, and ensure the reproducibility and validation of results among researchers and practitioners.

Moreover, achieving consensus on error metrics for evaluating energy savings estimation models is crucial. While this study utilized metrics such as CV(RMSE), NMBE, and MAE, establishing a standardized set of metrics specifically tailored for energy savings estimation would enable meaningful comparisons and benchmarking. This common framework for assessing model accuracy would support the development of best practices and foster progress in the field.

References

1. M. Agenis-Nevers, Y. Wang, M. Dugachard, R. Salvazet, G. Becker, D. Chenu, Measurement and verification for multiple buildings: an innovative baseline model selection framework applied to real energy performance contracts. Energy Build. **249**, 111183 (2021)
2. R. Agrawal, L. De Tommasi, P. Lyons, S. Zanoni, G. Papagiannis, C. Karakosta, A. Papapostolou, A. Durand, L. Martinez, G. Fragidis et al., Challenges and opportunities for improving energy efficiency in smes: learnings from seven european projects. Energ. Effi. **16**(3), 17 (2023)
3. C.A. Balaras, E.G. Dascalaki, K.G. Droutsa, S. Kontoyiannidis, Empirical assessment of calculated and actual heating energy use in Hellenic residential buildings. Appl. Energy **164**, 115–132 (2016)
4. P. Bertoldi, M. Economidou, V. Palermo, B. Boza-Kiss, V. Todeschi, How to finance energy renovation of residential buildings: review of current and emerging financing instruments in the eu. Wiley Interdiscip. Rev.: Energy Environ. **10**(1), e384 (2021)
5. L. Breiman, Bagging predictors. Mach. Learn. **24**(2), 123–140 (1996)
6. L. Breiman, Bagging predictors. Mach. Learn. **45**(1), 5–32 (2001)
7. T. Catalina, V. Iordache, B. Caracaleanu, Multiple regression model for fast prediction of the heating energy demand. Energy Build. **57**, 302–312 (2013)
8. T. Chen, C. Guestrin, XGBoost: a scalable tree boosting system, in *Proceedings of the 22nd ACM SIGKDD International Conference on Knowledge Discovery and Data Mining*, KDD '16, pp. 785–794, New York, NY, USA (ACM, 2016)
9. J.A. Díaz, J.S. Ramos, M.C.G. Delgado, D.H. García, F.G. Montoya, S.Á. Domínguez, A daily baseline model based on transfer functions for the verification of energy saving. A case study of the administration room at the palacio de la madraza, granada. Appl. Energy **224**, 538–549 (2018)
10. R.E. Edwards, J. New, L.E. Parker, Predicting future hourly residential electrical consumption: a machine learning case study. Energy Build. **49**, 591–603 (2012)
11. B.B. Ekici, U.T. Aksoy, Prediction of building energy consumption by using artificial neural networks. Adv. Eng. Softw. **40**(5), 356–362 (2009)
12. C.V. Gallagher, K. Bruton, K. Leahy, D.T. O'Sullivan, The suitability of machine learning to minimise uncertainty in the measurement and verification of energy savings. Energy Build. **158**, 647–655 (2018)

13. J. Granderson, S. Touzani, C. Custodio, M.D. Sohn, D. Jump, S. Fernandes, Accuracy of automated measurement and verification (m&v) techniques for energy savings in commercial buildings. Appl. Energy **173**, 296–308 (2016)
14. C. Granell, A. Kamilaris, A. Kotsev, F.O. Ostermann, S. Trilles, Internet of things, in Manual of Digital Earth, pp. 387–423 (2020)
15. J.L. Hernández, S. Serroni, S. Stamatopoulos, E. Sarmas, G.M. Revel, F. Vélez, Multi-criteria building performance assessment, in *2024 15th International Conference on Information, Intelligence, Systems & Applications (IISA)* (IEEE, 2024), pp. 1–7
16. I. IEA, Energy technology perspectives 2020. International Energy Agency (2020)
17. R.K. Jain, K.M. Smith, P.J. Culligan, J.E. Taylor, Forecasting energy consumption of multi-family residential buildings using support vector regression: investigating the impact of temporal and spatial monitoring granularity on performance accuracy. Appl. Energy **123**, 168–178 (2014)
18. D. Jin, Y. Lu, J. Qin, Z. Cheng, Z. Mao, Swiftids: real-time intrusion detection system based on lightgbm and parallel intrusion detection mechanism. Comput. & Secur. **97**, 101984 (2020)
19. S. Kapp, J.-K. Choi, T. Hong, Predicting industrial building energy consumption with statistical and machine-learning models informed by physical system parameters. Renew. Sustain. Energy Rev. **172**, 113045 (2023)
20. S. Karatasou, M. Santamouris, V. Geros, Modeling and predicting building's energy use with artificial neural networks: methods and results. Energy Build. **38**(8), 949–958 (2006)
21. Ke, G., Meng, Q., Finley, T., Wang, T., Chen, W., Ma, W., Ye, Q., and Liu, T.-Y. (2017). Lightgbm: A highly efficient gradient boosting decision tree. *Advances in neural information processing systems*, 30
22. J.K. Kissock, C. Eger, Measuring industrial energy savings. Appl. Energy **85**(5), 347–361 (2008)
23. Q. Li, Q. Meng, J. Cai, H. Yoshino, A. Mochida, Predicting hourly cooling load in the building: a comparison of support vector machine and different artificial neural networks. Energy Convers. Manag. **50**(1), 90–96 (2009)
24. L. Lundström, F. Wallin, Heat demand profiles of energy conservation measures in buildings and their impact on a district heating system. Appl. Energy **161**, 290–299 (2016)
25. S. Makridakis, E. Spiliotis, V. Assimakopoulos, M5 accuracy competition: results, findings, and conclusions. Int. J. Forecast. (2022)
26. R.F. Mustapa, N.Y. Dahlan, A.I.M. Yassin, A.H.M. Nordin, Quantification of energy savings from an awareness program using narx-ann in an educational building. Energy Build. **215**, 109899 (2020)
27. A. Natekin, A. Knoll, Gradient boosting machines, a tutorial. Front. Neurorobot. **7**, 21 (2013)
28. A.H. Neto, F.A.S. Fiorelli, Comparison between detailed model simulation and artificial neural network for forecasting building energy consumption. Energy Build. **40**(12), 2169–2176 (2008)
29. N. Oses, A. Legarretaetxebarria, M. Quartulli, I. García, M. Serrano, Uncertainty reduction in measuring and verification of energy savings by statistical learning in manufacturing environments. Int. J. Interact. Design Manuf. (IJIDeM) **10**, 291–299 (2016)
30. A.M. Prasad, L.R. Iverson, A. Liaw, Newer classification and regression tree techniques: bagging and random forests for ecological prediction. Ecosystems **9**(2), 181–199 (2006)
31. P. Price, Methods for analyzing electric load shape and its variability. Technical report, Lawrence Berkeley National Lab. (LBNL), Berkeley, CA (United States) (2010)
32. F. Rossi, D. Velázquez, A methodology for energy savings verification in industry with application for a chp (combined heat and power) plant. Energy **89**, 528–544 (2015)
33. P. Samaras, E. Stamatopoulos, A. Arsenopoulos, E. Sarmas, E. Marinakis, Readiness to adopt the smart readiness indicator scheme across Europe: a multi-criteria decision analysis approach, in *2024 IEEE International Workshop on Metrology for Living Environment (MetroLivEnv)* (IEEE, 2024), pp. 268–273
34. E. Sarmas, A. Forouli, V. Marinakis, H. Doukas, Baseline energy modeling for improved measurement and verification through the use of ensemble artificial intelligence models. Inf. Sci. **654**, 119879 (2024)

35. E. Sarmas, E. Spiliotis, N. Dimitropoulos, V. Marinakis, H. Doukas, Estimating the energy savings of energy efficiency actions with ensemble machine learning models. Appl. Sci. **13**(4), 2749 (2023)
36. E. Sarmas, E. Spiliotis, V. Marinakis, T. Koutselis, H. Doukas, A meta-learning classification model for supporting decisions on energy efficiency investments. Energy Build. **258**, 111836 (2022)
37. R.E. Schapire, The boosting approach to machine learning: an overview. *Nonlinear Estimation and Classification* (2003), pp. 149–171
38. P. Skaloumpakas, E. Sarmas, Z. Mylona, A. Cavadenti, F. Santori, V. Marinakis, Predicting thermal comfort in buildings with machine learning and occupant feedback, in *2023 IEEE International Workshop on Metrology for Living Environment (MetroLivEnv)* (IEEE, 2023), pp. 34–39
39. E. Stamatopoulos, A. Forouli, D. Stoian, P. Kouloukakis, E. Sarmas, V. Marinakis, An adaptive framework for assessing climate resilience in buildings. Build. Environ. **264**, 111869 (2024)
40. D. Stoian, E. Spiliotis, E. Stamatopoulos, E. Sarmas, P. Rekomaa, V. Marinakis, A neural network-based approach for hierarchical electricity consumption forecasting in buildings, in *2024 IEEE International Workshop on Metrology for Living Environment (MetroLivEnv)* (IEEE, 2024), pp. 274–279
41. C. Strobl, J. Malley, G. Tutz, An introduction to recursive partitioning: rationale, application, and characteristics of classification and regression trees, bagging, and random forests. Psychol. Methods **14**(4), 323 (2009)
42. S. Touzani, J. Granderson, S. Fernandes, Gradient boosting machine for modeling the energy consumption of commercial buildings. Energy Build. **158**, 1533–1543 (2018)
43. V. Tsiatsis, S. Karnouskos, J. Holler, D. Boyle, C. Mulligan, *Internet of Things: Technologies and Applications for a New Age of Intelligence* (Academic, 2018)
44. C. Tsolkas, E. Spiliotis, E. Sarmas, V. Marinakis, H. Doukas, Dynamic energy management with thermal comfort forecasting. Build. Environ. **237**, 110341 (2023)
45. D. Ürge-Vorsatz, L. Danny Harvey, S. Mirasgedis, M.D. Levine, Mitigating co2 emissions from energy use in the world's buildings. Buildi. Res. & Inf. **35**(4), 379–398 (2007)
46. C. Wade, *Hands-On Gradient Boosting with XGBoost and Scikit-learn: Perform Accessible Machine Learning and Extreme Gradient Boosting with Python.* (Packt Publishing Ltd, 2020)
47. Z. Wen, J. Shi, B. He, J. Chen, K. Ramamohanarao, Q. Li, Exploiting gpus for efficient gradient boosting decision tree training. IEEE Trans. Parallel Distrib. Syst. **30**(12), 2706–2717 (2019)
48. S.L. Wong, K.K. Wan, T.N. Lam, Artificial neural networks for energy analysis of office buildings with daylighting. Appl. Energy **87**(2), 551–557 (2010)
49. K. Ye, G. Demirezen, A. Fung, E. Janssen, The use of artificial neural networks (ann) in the prediction of energy consumption of air-source heat pump in retrofit residential housing, in *IOP Conference Series: Earth and Environmental Science*, vol. 463 (IOP Publishing, 2020), p. 012165

Part III
Advances in Energy Efficiency

Chapter 6
Meta-Learning Approaches for Assessing Energy Efficiency Investments in Buildings

Abstract This chapter presents a methodology aimed at bridging the financing gap for energy efficiency projects. The energy efficiency of buildings is critical for achieving global energy and climate goals; however, it necessitates substantial investments. The absence of mature decision-support systems and the reliance on traditional investment mechanisms, which emphasize the economic dimensions of energy efficiency projects while neglecting their environmental impacts, can impede the ability of such projects to secure funding. In this chapter, various classification methods are evaluated and subsequently integrated through a meta-learning model designed to enhance overall classification performance and allocate funding to each investment based on its specific characteristics. The proposed methodology is assessed using a dataset comprising more than 300 completed renovation projects in buildings located in Latvia. The results demonstrate that the meta-learning model achieves superior accuracy compared to all baseline machine learning models, effectively identifying high- and medium-potential projects while distinguishing low-potential projects.

Keywords Artificial intelligence · Energy efficiency · Machine learning · Meta-learning · Building renovation

6.1 Introduction

This chapter explores the challenges associated with evaluating the dynamics of energy efficiency (EE) investments, particularly concerning renovation costs and the energy savings they generate. The lack of advanced information systems to facilitate decision-making frequently leads to renovation projects failing to obtain financial support from funding institutions, thereby jeopardizing global environmental objectives for the coming decades. To address this issue, the chapter proposes a methodology designed to accurately identify promising EE projects and estimate the required capital investments on an individual basis. Furthermore, it analyzes critical investment characteristics of completed projects and employs meta-learning (ML) techniques to predict the dynamics of future initiatives.

E. Sarmas et al., *Artificial Intelligence for Energy Systems*, Learning and Analytics in Intelligent Systems 46, https://doi.org/10.1007/978-3-031-85209-1_6

Climate change is already impacting daily life and threatening Earth's ecosystems, prompting urgent warnings from scientists, governments, and institutions regarding navigation through a future fraught with uncertainty. The Paris Agreement represents the latest effort to curb CO_2 emissions and limit temperature increases, mandating that the global temperature rise must remain below 2 °C in the coming decades [19]. Achieving this goal necessitates significant measures, such as the complete decarbonization of the energy sector by 2050, as recommended by the International Renewable Energy Agency (IRENA) [31]. However, the global transition to green energy is uneven, shaped by differing developmental stages, historical and cultural contexts, and varying economic, social, political, and environmental conditions across countries [56]. Consequently, meeting these global targets relies heavily on the actions of individual countries and sectors, which frequently operate in discord with respect to greenhouse gas emissions.

These challenges underscore the necessity for complementary strategies aimed at mitigating energy footprints, particularly through the reduction of energy consumption. One such strategy involves promoting EE investments, particularly in building renovations, while also encouraging citizen participation in community renewable energy projects. Factors driving this participation include environmental awareness and aspirations for energy self-sufficiency [7], though social influences such as trust and social norms also play a role [64]. However, the resistance of older generations to adopting climate-friendly habits [2] underscores the importance of deploying more efficient technologies. In this context, smart cities and building renovation initiatives emerge as pivotal for advancing the green energy transition.

The International Energy Agency estimates that implementing energy efficiency (EE) policies could account for nearly 36% of the avoided greenhouse gas emissions by 2050 [28, 29], with the majority of this reduction anticipated to stem from building renovation projects [62]. However, while these projections are encouraging, the practical implementation of EE investments faces significant obstacles. EE projects are often fragmented, and their costs and energy savings may be unknown at the outset or challenging to estimate accurately. This lack of clarity leaves stakeholders-such as private financial institutions, investment funds, national and regional authorities, and energy solution providers-without reliable decision-making tools to forecast the effectiveness of future EE projects [4]. Developing countries, despite their considerable potential for energy savings, encounter additional challenges such as limited access to appropriate financial mechanisms [46].

Financing large-scale EE investments is further hindered by a potential scarcity of sufficient data[1] to evaluate their utility. Even when such data is available, it is not consistently utilized to inform decision-making effectively. Funding entities employ a variety of methods, standards, and data sources to assess the risks and benefits associated with future investments [25], applying diverse indicators that account for factors such as economic structure, climate, and geography. The barriers to renewable energy investments vary across technologies and countries [45]. In recognition of

[1] For example, many current projects are small-scale or at a pilot level, complicating efforts to benchmark effectively.

these challenges, the European Union has supported research initiatives such as EEnvest, Triple-A, and Quest: 2050 to mitigate the risks linked to renewable energy investments [38]. Despite these initiatives, EE projects are frequently viewed as higher-risk ventures, leading financial institutions to favor other types of investments or traditional projects [52].

This study proposes a data-driven methodology to tackle the challenges of EE investment evaluation. Leveraging advancements in machine learning (ML) and its success in classification tasks, we introduce a meta-learning model[2] to predict the utility of future EE projects in terms of costs and energy savings. Using an extensive dataset that includes (i) renovation costs, (ii) current energy consumption, (iii) specific building characteristics, and (iv) both expected and (v) realized energy savings from multiple past EE projects, we train various classification models to evaluate the attractiveness of future investments (categorized as high, medium, or low). A meta-learner then determines the most suitable classifier for making predictions based on the unique characteristics of each project, further enhancing prediction accuracy.

We demonstrate the advantages of this approach and illustrate its practical applications, showing how stakeholders can diversify their EE investments based on expected utility, thereby improving decision-making and mitigating perceived risks.

The key contributions of this study are as follows:

- We introduce an innovative meta-learning model designed to systematically assess the utility of future EE projects based on costs and actual energy savings. This model supports financial institutions in making informed decisions and is straightforward to replicate, allowing seamless integration into existing institutional frameworks at minimal cost.
- Rather than classifying future EE investments based on theoretical indicators that estimate their efficiency, we derive our recommendations from the actual utility realized by similar past EE projects. This empirical approach enables us to learn from both their successes and limitations, providing tailored recommendations that account for the unique characteristics of each new investment.
- To address the risks associated with relying on a single classifier-such as reduced accuracy and robustness when handling small, imbalanced, or noisy datasets-for evaluating future EE investments, we propose employing a meta-learner. This meta-learner selects the classifier most likely to provide reliable and accurate predictions for each specific scenario.
- We present our classification predictions as probabilities to account for uncertainty, leveraging these results to dynamically generate recommendations on the appropriate percentage of subsidized funding that future EE investments should receive.

[2] Meta-learning uses ML techniques to learn how to optimize or combine predictions from other ML models.

The chapter is structured as follows: Sect. 6.2 outlines the problem framework and reviews the relevant literature. Section 6.3 details the proposed methodology. Section 6.4 presents an experimental application based on a real case study conducted in Latvia. Finally, Sect. 6.5 offers concluding remarks.

6.2 Problem Formulation and Related Work

It is widely recognized that energy consumption in buildings can be significantly reduced through the implementation of existing energy-saving technologies. However, EE projects often encounter challenges in securing funding due to uncertainties and the absence of a comprehensive methodology for evaluating and comparing their potential [34]. For instance, Intrachooto and Horayangkura [30] identified factors such as the interdependence of renovation components, speculative financial performance, and a low-cost mindset as key contributors to financial barriers. Currently, building EE renovations are primarily evaluated using traditional investment mechanisms established by large financial institutions and banks. While numerous projects have been financed through these methods, a considerable amount of potential remains untapped [63].

In recent years, advancements in information and communication technologies, including the Internet of Things and artificial intelligence, have facilitated the collection of large volumes of diverse data, leading to innovative solutions and analyses [51, 58]. Specifically, in EE financing, data gathered from smart meters post-renovation now provides valuable insights into actual project performance. Despite these technological advances, EE financing for buildings remains relatively unexplored and underfunded, as noted by Zhang et al. [70]. This underscores the need for future research to focus on the application of artificial intelligence methods [16, 60] and the optimization of available data to develop tools capable of accurately evaluating EE investments [33].

Few studies have proposed a classification framework[3] to assist the financial sector in identifying promising renovation projects [53, 54]. A review of EE assessment methodologies is provided in Tuominen et al. [66], which highlights that economic indicators are often neglected or superficially addressed in EE research. According to this study, the most commonly considered economic objectives are payback periods and cost mitigation. Research focused on optimal cost analysis for energy renovations can be found in Stocker et al. [59], de Vasconcelos et al. [15], and Ortiz et al. [44], which examine projects in the Alps, Portugal, and Catalonia, respectively. Beyond cost and payback periods, financial risk is also employed as a complementary objective [22]. However, economic efficiency is not an isolated consideration but rather a subset of the broader EE concept, which also encompasses environmental and social dimensions [4].

[3] These frameworks may involve binary predictions (investment vs. no investment) or multi-category classifications (e.g., low, medium, or high investment potential).

In this context, our study incorporates additional parameters into the EE evaluation process, including current energy consumption, building characteristics (e.g., construction year, heated area, and number of apartments), and the CO_2 reductions expected from EE investments. These factors are treated as fundamental components of a holistic approach to evaluating EE projects [26].

Our study introduces a novel application of machine learning (ML) and meta-learning methods [39] to the evaluation of energy efficiency (EE) investments. Most prior research has relied on frameworks such as multi-criteria decision analysis [55]-using either discrete methods [47, 68] or continuous multi-objective optimization approaches [1]-as well as cost-benefit analysis [42] and cost-effectiveness analysis [21, 66]. While these approaches provide valuable insights, they require a comprehensive understanding of the economic, social, and climatic factors influencing each investment, and their performance heavily depends on the assumptions underlying their methodologies. In contrast, ML methods offer data-driven alternatives to traditional EE assessment frameworks [69]. A recent review of ML applications in energy economics and finance [24] revealed that ML has primarily been used for predicting crude oil and electricity prices. However, applying ML to classify EE investments based on their expected utility remains an underexplored area.

This study seeks to advance ML methods and data analysis techniques for assessing EE investments. In Doukas et al. [20], traditional statistical classification methods, such as ordinal logit, ordinal probit, and linear discriminant analysis, were integrated with a limited number of ML techniques, including k-nearest neighbors and support vector machines. Building on this foundation, our research further explores advanced methods such as Gaussian naive Bayes, extreme gradient boosting trees, and random forests, with the objective of enhancing the predictive accuracy of higher-capacity ML models.

A key innovation of this research is the integration of a meta-learner within the proposed framework. The meta-learner determines the most appropriate classification method for each investment, taking into account its specific characteristics. This optimization of the classification process enhances its practical utility. The primary contribution of this study lies in its utilization of a meta-learning model to mitigate the risks associated with reliance on a single classifier, particularly in scenarios involving small or imbalanced datasets. In contrast to prior research, which primarily tailored models to specific datasets, this study introduces a generalized framework capable of delivering accurate and robust predictions across diverse datasets, independent of feature selection or data volume. This approach enhances reliability and broadens the applicability of ML techniques in evaluating EE investments.

6.3 Proposed Methodology

The methodology we present aims to offer grant funding recommendations for future EE projects by leveraging insights from a dataset of previously implemented projects with similar characteristics. The core of our approach is a stacking ensemble classification model that integrates five key classification methods: *k-Nearest Neighbors*, *Gaussian Naive Bayes*, *Extreme Gradient Boosted Trees*, *Random Forest*, and *Support Vector Machine*.

The classifiers are trained on a dataset comprising information about past renovations conducted on various buildings. This dataset includes data initially available before the renovation (used as features) and recorded data after the investment's completion (used as labels).

In practice, to facilitate automation, this information should be stored in the database of the information system utilized by funding institutions for evaluating future EE investments.

6.3.1 Model Architecture

The investment features pertain to specific attributes of the projects, which in this study include the building's age, the number of apartments, the total heated area, current energy consumption, anticipated CO_2 reduction, and renovation costs. The outputs of the base classifiers are the probabilities that an investment falls into one of the following three distinct classes:

- *Class A*: The project should be funded.
- *Class B*: The project should be partially funded.
- *Class C*: The project should not be funded.

To label past investments, the realized utility of each project is determined using a specific selection metric. In this study, we assess the efficiency ranking of each investment, as detailed in Sect. 6.4.1. Based on this ranking, the top-performing projects are labeled as *Class A*, the lowest-performing ones as *Class C*, and the remainder as *Class B*.

Once the labeling process is complete, the base classification methods are employed to classify past projects, and their predictions are documented. These predictions, in conjunction with the actual labels of the investments and the features initially utilized by the base classifiers, are subsequently used to train a meta-learner-a logistic regression classifier. This meta-learner discerns the most suitable base method for predicting the class of each project.

Using this trained meta-learner, the "optimal" classifier is selected for each investment to calculate the probabilities for each class. These probabilities are subsequently combined to generate a recommendation on the appropriate grant percentage for future investments, as described in Sect. 6.3.4. The complete recommendation process is illustrated in Fig. 6.1.

The proposed methodology framework is highly adaptable with respect to investment features, classes, methods, and utility metrics. Decision-makers can customize the input variables used by the classifiers based on their preferences or the availability of data. Similarly, the set of base classifiers can be modified to include different methods or a varying number of classifiers. Furthermore, the number of classification labels and the metrics or approaches used to define them can be adjusted. Consequently, the specific choices made in this study to construct and demonstrate the framework should be regarded as indicative, though they incorporate a broad range of parameters.

This flexibility is complemented by the use of a meta-learner, which enhances classification performance. Different classification methods exhibit varying strengths based on their algorithmic characteristics. For instance, tree-based methods like XGBoost and Random Forest (RF) excel at identifying the most relevant features by evaluating their contribution to reducing prediction error during model training. These methods also perform well when handling a large number of features. On the other hand, methods such as RF and Support Vector Machines (SVM) are robust to data randomness, while simpler methods like k-Nearest Neighbors (k-NN) and Gaussian Naive Bayes are effective with smaller training datasets due to their lower data requirements. However, the latter methods are less capable of automatically selecting the most relevant features and may be negatively impacted when multiple input variables are provided.

The meta-learner is designed to address these nuances and challenges. By accounting for the specific strengths and weaknesses of individual classifiers, it mitigates the risks associated with relying on a single method for evaluating future EE investments. This ensures the most reliable predictions possible, tailored to the dataset and classification problem at hand.

6.3.2 Baseline Classification Models

The proposed methodology incorporates five base classification methods, chosen to provide the meta-learner with a diverse set of options for selecting the most suitable classifier. This diversity[4] arises from the varying assumptions each method makes

---

[4] Diversity refers to the distinct algorithmic characteristics of the selected ML methods, including their strengths and weaknesses in terms of accuracy and robustness. These are particularly relevant when dealing with small, imbalanced, or noisy datasets, as well as managing multiple features. These properties are summarized in the introductory part of Sect. 6.3 and further detailed in the following subsections.

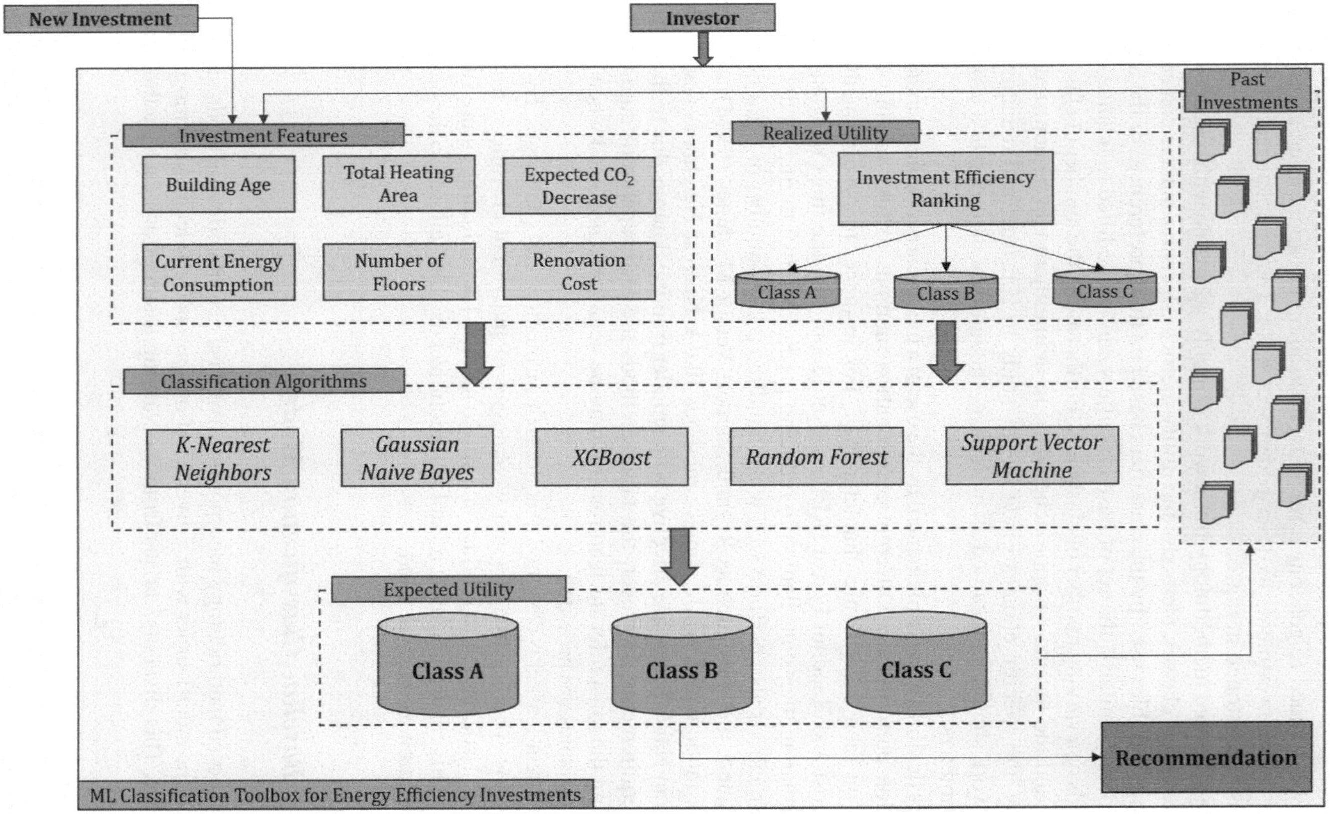

Fig. 6.1 Overview of the proposed methodology. The recommendation on the grant financing percentage of future investments is based on a data-driven approach that exploits useful insights from past projects, labeled based on the achieved energy consumption reduction and the investment cost (realized utility). The output of the baseline classification methods, as well as the meta-learner, provides the percentage of recommended grant financing

about the prediction task. This diversity has been shown to effectively enhance overall accuracy [61]. A brief overview of the base classification methods is provided below.

k-Nearest Neighbors

The k-Nearest Neighbors (k-NN) algorithm is a non-parametric classification method [14]. It uses n features to characterize each observation (in this case, an EE investment) and assigns a label indicating its class (the realized utility of the investment). The algorithm classifies new, unlabeled observations by identifying the k most similar labeled observations (nearest neighbors) and assigning the majority class among these neighbors.

The method operates in two phases: training and testing. During the training phase, labeled observations-represented as vectors in a multidimensional space-are stored with their respective class labels, and the number of neighbors (k) is specified. In the testing phase, unlabeled vectors are classified by associating them with the class most frequently occurring among the k closest labeled observations. The notion of "closest" is determined using various distance metrics, depending on the nature of the input features. For continuous variables, the Euclidean distance is most commonly used, while for discrete variables, the Hamming distance is prevalent [43]. Additionally, correlation-based measures, such as Pearson correlation, can be applied in some cases [36]. To enhance accuracy, normalizing feature values is recommended since the method is sensitive to variations in input scales [65].

The selection of k-the algorithm's primary hyperparameter-is critical to classification performance. A value of $k = 1$ classifies observations based on their single closest neighbor, while k set to the total number of observations in the training set results in classification by the majority label across all training data. The "optima" k is typically determined through heuristic methods, such as cross-validation, which are widely used in ML literature for hyperparameter tuning.

In this study, the features under consideration are continuous variables, and the Euclidean distance is used to identify the nearest neighbors. For two points p and q in an n-dimensional Euclidean space, the distance is calculated as:

$$d(p, q) = \sqrt{(p_1 - q_1)^2 + (p_2 - q_2)^2 + \cdots + (p_i - q_i)^2 + \cdots + (p_n - q_n)^2}. \tag{6.1}$$

Gaussian Naive Bayes

The family of probabilistic classifiers includes classifiers capable of predicting the probability distribution over a given set of classes, rather than providing a single class prediction to which the observation should belong [41]. Naive Bayes classifiers are a subset of probabilistic classifiers that rely on Bayes' theorem, assuming strong independence between features. The Gaussian naive Bayes classifier has been applied in various fields, including tasks related to evaluating investment performance [13], forecasting photovoltaic production [5], and analyzing the energy efficiency of buildings [50].

According to Zhang [71], given a feature vector x_1 to x_n and a class variable y, the following relationship is expressed by Bayes' theorem.

$$p(y \mid x_1, \ldots, x_n) = \frac{p(y)\, p(x_1, \ldots, x_n \mid y)}{p(x_1, \ldots, x_n)}. \tag{6.2}$$

Using the naive assumption of conditional independence for all i, this relationship can be transformed into the following equation:

$$p(y \mid x_1, \ldots, x_n) = \frac{p(y) \prod_{i=1}^{n} p(x_i \mid y)}{p(x_1, \ldots, x_n)}. \tag{6.3}$$

The classification rule given by the following equation is derived since $p(x_1, \ldots, x_n)$ is constant for a given input:

$$\hat{y} = \underset{y}{\operatorname{argmax}}\; p(y) \prod_{i=1}^{n} p(x_i \mid y). \tag{6.4}$$

There are different Naive Bayes classifiers depending on the assumptions made about the distribution of $p(x_i \mid y)$. The problem under consideration involves continuous data, so we can assume that the continuous values associated with each class follow a Gaussian distribution [32]. Consequently, the Gaussian Naive Bayes classifier is used, where the probability of the features is assumed to be Gaussian:

$$p(x_i \mid y) = \frac{1}{\sqrt{2\pi\sigma_y^2}} e^{-\frac{(x_i - \mu_y)^2}{2\sigma_y^2}}. \tag{6.5}$$

Extreme Gradient Boosted Trees

Classification trees are a machine learning (ML) method that uses a series of rules to recursively partition the data space and classify observations [8]. The tree starts with a root node containing the entire training dataset, and sequentially splits the data into branches based on the first rule. Each branch contains observations that satisfy this rule and can be further divided using additional rules. Branches that are not split further are called leaves and represent the final predictions of the tree. The path to the appropriate leaf for each observation is determined by following the sequence of rules. These rules are automatically generated using the features provided for training and are based on specific criteria. To avoid overfitting and reduce computational costs, additional criteria such as learning rate, maximum tree depth, maximum number of leaves, minimum sum of instance weights in a child, and minimum loss reduction for further splits are often employed as hyperparameters to regulate tree growth.

Gradient boosting (GB) is an advanced technique designed to improve ML methods by reducing prediction bias and variance, effectively transforming weak learners into strong ones. It achieves this by sequentially combining multiple weak learners to form a single, robust model with greater predictive power and accuracy [23]. Each new weak learner focuses on correcting the errors of its predecessors, thereby progressively improving overall performance. GB can also enhance the capabilities of classification trees, creating stronger and more effective classifiers.

Extreme gradient boosting (XGBoost) is a widely recognized implementation of gradient-boosted trees, known for its superior speed and performance due to its efficient use of memory and hardware resources [11]. XGBoost incorporates three core gradient boosting techniques: gradient boosting, regularized boosting, and stochastic boosting [17]. In this study, XGBoost was implemented using the gradient boosting approach to construct a powerful classifier.

Random Forest

Random Forest (RF) is a machine learning classification method that builds multiple classification trees and aggregates their votes to produce the final prediction [9]. To leverage the benefits of ensemble learning-reducing both prediction bias and variance-each tree in the forest is trained on a different subset of observations and features, randomly selected from the full dataset. This technique, known as bagging, ensures diversity among the trees by exposing them to different portions of the data, making the model less sensitive to variations in the training set.

In the energy and building sectors, RF has been widely applied in various contexts, such as predicting energy consumption in buildings [3], developing personalized air conditioning systems for offices [37], and identifying faults in distribution networks [10], among other use cases.

Support Vector Machines

Support Vector Machines (SVMs) are a machine learning method that constructs a hyperplane, or a set of hyperplanes, in a high-dimensional or infinite-dimensional space to solve classification problems, including binary and multi-class tasks [57]. The hyperplane that maximizes the functional margin-the distance between the hyperplane and the nearest data points in the training set-minimizes generalization errors and achieves better separation of the data. In this context, SVMs are often referred to as Support Vector Classifiers (SVC) [6, 40].

Given vectors $x_i \in \mathbb{R}^p,\ i = 1, \ldots, n$ and a vector $y \in 1, -1$, the SVC computes values $w \in \mathbb{R}^p$ and $b \in \mathbb{R}$ to make correct predictions for each observation using the value of $\text{sign}(w^T \phi(x) + b)$. The primary problem solved by SVCs can be formulated as follows:

$$\begin{aligned} \min_{w,b,z} \quad & \frac{1}{2} w^T w + C \sum_{i=1}^{N} z_i, \\ \text{s.t.} \quad & y_i (w^T \phi(x_i) + b) \geq 1 - z_i, \\ & z_i \geq 0. \end{aligned} \tag{6.6}$$

This mathematical formula describes how SVCs aim to maximize the margin while imposing a penalty C for observations that are misclassified or fall within the margin boundary. Finally, the slack variable z_i for each vector X_i allows an observation to be at a certain distance from the correct margin boundary, acknowledging that most problems are not separable by a single hyperplane.

For multi-class classification, the SVC problem environment differs significantly from the binary classification approach [18, 27]. There are two approaches used to generalize from binary to multi-class classification. The direct method proposes generalizing a binary classifier to produce multi-class predictions, while the more common approach, the indirect one, combines N independent binary classifiers to generate a single multi-class model. In the latter approach, binary tasks can be defined in the form of a matrix of size $M \times N$, where M represents the number of classes and N the number of tasks. The values of each element in the matrix $R_{i,j} \in -1, 0, 1$ produced by the binary classifiers f are used to predict the class label according to the following equation:

$$\hat{y} = \arg \min_{y \in 1, \ldots, n} \left\{ \sum_{k=1}^{N} R_{yk} f^k(x) \right\}. \tag{6.7}$$

6.3.3 *Meta-Learning Model*

Stacking (or stacked) ensemble modeling is a technique designed to minimize the generalization error by combining predictions from multiple methods within an ensemble [67]. This approach typically employs a meta-learner, which is responsible for either optimally combining the predictions of various base ML methods or selecting the most accurate prediction for a specific instance. This allows the stacked ensemble to harness the strengths of diverse high-performing methods, each with unique characteristics, structures, and assumptions about the data.

The effectiveness of stacked ensemble modeling is based on the principle that prediction errors from different base methods are often uncorrelated. When this assumption holds-where each method captures distinct aspects of the problem-the ensemble can provide more accurate predictions than any individual method. By integrating these diverse perspectives, stacking reduces biases inherent in individual models relative to the training data.

The stacked ensemble model developed in this study follows a simple two-level architecture. The first level, or "level-0," consists of the five base classifiers described in Sect. 6.3.2. The second level, or "level-1," is a logistic regression model chosen as the meta-learner. This selection was made due to the interpretability of logistic regression, which provides clear insights into why the meta-learner favors one base classifier over another. Research has shown that using relatively simple classifiers as meta-learners is effective for integrating predictions from base methods [12, 48]. The overall architecture of the stacked ensemble model is depicted in Fig. 6.2.

The process of constructing the stacking model consists of three main phases: ensemble setup, training, and prediction on new data. These phases are outlined below:

- **Phase 1: Ensemble Setup** In this initial phase, the base classifiers and the meta-learning classifier are selected. The base methods should represent a diverse range of algorithms with different characteristics (e.g., not solely tree-based methods) to maximize the ensemble's effectiveness. The meta-learner should be a relatively simple and interpretable classifier to ensure clarity in understanding its decisions.
- **Phase 2: Training** First, each base classifier is trained using optimized hyperparameters determined through cross-validation (e.g., k-fold cross-validation). Next, the meta-learner is trained by using the predictions from the base classifiers (level-0 methods) and their associated input vectors as features, while the true labels of the observations serve as the target output.
- **Phase 3: Prediction** The stacked ensemble generates predictions for new test data. Each base classifier first makes its individual prediction for the test data. These predictions are then input into the meta-learner, which determines how to combine or select the baseline predictions to produce the final output, based on its trained model.

6.3.4 Funding Recommendation

The proposed methodology is designed to provide recommendations on the percentage of subsidized funding that future investments should receive. If the predicted class from the meta-learner were directly used to determine this percentage, the recommendations would result in discrete funding levels corresponding to the three possible classes. For example, if *Class A* suggests 100% funding, *Class B* suggests 50%, and *Class C* suggests 0%, all projects would be funded at levels represented by the coefficients $f_A = 1$, $f_B = 0.5$, or $f_C = 0$, depending on their assigned class. However, this approach significantly limits the methodology's flexibility and disregards the uncertainty inherent in the predictions.

For instance, consider a case where an investment is assigned probabilities 0.55 for *Class A*, 0.35 for *Class B*, and 0.10 for *Class C*. Using the discrete approach,

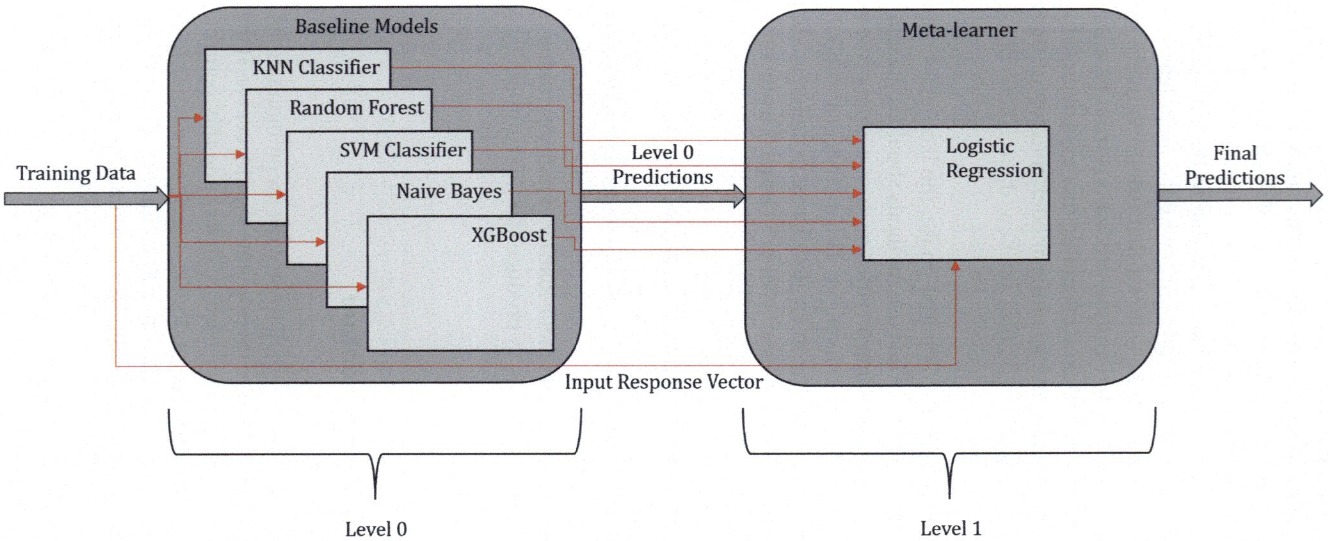

Fig. 6.2 Overview of the proposed stacked ensemble model. The baseline classification methods (level-0) are trained following the process described in Fig. 6.1 and provide predictions for future investments. These predictions are then fed to the meta-learner along with the input vectors of the baseline methods. The output of the meta-learner is the final prediction of the stacked ensemble model, i.e. the prediction of the most appropriate baseline classifier

the investment would receive 100% funding, eventhough the prediction probabilities suggest considerable uncertainty about the decision.

To address these limitations and incorporate uncertainty, we propose basing funding recommendations on the probabilities of each class rather than solely the dominant class. This approach provides a nuanced recommendation by summarizing the likelihood that a future investment belongs to each class. Using p_A, p_B, and p_C to represent the probabilities of an investment being classified as A, B, and C, respectively, and f_A, f_B, and f_C, as the corresponding funding coefficients, the funding recommendation for a future investment can be calculated using the following formula:

$$\text{Financing Percentage} = \sum_{i \in A,B,C} f_i \times p_i. \tag{6.8}$$

Using the above equation, in the earlier example, the decision support system would recommend funding the new project at $0.55 \times 1.00 + 0.35 \times 0.50 + 0.10 \times 0.00 = 72.5\%$. This recommendation effectively incorporates the insights provided by the proposed data-driven methodology.

6.4 Case Study

This section presents a comprehensive experimental application of the proposed methodology using data from real-world renovation projects. It details the training process for the five base classification methods, including the fine-tuning of their hyperparameters, and outlines the subsequent training of the meta-learning model. The results of the application are then presented and analyzed.

6.4.1 Dataset

The experimental application of the proposed methodology was carried out on data collected from the Latvian Environmental Investment Fund (LEIF), an organization under the Ministry of Environmental Protection and Regional Development of Latvia, established in 1997. Since 2009, LEIF has overseen the implementation and post-implementation monitoring of various projects co-financed by climate change financial instruments, making it the sole institution in Latvia with reliable data on the actual performance of various EE investments in terms of energy savings.

Although our approach is based on the evaluation of investments in a single building, many of the collected records involved investments applied to more than one building. Consequently, these records were excluded from the dataset, resulting in a final sample of 312 EE projects. The selected ML methods were trained using a randomly chosen 80% of the total sample, i.e., 249 investments, while the remaining 63 investments were used to test the performance of our methodology.

The aforementioned investments were classified using three labels:

- *Class A*, which includes high-potential investments;
- *Class B*, which includes medium-potential investments; and
- *Class C*, which includes low-potential investments.

The labeling of the investments was based on the ranking of investment efficiency. This is a simple ranking method using two criteria: the project's investment cost and the reduction in energy consumption in the building after renovation. These criteria were combined using the following steps: First, the values for each criterion were normalized on a scale of (0, 1). Second, the weighted average of the normalized values was calculated for all available investments. This process was conducted using equal weights for the selected criteria; however, the choice of weighting factors can be adjusted based on investor preferences. Finally, the values were ranked from best to worst. The top 20% of the projects were classified as *Class A* investments, while the bottom 20% were classified as *Class C* investments. The remaining 60% of investments were classified as *Class B*.

The labeling system used in this study inherently produces an imbalanced training set, with *Class B* investments being more likely to be identified as the correct class due to their larger representation. While this issue could be mitigated through resampling techniques-such as randomly undersampling the majority class (*Class B*) or generating synthetic data to oversample the minority classes (*Class A* and *Class C*) [35]-the meta-learner is expected to address class imbalance by selecting the base classification method that best predicts each investment class. Moreover, because imbalanced datasets are common in real-world scenarios, where some classes naturally occur more frequently than others, this imbalance provides an additional stress test to assess the robustness and added value of the proposed approach.

The investment features used in the case study were selected based on data availability. Six key features were chosen as input for the classification methods:

1. The building's construction year,
2. The total heating area (m^2),
3. The expected reduction in CO_2 emissions due to renovation actions ($kgCO_2$),
4. The building's current energy consumption (MWh),
5. The number of floors to be renovated, and
6. The grant amount requested for the renovation (€).

Since all investments were conducted within the same country, geographical variables (e.g., city identifiers) were not included as additional features.

6.4.2 Training and Hyperparameter Tuning

The optimal hyperparameters for the five base classification methods were determined using a stratified k-fold cross-validation process. Unlike traditional k-fold cross-validation, the stratified approach ensures that the proportion of samples in each class is preserved within each fold. This method was chosen because the classification task involves imbalanced class distributions, with 20% of investments in *Class A*, 60% in *Class B*, and 20% in *Class C*.

The process was configured with 10 splits and repeated 3 times to ensure robust evaluation. The selected hyperparameter values for each classification method and the functions used for their implementation are summarized below.

- *k-Nearest Neighbors*: The most critical hyperparameters for this method are the number of neighbors used to perform the queries, which is set to 6, and the weighting function used for predictions, which is set to uniform. The method was implemented using the *neighbors* function from the *sklearn* library in Python [49].
- *Gaussian Naive Bayes*: The method does not require extensive tuning of its hyperparameters. The implementation of the method was carried out using the *naivebayes* function from the *sklearn* library in Python.
- *Extreme Gradient Boosted Trees*: XGBoost includes several hyperparameters that are critical for its performance. The most important ones are the *learning rate*, the *number of leaves* in the decision tree, the *bagging fraction*, which reduces prediction variance, and the *feature fraction*, which allows the random selection of a subset of features at each iteration. After applying the cross-validation process, the following values were selected: Learning rate = 0.95; maximum tree depth = 10; minimum sum of instance weight needed in a child = 1; number of gradient-boosted trees = 1000; all other hyperparameters were set to their default values. The method was implemented using the *Classifier* function from the *XGBoost* library for Python [11].
- *Random Forest*: The number of trees (number of estimators) in the RF method was set to 500, the class weight factor was set to balanced (using the values of the target variable to adjust weights inversely proportional to the class frequencies in the feature variables), and the maximum number of features used in each iteration was set to log_2. The method was implemented using the *ensemble* function from the *sklearn* library in Python.
- *Support Vector Machine*: The SVM method used the linear kernel, and the regularization parameter C was set to 1. The method was implemented using the *svm* function from the *sklearn* library in Python.

After configuring the hyperparameters for the base classification methods, the final step in the training process involves training the level-1 model. In this study, a logistic regression classifier was used as the meta-learner due to its simplicity, which is commonly recommended for level-1 predictors. Since logistic regression does not require hyperparameter tuning, it was directly trained on the level-1 training dataset. This dataset comprises the predictions from the base methods (level-0 models) and

Table 6.1 Classification performance (mean and standard deviation) of the baseline methods and the stacking model on the stratified 10-fold validation process. Column-wise best values are displayed in bold

Classifier	Accuracy	Precision	F1 score
k-nearest neighbors	0.77 (0.08)	0.78 (0.10)	0.74 (0.09)
Gaussian Naive Bayes	0.70 (0.10)	0.66 (0.14)	0.67 (0.12)
XGBoost	0.78 (0.10)	0.79 (0.10)	0.78 (0.10)
Random forest	0.80 (0.09)	0.82 (0.08)	0.79 (0.09)
Support vector Machine	0.82 (0.09)	**0.85 (0.07)**	0.82 (0.09)
Stacking Model	**0.83 (0.09)**	0.84 (0.09)	**0.83 (0.09)**

the actual labels, generated through a 3-fold cross-validation process on the training set.

The performance of the base methods and the stacking classifier during the stratified k-fold validation process (10 folds $\times$ 3 repetitions = 30 validations) is summarized in Table 6.1. Among the base classifiers, the SVM method achieved the highest performance based on the F1 score, followed by RF and XGBoost. As noted in Sect. 6.3, the strong performance of XGBoost and RF can be attributed to their ability to identify the most relevant features from a large set of explanatory variables, as well as the robustness of RF and SVM in handling data randomness.

Additionally, the stacking model outperformed all the base classifiers in terms of accuracy and F1 score metrics, though its precision was slightly lower than that of the SVM. While the differences between the meta-learner and the SVM are relatively minor according to the classification metrics, the meta-learner is more robust. It consistently classifies most investments more accurately and avoids extreme errors. This robustness is evident in Fig. 6.3, which visually compares classifier performances using box plots. The median values for F1 score and accuracy are higher for the meta-classifier than for the SVM, indicating a greater proportion of correctly classified investments. Furthermore, the interquartile range for the meta-classifier is significantly smaller, reflecting better consistency and robustness across different datasets.

6.4.3 *Results*

Following the training of the stacking model, the performance of the proposed methodology was evaluated on the test dataset, which comprises 63 randomly selected EE investments. The results are presented in the confusion matrix in Fig. 6.4, comparing the meta-learner's predicted classes with the actual labels.

The analysis reveals that 32 out of 36 *Class B* investments (88.8%) were correctly classified. For *Class A*, 10 out of 14 investments (71.4%) were accurately predicted,

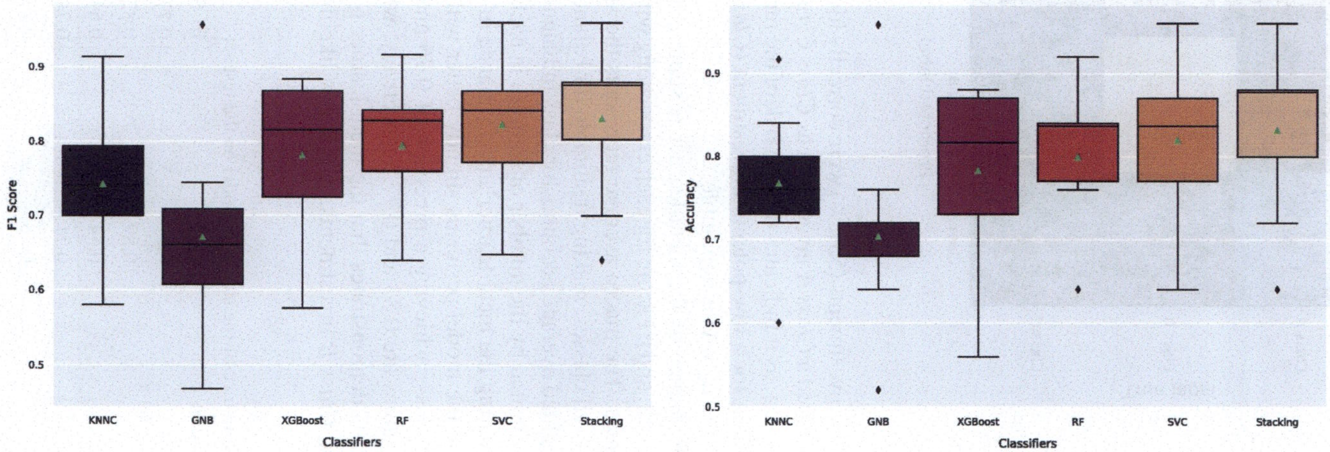

(a) F1 Score

(b) Accuracy

Fig. 6.3 Comparison of the baseline classifiers and the stacking on the stratified 10-fold validation process. The left sub-figure shows the F1 score of the classifiers, while the right one their accuracy

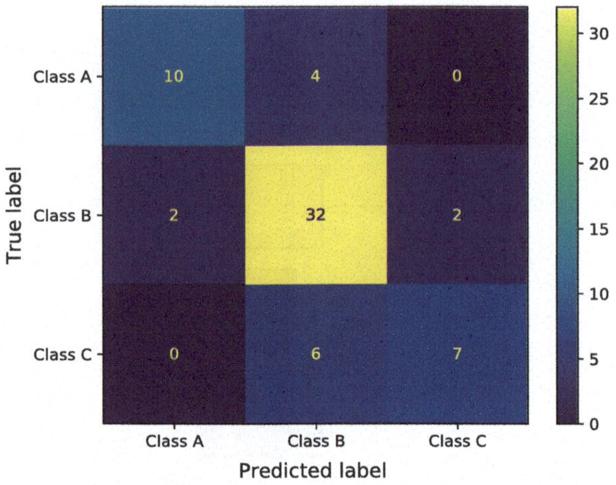

Fig. 6.4 Confusion matrix for the stacking model on the test set

while for *Class C*, 7 out of 13 investments (53.2%) were correctly identified. Notably, the classifier never misclassified a *Class A* investment as *Class C*, or vice versa. This result is particularly promising, as it indicates that the risk of the meta-learner incorrectly recommending funding for a low-potential investment (*Class C*) or rejecting a high-potential investment (*Class A*) is negligible.

The classification performance of the meta-learner is summarized in Table 6.2, which reports precision, recall, and F1 score for each class, along with macro-average and weighted average values. The macro-average represents the simple arithmetic mean of each metric across all classes, making it suitable for balanced classification tasks. Conversely, the weighted average accounts for class imbalance by calculating the mean of the metrics weighted by the number of samples in each class, making it more appropriate for evaluating the meta-learner's performance in this case study.

The results indicate that the model performs slightly better with *Class B* investments, the most frequent class, achieving a precision of 0.76 and an F1 score of 0.82. For *Class A*, these values are 0.83 and 0.77, respectively, while for *Class C*, they are 0.78 and 0.64. This demonstrates that the meta-learner is highly effective in identifying *Class A* and *Class B* investments, with a low likelihood of failing to

Table 6.2 Classification performance of the stacking model on the test set

	Precision	Recall	F1 Score
Class A	0.83	0.71	0.77
Class B	0.76	0.89	0.82
Class C	0.78	0.54	0.64
Macro Average	0.79	0.71	0.74
Weighted Average	0.78	0.78	0.77

correctly classify these classes. In contrast, *Class C* investments are more prone to being misclassified as *Class B*.

These findings suggest that the meta-learner is well-suited for practical application in identifying high- and medium-potential investments, thereby supporting the funding of large-scale EE projects. The model's overall accuracy across all classes is 0.78, further validating its effectiveness. Additionally, despite the imbalanced dataset used to train the base classification methods, the recall results in Table 6.2 indicate that the meta-learner successfully mitigates this issue, delivering unbiased predictions, particularly for *Class A* and *Class B* investments.

Given that the accuracy metric does not account for class imbalance-a critical factor in this problem, as 60% of the observations belong to the same class-we employ Cohen's kappa coefficient (κ) as an alternative. This statistical measure evaluates the reliability of agreement between raters for categorical data, offering a more robust metric than simple percentage agreement by accounting for the probability of agreement occurring by chance through the κ parameter. Cohen's kappa coefficient is calculated using the following formula:

$$\kappa = \frac{p_0 - p_e}{1 - p_e}, \tag{6.9}$$

where p_0 represents the observed proportional agreement between actual and predicted values, while p_e denotes the expected agreement if both raters were to assign labels randomly. For the test set, Cohen's kappa score is calculated as 0.594, indicating substantial agreement for the stacking classifier.

However, because the proposed methodology relies on classifiers that produce class membership probabilities, these metrics alone do not provide a complete assessment. To evaluate the model's ability to distinguish between classes, the Area Under the Receiver Operating Characteristic curve (ROC AUC) is used. This metric is calculated by generating predictions across a range of decision thresholds [0, 1] and computing the true positive rate (TPR, or recall) and the false positive rate (FPR) for each threshold.

For multi-class classification, the ROC AUC is calculated separately for each class using a one-vs-rest (OVR) strategy, treating each class as a positive class against the rest. The ROC curve for each class is shown in Fig. 6.5. A larger area under the curve signifies better discrimination among classes. The overall ROC AUC value for the stacking classifier is 0.883, which is a strong result given that ROC AUC values range from 0.5 (indicating a poor classifier) to 1 (indicating an excellent classifier).

6.5 Conclusions

This study addresses the challenge of evaluating the potential of future energy efficiency (EE) investments, with a focus on renovation costs and realized energy savings. The absence of robust information systems to support these evaluations

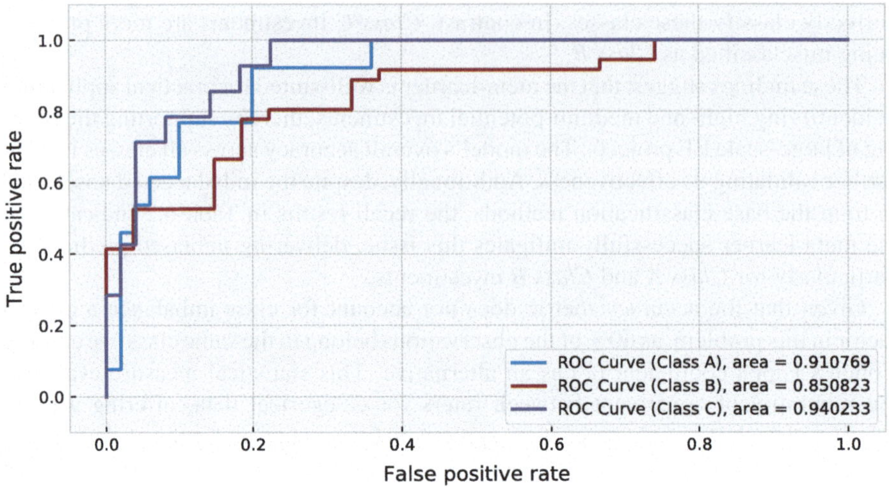

Fig. 6.5 ROC curve for the three classes based on the stacking classifier's generated probabilistic predictions

has hindered many renovation projects from securing financial backing, ultimately threatening the global environmental targets established for the coming decades. To address this gap, a data-driven methodology is proposed to identify promising EE projects and determine the appropriate capital investments. By analyzing critical characteristics of past projects and leveraging machine learning (ML) methods, the approach learns from prior successes and shortcomings. This methodology aims to enhance EE financing processes and provide stakeholders-including funders, bankers, and investors-with a standardized and less uncertain framework for comparing and labeling EE projects.

The core of the methodology is a stacking ensemble model that integrates multiple ML classification methods to optimize performance. Key characteristics of EE investments, such as building age, total heating area, expected CO_2 emissions reductions, and renovation costs, are collected and used to train the model. The approach is evaluated through a real-world case study in Latvia, where the results demonstrate that the stacking ensemble model outperforms all individual ML classification methods, achieving high accuracy in assessing future EE investments. Additionally, the model effectively identifies high- and medium-potential projects and excels in distinguishing between low- and high-potential investments. These findings suggest that the methodology can assist stakeholders in reducing investment risks and maximizing returns.

Despite its contributions, this study has limitations that provide opportunities for future research. Expanding the methodology to encompass other types of EE investments, such as those in manufacturing, transportation, and outdoor lighting, would broaden its scope and enhance its economic and environmental impacts. The developed meta-learning model is adaptable for these applications, provided that sufficient data are available for training. Another promising direction involves exploring the influence of geographical and climatic factors on the effectiveness of EE investments. Incorporating data from projects across multiple countries with diverse characteristics could improve the model's generalizability and attract a wider range of stakeholders.

Future studies could also explore alternative measures for evaluating the potential of EE projects to better reflect real-world conditions. For large-scale applications, it is essential to identify accessible sources of energy- and investment-related data for training the classifiers. One potential source is energy performance certificates for renovated buildings, which supply standardized and interpretable information on building energy efficiency before and after renovations, along with associated costs. Such data could be easily shared with public authorities and financial institutions, facilitating the widespread adoption of the proposed methodology and enabling its application on a larger scale.

References

1. N. Abdou, Y.E. Mghouchi, S. Hamdaoui, N.E. Asri, M. Mouqallid, Multi-objective optimization of passive energy efficiency measures for net-zero energy building in morocco. Build. Envir. **204**, 108141 (2021)
2. M.I. Abreu, R.A. de Oliveira, J. Lopes, Younger vs. older homeowners in building energy-related renovations: Learning from the portuguese case. Energy Rep. **6**, 159–164 (2020)
3. M.W. Ahmad, M. Mourshed, Y. Rezgui, Trees vs neurons: Comparison between random forest and ann for high-resolution prediction of building energy consumption. Energy Build. **147**, 77–89 (2017)
4. A. Arsenopoulos, E. Sarmas, A. Stavrakaki, I. Giannouli, J. Psarras, A data-driven decision support tool at the service of energy suppliers and utilities for tackling energy poverty: a case study in Greece, in *2021 12th International Conference on Information, Intelligence, Systems & Applications (IISA)* (IEEE, 2021), pp. 1–6
5. R. Bayindir, M. Yesilbudak, M. Colak, N. Genc, A novel application of Naive Bayes classifier in photovoltaic energy prediction, in *2017 16th IEEE International Conference on Machine Learning and Applications (ICMLA)* (IEEE, 2017), pp. 523–527
6. G. Bo, H. Xianwu, Svm multi-class classification. J. Data Acquis. & Proc. **21**(3), 334–339 (2006)
7. F.P. Boon, C. Dieperink, Local civil society based renewable energy organisations in the Netherlands: exploring the factors that stimulate their emergence and development. Energy Policy **69**, 297–307 (2014)
8. L. Breiman, *Classification and Regression Trees* (Chapman & Hall, Boca Raton, FL, 1993)
9. L. Breiman, Bagging predictors. Mach. Learn. **24**(2), 123–140 (1996)
10. D. Chakraborty, U. Sur, P.K. Banerjee, Random forest based fault classification technique for active power system networks, in *2019 IEEE International WIE Conference on Electrical and Computer Engineering (WIECON-ECE)* (IEEE, 2019), pp. 1–4
11. T. Chen, C. Guestrin, Xgboost: a scalable tree boosting system, in *Proceedings of the 22nd ACM Sigkdd International Conference on Knowledge Discovery and Data Mining* (2016), pp. 785–794
12. Z. Chen, P. Zhao, F. Li, T.T. Marquez-Lago, A. Leier, J. Revote, Y. Zhu, D.R. Powell, T. Akutsu, G.I. Webb et al., Ilearn: an integrated platform and meta-learner for feature engineering, machine-learning analysis and modeling of dna, rna and protein sequence data. Brief. Bioinf. **21**(3), 1047–1057 (2020)

13. C. Chia-Cheng, Y. Liu, T.-H. Hsu, An analysis on investment performance of machine learning: an empirical examination on Taiwan stock market. Int. J. Econ. Financ. Issues **9**(4), 1 (2019)
14. T. Cover, P. Hart, Nearest neighbor pattern classification. IEEE Trans. Inf. Theory **13**(1), 21–27 (1967)
15. A.B. de Vasconcelos, M.D. Pinheiro, A. Manso, A. Cabaço, Epbd cost-optimal methodology: Application to the thermal rehabilitation of the building envelope of a Portuguese residential reference building. Energy Build. **111**, 12–25 (2016)
16. C. Debrah, A.P.C. Chan, A. Darko, Green finance gap in green buildings: a scoping review and future research needs. Build. Environ. 108443 (2021)
17. S.S. Dhaliwal, A.-A. Nahid, R. Abbas, Effective intrusion detection system using xgboost. Information **9**(7), 149 (2018)
18. D.R. Don, I.E. Iacob, Dcsvm: fast multi-class classification using support vector machines. Int. J. Mach. Learn. Cybern. **11**(2), 433–447 (2020)
19. H. Doukas, A. Nikas, M. González-Eguino, I. Arto, A. Anger-Kraavi, From integrated to integrative: delivering on the Paris agreement. Sustainability **10**(7), 2299 (2018)
20. H. Doukas, P. Xidonas, N. Mastromichalakis, How successful are energy efficiency investments? a comparative analysis for classification & performance prediction. Comput. Econ. 1–20 (2021)
21. E. Foda, A. El-Hamalawi, J. Le Dréau, Computational analysis of energy and cost efficient retrofitting measures for the french house. Build. Environ. **175**, 106792 (2020)
22. A. Forouli, N. Gkonis, A. Nikas, E. Siskos, H. Doukas, C. Tourkolias, Energy efficiency promotion in Greece in light of risk: Evaluating policies as portfolio assets. Energy **170**, 818–831 (2019)
23. J.H. Friedman, Stochastic gradient boosting. Comput. Stat. & Data Anal. **38**(4), 367–378 (2002)
24. H. Ghoddusi, G.G. Creamer, N. Rafizadeh, Machine learning in energy economics and finance: a review. Energy Econ. **81**, 709–727 (2019)
25. S. Hayes, S. Nadel, C. Granda, K. Hottel, What have we learned from energy efficiency financing programs, in *American Council for an Energy-Efficient Economy* (2011)
26. J.L. Hernández, S. Serroni, S. Stamatopoulos, E. Sarmas, G.M. Revel, F. Vélez, Multi-criteria building performance assessment, in *2024 15th International Conference on Information, Intelligence, Systems & Applications (IISA)* (IEEE, 2024), pp. 1–7
27. S. Herrero-Lopez, J.R. Williams, A. Sanchez, Parallel multiclass classification using svms on gpus, in *Proceedings of the 3rd Workshop on General-Purpose Computation on Graphics Processing Units* (2010), pp. 2–11
28. IEA, Implementing Energy Efficiency Policies: are IEA Member Countries on Track? (2009). https://www.oecd.org/publications/implementing-energy-efficiency-policies-are-iea-member-countries-on-track-9789264075696-en.htm
29. IEA, Net Zero by 2050: A Roadmap for the Global Energy Sector (2021). https://iea.blob.core.windows.net/assets/4719e321-6d3d-41a2-bd6b-461ad2f850a8/NetZeroby2050-ARoadmapfortheGlobalEnergySector.pdf
30. S. Intrachooto, V. Horayangkura, Energy efficient innovation: overcoming financial barriers. Build. Environ. **42**(2), 599–604 (2007)
31. IRENA, Renewable energy statistics (2015). https://www.irena.org/publications/2015/Jun/Renewable-Energy-Target-Setting
32. A.H. Jahromi, M. Taheri, A non-parametric mixture of gaussian Naive Bayes classifiers based on local independent features, in *2017 Artificial Intelligence and Signal Processing Conference (AISP)*. (IEEE, 2017), pp. 209–212
33. P.A. Jensen, E. Maslesa, Value based building renovation-a tool for decision-making and evaluation. Build. Environ. **92**, 1–9 (2015)
34. P.A. Jensen, E. Maslesa, J.B. Berg, C. Thuesen, 10 questions concerning sustainable building renovation. Build. Environ. **143**, 130–137 (2018)
35. B. Krawczyk, Learning from imbalanced data: open challenges and future directions. Progr. Artif. Intell. **5**(4), 221–232 (2016)

36. J.Y. Lee, M.P. Styczynski, Ns-knn: a modified k-nearest neighbors approach for imputing metabolomics data. Metabolomics **14**(12), 1–12 (2018)
37. Q.Y. Li, J. Han, L. Lu, A random forest classification algorithm based personal thermal sensation model for personalized conditioning system in office buildings. Comput. J. **64**(3), 500–508 (2021)
38. T. Loureiro, M. Gil, R. Desmaris, A. Andaloro, C. Karakosta, S. Plesser, De-risking energy efficiency investments through innovation, in *Multidisciplinary Digital Publishing Institute Proceedings*, vol. 65 (2020), p. 3
39. S. Makridakis, E. Spiliotis, V. Assimakopoulos, Statistical and machine learning forecasting methods: concerns and ways forward. PLoS ONE **13**(3), 1–26 (2018)
40. A. Mathur, G.M. Foody, Multiclass and binary svm classification: implications for training and classification users. IEEE Geosci. Remote Sens. Lett. **5**(2), 241–245 (2008)
41. D. Mena, E. Montañés, J.R. Quevedo, J.J. del Coz, An overview of inference methods in probabilistic classifier chains for multilabel classification. Wiley Interdiscip. Rev.: Data Mining Knowl. Discov. **6**(6), 215–230 (2016)
42. J. Morrissey, B. Meyrick, D. Sivaraman, R. Horne, M. Berry, Cost-benefit assessment of energy efficiency investments: accounting for future resources, savings and risks in the australian residential sector. Energy Policy **54**, 148–159 (2013)
43. M. Norouzi, D.J. Fleet, R.R. Salakhutdinov, Hamming distance metric learning, in *Advances in Neural Information Processing Systems* (2012), pp. 1061–1069
44. J. Ortiz, A.F. i Casas, J. Salom, N.G. Soriano, P.F. i Casas, Cost-effective analysis for selecting energy efficiency measures for refurbishment of residential buildings in Catalonia. Energy Build. **128**, 442–457 (2016)
45. J.P. Painuly, Barriers to renewable energy penetration; a framework for analysis. Renew. Energy **24**(1), 73–89 (2001)
46. J.P. Painuly, H. Park, M.-K. Lee, J. Noh, Promoting energy efficiency financing and escos in developing countries: mechanisms and barriers. J. Clean. Prod. **11**(6), 659–665 (2003)
47. A. Papapostolou, F.D. Mexis, E. Sarmas, C. Karakosta, J. Psarras, Web-based application for screening energy efficiency investments: a mcda approach, in *2020 11th International Conference on Information, Intelligence, Systems and Applications (IISA* (IEEE, 2020), pp. 1–7
48. B. Pavlyshenko, Using stacking approaches for machine learning models, in *2018 IEEE Second International Conference on Data Stream Mining & Processing (DSMP)* (IEEE, 2018), pp. 255–258
49. F. Pedregosa, G. Varoquaux, A. Gramfort, V. Michel, B. Thirion, O. Grisel, M. Blondel, P. Prettenhofer, R. Weiss, V. Dubourg, J. Vanderplas, A. Passos, D. Cournapeau, M. Brucher, M. Perrot, E. Duchesnay, Scikit-learn: machine learning in Python. J. Mach. Learn. Res. **12**, 2825–2830 (2011)
50. B. Prasetiyo, M. Muslim, et al., Analysis of building energy efficiency dataset using Naive Bayes classification classifier. J. Phys.: Conf. Ser. **1321**, 032016 (2019). IOP Publishing
51. P. Samaras, E. Stamatopoulos, A. Arsenopoulos, E. Sarmas, E. Marinakis, Readiness to adopt the smart readiness indicator scheme across Europe: a multi-criteria decision analysis approach, in *2024 IEEE International Workshop on Metrology for Living Environment (MetroLivEnv)* (IEEE, 2024), pp. 268–273
52. A. Sarkar, J. Singh, Financing energy efficiency in developing countries-lessons learned and remaining challenges. Energy Policy **38**(10), 5560–5571 (2010)
53. E. Sarmas, A. Forouli, V. Marinakis, H. Doukas, Baseline energy modeling for improved measurement and verification through the use of ensemble artificial intelligence models. Inf. Sci. **654**, 119879 (2024)
54. E. Sarmas, A. Fragkiadaki, V. Marinakis, Explainable ai-based ensemble clustering for load profiling and demand response. Energies **17**(22), 5559 (2024)
55. E. Sarmas, P. Xidonas, H. Doukas, *Multicriteria Portfolio Construction with Python* (Springer, 2020)

56. M. Sarrica, S. Brondi, P. Cottone, B.M. Mazzara, One, no one, one hundred thousand energy transitions in Europe: the quest for a cultural approach. Energy Res. & Soc. Sci. **13**, 1–14 (2016)
57. A.J. Smola, B. Schölkopf, A tutorial on support vector regression. Stat. Comput. **14**(3), 199–222 (2004)
58. E. Stamatopoulos, A. Forouli, D. Stoian, P. Kouloukakis, E. Sarmas, V. Marinakis, An adaptive framework for assessing climate resilience in buildings. Build. Environ. **264**, 111869 (2024)
59. E. Stocker, M. Tschurtschenthaler, L. Schrott, Cost-optimal renovation and energy performance: evidence from existing school buildings in the alps. Energy Build. **100**, 20–26 (2015)
60. D. Stoian, E. Spiliotis, E. Stamatopoulos, E. Sarmas, P. Rekomaa, V. Marinakis, A neural network-based approach for hierarchical electricity consumption forecasting in buildings, in *2024 IEEE International Workshop on Metrology for Living Environment (MetroLivEnv)* (IEEE, 2024), pp. 274–279
61. I. Syarif, E. Zaluska, A. Prugel-Bennett, G. Wills, Application of bagging, boosting and stacking to intrusion detection, in *International Workshop on Machine Learning and Data Mining in Pattern Recognition* (Springer, 2012), pp. 593–602
62. P.G. Taylor, O.L. d'Ortigue, M. Francoeur, N. Trudeau, Final energy use in iea countries: The role of energy efficiency. Energy Policy **38**(11), 6463–6474 (2010)
63. R.P. Taylor, C. Govindarajalu, J. Levin, A.S. Meyer, W.A. Ward, *Financing Energy Efficiency: Lessons from Brazil, China, India, and beyond* (World Bank Publications, 2008)
64. J. Thøgersen, A. Grønhøj, Electricity saving in households-a social cognitive approach. Energy Policy **38**(12), 7732–7743 (2010)
65. G. Toussaint, Geometric proximity graphs for improving nearest neighbor methods in instance-based learning and data mining. Int. J. Comput. Geom. & Appl. **15**(02), 101–150 (2005)
66. P. Tuominen, F. Reda, W. Dawoud, B. Elboshy, G. Elshafei, A. Negm, Economic appraisal of energy efficiency in buildings using cost-effectiveness assessment. Procedia Econ. Finan. **21**, 422–430 (2015)
67. D.H. Wolpert, Stacked generalization. Neural Netw. **5**(2), 241–259 (1992)
68. P. Xidonas, H. Doukas, E. Sarmas, A python-based multicriteria portfolio selection dss. RAIRO-Oper. Res. **55**, S3009–S3034 (2021)
69. M. Zekić-Sušac, S. Mitrović, A. Has, Machine learning based system for managing energy efficiency of public sector as an approach towards smart cities. Int. J. Inf. Manage. **58**, 102074 (2021)
70. D. Zhang, Z. Zhang, S. Managi, A bibliometric analysis on green finance: current status, development, and future directions. Financ. Res. Lett. **29**, 425–430 (2019)
71. H. Zhang, The optimality of Naive Bayes. AA **1**(2), 3 (2004)

Chapter 7
Ensemble Machine Learning Models for Estimating Energy Savings from Efficiency Measures in Buildings

Abstract This chapter presents a Machine Learning (ML)-based framework for predicting energy savings resulting from various Energy Efficiency (EE) renovations. The framework employs three widely recognized tree-based ensemble algorithms: Random Forest (RF), XGBoost, and LightGBM. Predictions generated by these models are aggregated through an ensembling layer with equal weights to mitigate uncertainty and enhance forecasting accuracy. The performance of both the individual models and the ensemble is assessed using a comprehensive database of EE renovation investments. The findings indicate that the ensemble model consistently outperforms the individual ML models across four accuracy metrics. Furthermore, the analysis demonstrates that certain types of EE measures exhibit greater predictability than others, providing valuable insights for stakeholders evaluating investment risks. Importantly, none of the EE categories displayed significantly poor predictive performance.

Keywords Artificial intelligence · Machine learning · Energy efficiency · Building renovations · Energy sector · Digitization

7.1 Introduction

As of 2020, the building sector accounts for 35% of global energy consumption and 38% of greenhouse gas (GHG) emissions [GHGs; 23]. Achieving climate goals requires significant efforts to reduce energy use in this sector, impacting both new constructions and the retrofitting of existing building stock [57]. The European Commission (EC) has recently revised the Energy Performance of Buildings Directive [15], advancing from the nearly zero-emission buildings (NZEB) standard to the zero-emission buildings (ZEB) initiative. This directive mandates that all new buildings in EU member states must meet ZEB standards by 2030. Additionally, the most recent update of the Directive [16] requires EU countries to establish long-term renovation strategies to transform their building stock into nearly zero-energy buildings by 2050. Despite these initiatives and recent improvements in energy efficiency (EE), the International Energy Agency (IEA) reports that progress remains inadequate to

E. Sarmas et al., *Artificial Intelligence for Energy Systems*, Learning and Analytics in Intelligent Systems 46, https://doi.org/10.1007/978-3-031-85209-1_7

achieve net-zero targets. The energy intensity of the building sector must decrease nearly five times faster over the next decade than it did in the previous five years to align with the Net Zero Emissions by 2050 Scenario [26]. This implies a 45% reduction in energy consumption per square meter by 2030 compared to 2020 levels.

In this context, EE measures are vital for reducing building energy consumption and GHG emissions while supporting global economic growth [47]. However, retrofits and deep renovations are expensive and involve uncertainties [25]. Therefore, it is essential to ensure that funded EE activities effectively reduce energy intensity while optimizing the use of financial resources. Several studies have sought to address this issue, primarily focusing on identifying the EE measures with the highest financial returns [24, 32, 50, 52]. However, beyond financial considerations, the environmental impact of EE investments introduces significant uncertainty in estimating the energy savings of renovation projects.

This chapter emphasizes the importance of quantifying the expected savings of various EE measures to objectively evaluate their potential impact. It proposes leveraging advancements in machine learning (ML) and open-source data to support stakeholder decision-making in financing EE projects. To this end, we present an ML-based methodological framework for predicting energy savings from EE renovations. The framework combines advanced bagging and boosting prediction algorithms with a comprehensive dataset of previously funded EE measures. The key contributions of this study are summarized as follows:

- Rather than relying on simulation tools or physical models to estimate the potential impact of EE measures, we propose a data-driven approach utilizing advanced machine learning ensemble algorithms, including Random Forest, Extreme Gradient Boosting, and Light Gradient Boosting Machine.
- Unlike similar frameworks that primarily evaluate EE measures from a financial perspective-emphasizing return on investment and other financial metrics-our approach focuses on the environmental impact. It estimates the energy savings achievable based on the specific characteristics of the EE action, including the type and estimated cost of the investment, as well as the country and sector in which it is implemented.
- We demonstrate the proposed framework using a comprehensive open-source database containing information on numerous EE investment projects. This allows us to empirically assess the performance of our approach while highlighting the value of leveraging and expanding such databases to improve the monitoring and benchmarking of EE measures.

The chapter is structured as follows: Sect. 7.2 presents a brief review of the literature on quantifying EE measures. Section 7.3 outlines the proposed ML-based framework. Section 7.4 details the empirical evaluation of the methodology, including information on the dataset, accuracy metrics, and a discussion of the results. Finally, Sect. 7.5 offers concluding remarks.

7.2 Problem Formulation and Related Work

Techniques for estimating the energy savings from an Energy Efficiency (EE) measure can be broadly categorized into two main approaches: a priori (or ex ante) and a posteriori (or ex post) methods. A posteriori estimation relies on Measurement and Verification (M&V) protocols to assess energy savings. These protocols use baseline models to compare a building's energy consumption before and after the implementation of an EE measure. M&V processes are guided by established standards, such as the International Performance Measurement and Verification Protocol (IPMVP) [13], the Uniform Methods Project developed by the US Department of Energy [41], and ASHRAE Guideline 14 for evaluating energy, demand, and water savings [7]. However, performance measured during implementation often deviates from design-phase expectations due to optimistic assumptions and model simplifications, necessitating validation and calibration with long-term monitoring data [35]. Traditionally, regression models have supported M&V efforts; however, recent studies are increasingly employing advanced models and algorithms to enhance accuracy [18].

Conversely, a priori estimation aims to predict energy savings before an EE measure is implemented. Reliable pre-implementation estimates are critical for guiding investment decisions and ensuring cost-effective measures, as energy consumption patterns often outweigh financial or social considerations in influencing investments [42]. The absence of robust methods for accurately predicting the impact of EE measures has been identified as a significant barrier to their widespread adoption [22].

In the context of a priori estimation, data-driven methods are gaining prominence. This is due to the increasing availability of building sub-metering data [59] and advancements in machine learning (ML) techniques, which transform raw data into actionable insights [34, 49]. Numerous studies have explored optimal EE measures across various building types and subsystems using both simulated data [3, 12, 55] and real-world data [40, 43]. These approaches include neural networks [21, 38, 43], support vector machines [12, 40], and ensemble models [3]. While these models typically focus on deriving insights rather than directly predicting energy savings, they can be integrated into a priori frameworks for estimating the impact of EE measures.

In addition, several studies have developed specialized data-driven frameworks to directly estimate energy savings before implementing EE measures. For instance, Marasco and Kontokosta [36] introduced a rule-based classifier to predict the responses of buildings to various EE measures, thereby enhancing energy audit outcomes and reducing retrofit costs and complexity. This approach utilized data from over 1,000 audited buildings in New York City, identifying potential improvements in systems such as heating, cooling, domestic hot water, ventilation, lighting, and building envelopes.

Similarly, Severinsen and Myrland [54] examined the accuracy of two statistical learning models-Broken Lines (BL) and Tao Vanilla Benchmarking (TVB)-for estimating energy savings in five Norwegian retail stores undergoing retrofits. The BL model provided weekly energy consumption estimates, simplifying the analysis of weather-related variations in energy usage. In contrast, the TVB model operated at an hourly level, offering more granular insights.

Additionally, Ali et al. [4] introduced a generalized data-driven approach to optimize urban-scale energy retrofit decisions in residential buildings. By leveraging energy performance certificate data, the study identified optimal retrofit actions for urban environments, offering a valuable knowledge base in scenarios with limited or no urban energy modeling data. Clustering techniques have also proven effective in pinpointing cost-efficient retrofit measures. For example, Geyer et al. [20] developed a method to create an effect matrix of retrofit actions, comparing hierarchical clustering methods, such as the Unweighted Pair Group Method with Arithmetic Mean and Shortest Distance, against type-age classifications.

Furthermore, Poortinga et al. [45] proposed an integrated building retrofit solution modeled as a nonlinear mixed-integer programming (NMIP) problem and solved using a genetic algorithm (GA). This approach considered both building envelope components and indoor appliances, aiming to enhance energy performance certificate (EPC) ratings through comprehensive renovation strategies.

As emphasized, estimating the energy savings of an EE measure is a complex and costly process that significantly impacts the feasibility of planned EE initiatives over time. Therefore, it is imperative to develop robust frameworks and methodologies for accurate a priori energy savings estimation. These tools are essential for meeting environmental objectives and reducing both energy intensity and greenhouse gas emissions associated with buildings.

7.3 Proposed Methodology

This section details the algorithms employed to address the regression problem described earlier using machine learning (ML). Various ML model families are available for regression tasks, including linear regression, decision trees, and neural networks (NNs). The selection of an appropriate algorithm is critical, as it is contingent upon the specific characteristics of the application and the nature of the dataset, which directly influences the accuracy of the predictions. In this study, all models are based on decision trees.

Before discussing the specific algorithms, it is essential to highlight why tree-based models were chosen over standard alternatives like linear regression and neural networks. Linear regression models are most effective when the relationships within the data are linear, making them unsuitable for the complexities of this application. Neural networks, especially deep learning models, are highly data-intensive and require substantial datasets for effective training and fine-tuning. The relatively small dataset of energy efficiency (EE) renovation action records limits the applicability of neural networks in this context. Conversely, decision trees excel at handling non-linear relationships, require less data for training, and involve fewer parameters that are straightforward to estimate [28, 56].

The subsections below outline the specific algorithms used in this study: Random Forest (RF), Extreme Gradient Boosting (XGBoost), and Light Gradient Boosting Machine (LightGBM). These tree-based methods employ ensemble techniques, such as bagging and boosting, to improve generalization and enhance predictive accuracy.

7.3.1 Random Forest

Random Forest (RF) is a supervised learning algorithm widely used for both regression and classification tasks [9]. This algorithm employs ensemble learning, specifically a bagging-based approach, to combine the predictions of multiple simple models into a final output that is often more accurate than those of individual models [19, 48]. The designation "Random Forest" signifies two fundamental aspects of the algorithm: randomization, which is accomplished through the bootstrapping of data and features, and the construction of multiple decision trees, thereby forming a "forest."

RF exemplifies bagging (bootstrap aggregating), a prominent ensemble learning technique. Bagging involves training multiple models on different subsets of the training dataset and combining their predictions to generate the final result [8]. In regression tasks, the final output is typically calculated using averaging techniques, such as mean or weighted average [46], while in classification tasks, the most common approach is majority voting, where the predicted class is determined by the majority of models [58].

The process of the RF algorithm for regression tasks is illustrated in Fig. 7.1. For a dataset D with N samples and F features, the algorithm builds K decision trees using the following steps:

1. *Subsampling with Replacement (Bootstrapping)*: The RF algorithm creates K subsets of samples from the original dataset D. Each subset contains N samples, randomly selected with replacement. As a result, some samples may appear multiple times within a subset, while others may not be included at all.
2. *Random Feature Selection*: For each decision tree, the algorithm randomly selects a subset of features ($f \leq F$ from the original set of features to use during training. This ensures diversity among the trees.
3. *Tree Construction*: Each subset of samples and selected features is used to train one decision tree. The trees operate independently, each making predictions based on its unique structure.

The use of bootstrapping reduces variance among tree predictions, helping to mitigate the influence of outliers and individual features that may disproportionately affect the training process. By averaging the predictions of the ensemble, RF produces robust and accurate results across a range of regression and classification problems.

RF models offer several advantages that have made them widely popular across various tasks. One key strength is their ability to handle features with non-linear

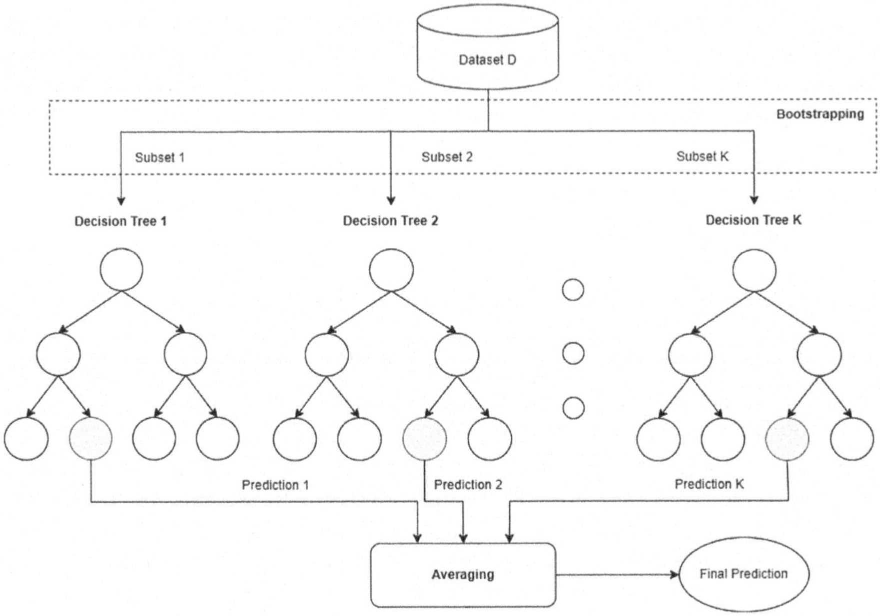

Fig. 7.1 The structure of the random forest algorithm which is based on bootstrap aggregating (or bagging) the predictions of multiple decision tree models

relationships and noisy datasets with high accuracy [5, 11]. Additionally, as a tree-based method, RF models typically require minimal to no data preprocessing, such as normalization or standardization.

However, RF models have some limitations. A significant drawback, as noted by Aria et al. [6], is their lack of interpretability. While RF provides better predictive accuracy compared to single decision trees, it sacrifices the inherent interpretability of individual trees, which allows for clearer insights into how the model captures patterns in the training data [17]. Another challenge is that RF models involve more parameters than single decision trees, requiring additional effort for proper tuning.

In this study, we utilized the Scikit-learn (Sklearn) implementation of the RF algorithm [44]. The key parameters fine-tuned for optimal performance include:

- *Number of estimators*: The number of decision trees utilized to generate individual predictions.
- *Maximum depth*: The maximum allowable depth for each decision tree. Deeper trees can model more complex functions, but excessive depth may lead to overfitting.
- *Maximum features*: The maximum size of the random subsets of features considered by the decision tree when determining a split at a node.
- *Bootstrap*: A Boolean parameter that determines whether bootstrapping will be applied by the model.

7.3.2 Extreme Gradient Boosting

Gradient Boosting (GB) is a highly effective ensemble learning algorithm widely used for both regression and classification tasks [39]. Similar to other boosting methods, GB seeks to create a "strong learner" by iteratively combining multiple "weak learners," each focused on correcting the prediction errors of its predecessor[1] [53]. By addressing the weaknesses of previous models, the algorithm significantly improves the overall performance of the final ensemble. After training all individual models, the ensemble is used to make predictions on new data.

The structure of the GB algorithm is depicted in Fig. 7.2. The ensemble consists of K decision trees. The first tree is trained using the input feature matrix X and the original labels y. Predictions are then generated, and the residual errors E_1 are computed. The second tree is trained using E_1 as the target labels rather than the original y. This process is repeated iteratively, with each subsequent tree trained on the residual errors of the previous tree. Once the final tree is trained, the ensemble is complete, and the entire collection of trees is utilized to make predictions for new samples.

XGBoost belongs to the Gradient Boosting (GB) family but stands out due to its incorporation of advanced regularization techniques, specifically L_1 and L_2, which improve the generalization capabilities of the models it produces. It is an optimized implementation of the GB algorithm, designed for superior computational efficiency

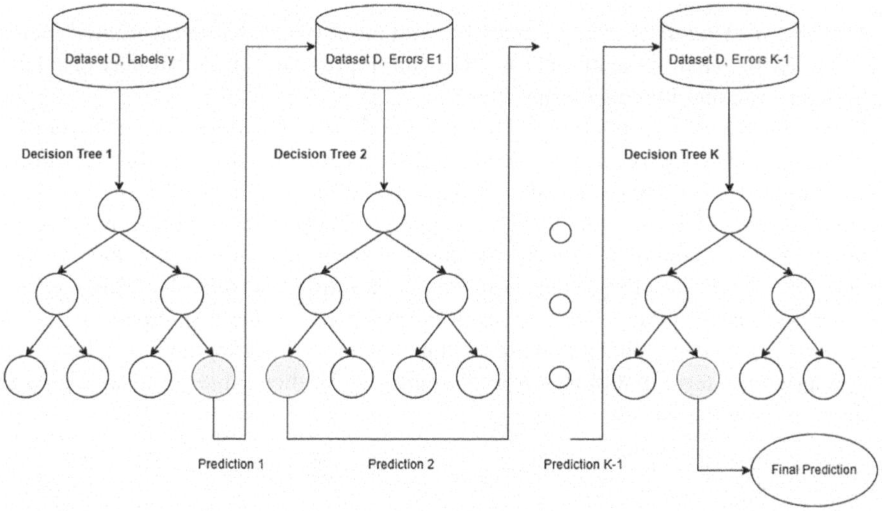

Fig. 7.2 The general structure of the gradient boosting algorithm which is based on boosting the predictions of multiple decision tree models

[1] At each iteration, a decision tree (either a regressor or classifier, depending on the task) is trained on the negative gradient of the loss function.

compared to traditional GB models. XGBoost achieves faster training times through parallelization across clusters [10]. Like Random Forest (RF), XGBoost creates ensembles of decision trees; however, it differs fundamentally in its sequential tree-building process, where each tree is trained to correct the prediction errors of its predecessors. This sequential method highlights the core distinction between bagging (used in RF) and boosting. The term "gradient boosting" refers to the use of gradient descent optimization to minimize prediction errors.

XGBoost offers several advantages that have established it as a leading GB model. One notable advantage is its unique tree pruning strategy, which speeds up training and enhances its scalability for large datasets [60]. Instead of exhaustively evaluating all potential splits, XGBoost uses an approximate greedy algorithm based on weighted quantiles for node splitting. It also partitions data into smaller chunks to enable parallel sample processing [61] and employs cache memory to store gradients, enabling faster computations [37]. These optimizations have made XGBoost a widely adopted choice for both classification and regression tasks across a range of applications over the past decade.

Like Random Forest (RF), XGBoost relies on a set of hyperparameters that are crucial for achieving high-accuracy predictions. These hyperparameters govern both the Gradient Boosting process-such as the number of trees trained-and the structure of individual trees. Important parameters include the maximum depth of the trees and the minimum loss reduction required to permit further splitting of a leaf node. In this study, the original implementation of the XGBoost library was employed [10]. The following key hyperparameters were fine-tuned to optimize performance:

- *Number of estimators*: The number of estimators (weak learners) trained incrementally. Increasing the number of trees can improve accuracy but may also lead to overfitting and longer training times.
- *Maximum depth*: The maximum allowable depth of a tree, representing the longest path from root to leaf. Generally, deeper trees increase the risk of overfitting, similar to what is observed in the Random Forest algorithm.
- *Learning rate*: This parameter scales the weight of each tree, controlling the magnitude of weight updates. Lower learning rates result in smaller, more incremental updates, leading to slower but more precise training. Conversely, higher values can cause faster updates but may result in divergence of the loss function.
- *Complexity control*: This parameter, often called the "Lagrangian multiplier", is a regularization term that ranges from 0 to infinity. Higher values indicate a greater degree of regularization.

7.3.3 Light Gradient Boosting Machine

LightGBM [30] is a Gradient Boosting (GB) framework similar to XGBoost, built on the GB algorithm and utilizing decision trees as weak learners. Developed by

Microsoft, LightGBM emphasizes high performance, scalability, and efficient memory usage, making it widely applicable to machine learning tasks such as ranking, classification, and regression [33].

Although LightGBM shares features with XGBoost-such as support for parallel processing, comparable loss functions, and regularization options-it takes a distinct approach to constructing decision trees. Unlike most GB algorithms, which grow trees depth-wise, LightGBM uses a leaf-wise growth strategy. This method selects the leaf that achieves the largest reduction in loss, irrespective of depth [29]. This approach often leads to lower loss and a more precise fit to the training data.

Another major distinction is in how optimal split points are determined for decision trees. While most GB frameworks, including XGBoost, identify split points by searching sorted feature values, LightGBM employs a histogram-based learning algorithm. This method is more memory-efficient and computationally faster [2].

LightGBM's efficiency is further enhanced by two advanced techniques: gradient-based one-side sampling and exclusive feature bundling. These innovations contribute to a highly efficient training framework that often delivers greater accuracy compared to many other GB algorithms.

Gradient-based one-side sampling is a sampling method that selects data instances based on their gradients. It leverages the observation that instances with smaller gradients are generally well-trained, while those with larger gradients are under-trained. By prioritizing samples with larger gradients, the algorithm achieves faster information gain, enhancing training efficiency.

Exclusive feature bundling, as described by Ke et al. [30], is a feature reduction technique that groups mutually exclusive features into bundles, treating them as a single feature. This approach is particularly effective in sparse feature spaces, where many features are nearly exclusive-meaning they are often zero simultaneously. A common example of such scenarios arises in tasks with numerous categorical variables, where one-hot encoding results in sparse and high-dimensional feature spaces.

The key parameters considered for fine-tuning the LightGBM model, implemented using the official LightGBM Python library Ke et al. [30], are as follows:

- *Number of estimators*: The total number of boosted decision trees trained during the model's learning process.
- *Maximum depth*:Similar to RF and XGBoost, this hyperparameter controls the maximum depth of the decision trees. Given LightGBM's leaf-wise growth structure, careful tuning of this parameter is essential to prevent overfitting.
- *Number of leaves*: This hyperparameter influences the complexity of the tree model. If the number of leaves is set to $2^{Maximum\ depth}$, the tree will have the same number of leaves as a depth-wise tree, which is generally discouraged. Conversely, leaving the number of leaves unrestricted can increase the risk of overfitting.
- *Learning Rate*: Similar to XGBoost, this hyperparameter controls the learning rate for the boosting process.

7.4 Case Study

7.4.1 Data Set

The proposed methodology was validated using data from multiple sources, including the De-risking Energy Efficiency Platform (DEEP) database. DEEP is an open-access online resource that provides comprehensive insights into a wide range of energy efficiency (EE) investment projects [51]. Its purpose is to enhance performance monitoring and benchmarking. By offering key indicators and statistics, this resource aids stakeholders in better comprehending the tangible benefits and potential risks associated with EE investments, based on empirical market data and performance records from thousands of projects across various building and industrial sectors. The dataset includes information such as the type and category of each EE project, the country and sector of implementation (building or industry), projected (a priori) net annual energy savings, investment costs, and various financial metrics, such as the payback period, internal rate of return, and net present value of the investment.

The selection of the most relevant features for predicting energy savings in EE renovation projects was guided by prior studies [1, 14, 27] and adapted so as to align with the available data. Consequently, the following input features were chosen:

- *Project sector* which refers to whether the investment pertains to a building or an industrial project.
- *Investment cost*, measured in Euros.
- *Projected net annual savings*, measured in Euros, as estimated at the time of making the investment.
- *Country* in which the EE project is developed.
- *Investment category* which indicates the main target of the EE renovation project in terms of energy usage or building fabric (e.g. heating, cooling, lighting, pumps, or building fabric).
- *Investment type* which provides a more detailed classification of the EE renovation project in relation to the investment category (e.g., distinguishing between technical and operational optimization for measures like pump or lighting upgrades).

The dataset comprises a fully anonymized sample of 4,183 EE investments from nine countries: Belgium, Bulgaria, Denmark, France, Germany, Latvia, Sweden, the United Kingdom, and the United States. The dataset was carefully curated to exclude EE projects with missing data in any feature. Of the original 4,183 projects, 412 were identified as outliers and removed, leaving a final dataset of 3,771 projects. The most challenging preprocessing task was the creation of investment categories and types, ensuring they are mutually exclusive and collectively exhaustive. As detailed in Table 7.1, the dataset includes 8 EE investment categories, subdivided into 27 distinct investment types.

Due to the inherent structural and technical differences among various types of investments, significant variability is expected across the categories (and even within

Table 7.1 The formulated investment categories and investment types per category for the EE investments used in the case study. For each type of investment, the number of available projects is also displayed

Category	Type	Projects
Heating	Oil/gas fired hot water/steam boilers—operational optimization (controls or automation)	26
	Oil/gas fired hot water/steam boilers—technical optimization (component renewal or supplement)	56
	Oil/gas fired hot water/steam boilers—complete renewal	12
	Heating distribution system—optimization (insulation of pipes/fittings/containment/other surfaces)	47
Cooling	Free cooling	35
Lighting	Re-lamp to LED	66
	Technical optimization (component renewal or supplement)	688
	Operational optimization (controls or automation)	400
Compressed Air	Compressor renewal	23
	Technical optimization (component renewal or supplement)	457
	Operational optimization (controls or automation)	653
	Distribution system optimization	20
Waste Heat	Heat from processes	97
	Heat from cross-cutting technologies	91
Pumps	Operational optimization (control or automation)	29
	Technical optimization (component renewal or supplement)	16
	Complete renewal	29
Building Fabric	Wall insulation	17
	Roof insulation	54
	Glazing upgrade	56
HVAC Plant	Boiler/furnace replacement or upgrade	287
	Chiller plant replacement or upgrade	2
	Pipe insulation	6
	Ventilation upgrade	950
	Installation or replacement of BEMS	1
	Drinking/tap water heating	59
	District heating supply	6

types) in terms of both investment cost and the energy savings achieved through the corresponding EE actions. A visual representation of these differences across investment categories is shown in Fig. 7.3.

In Fig. 7.3a, it is evident that building fabric measures and HVAC plant upgrades are among the most expensive investment categories, while heating, lighting, and compressed air measures are among the least costly. Regarding energy savings, Fig. 7.3b shows that building fabric measures, despite their high investment cost,

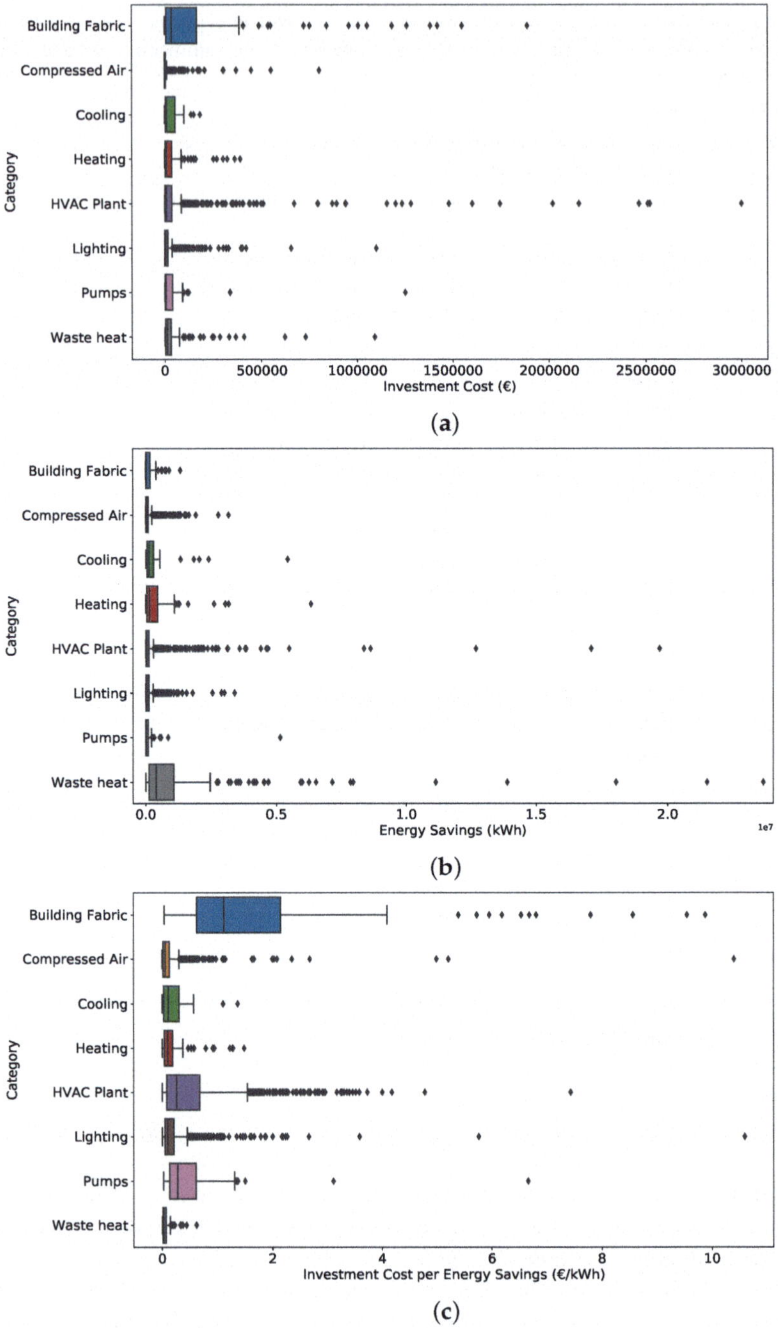

Fig. 7.3 Comparative analysis of the 8 investment categories in terms of cost, energy savings, and investment efficiency (defined as investment cost per energy savings)

deliver relatively low energy savings. In contrast, waste heat recovery actions achieve significant energy savings.

Finally, Fig. 7.3c illustrates the investment efficiency index, defined as the ratio of investment cost to energy savings (€/kWh). This index highlights which categories deliver greater energy savings relative to their cost. As anticipated, building fabric measures are the least efficient category, while investments in HVAC plants, lighting, and pumps also exhibit low efficiency. Conversely, compressed air, heating, and waste heat recovery investments rank among the most effective categories, offering substantial energy savings compared to their costs.

7.4.2 Experimental Setup

This subsection outlines the setup of the experimental application used to evaluate the proposed ML-based framework for estimating energy savings from EE renovation actions. The structure of the experimental workflow is summarized in Fig. 7.4.

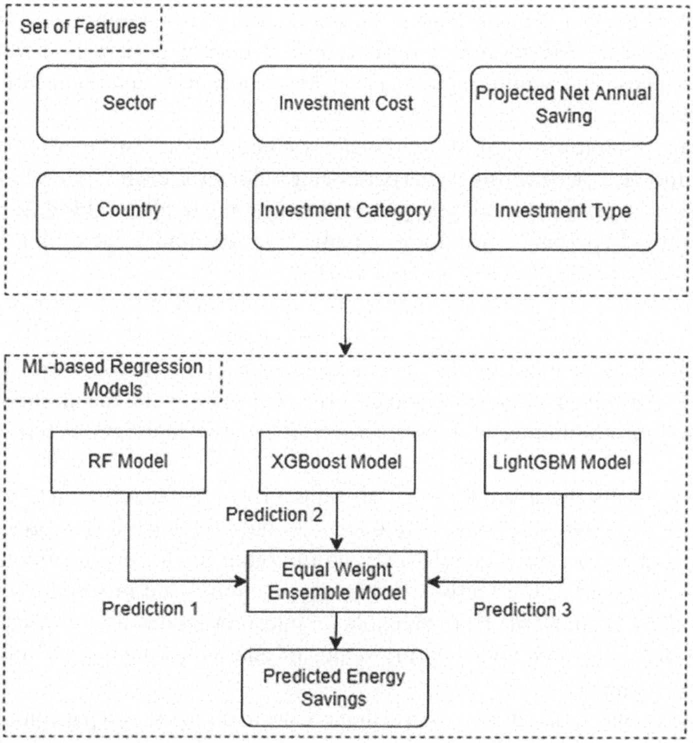

Fig. 7.4 Structure of the experimental application.

As shown, the input data includes the six features described in Sect. 7.4.1. While some features are used without additional processing, the categorical variables-sector, country, investment category, and investment type-are encoded using the label encoder function. For instance, in the sector feature, the category "Building" is encoded as 0, and "Industry" is encoded as 1. The dataset is then split into training and test sets through random sampling, with 80% allocated for training and 20% reserved for evaluating the framework.

At the core of the proposed ML framework are the tree-based algorithms introduced in Sect. 7.3. These algorithms utilize decision trees and leverage effective ensembling techniques, such as bagging and boosting. Specifically, the RF, XGBoost, and LightGBM models are employed to predict the energy savings of EE renovation actions. To enhance prediction accuracy and address model and parameter uncertainty, an additional ensembling layer is applied. In this layer, the predictions from the three tree-based models are averaged with equal weights to generate the final forecast, representing the energy savings from the EE actions. This additional ensembling step reduces the dispersion of the predictions, improving overall forecasting performance.

The final step of the experimental setup involves determining the optimal hyperparameters for the selected regression models. For each model, hyperparameter tuning was performed using a validation set comprising 20% of the training data. To ensure the robustness of the ML models across all investment categories, the validation set included samples representing diverse categories, measures, and countries from the original dataset.

The mean absolute error (MAE) and mean squared error (MSE) metrics were used to identify the best-performing hyperparameter values for each model. Grid search was employed to optimize the hyperparameters systematically. Table 7.2 provides a summary of the hyperparameter tuning results for each model, including the search space explored and the selected optimal values.

Selecting appropriate accuracy metrics is a critical step in evaluating any regression model. In this study, the accuracy of the four models (the three described in Sect. 7.3 and the ensemble model) is assessed using four metrics: mean absolute error (MAE), mean absolute percentage error (MAPE), root mean squared error (RMSE), and the coefficient of determination (R^2) between the predicted and actual values [31].

MAE represents the average of all absolute errors between the actual and predicted values. It is robust to outliers and easy to interpret since it is expressed in the same scale as the observed data. RMSE, on the other hand, assigns greater weight to larger errors, making it effective for identifying significant mispredictions by the models. Both MAE and RMSE are valuable for interpreting results, but when comparing investment categories with varying scales of energy savings, scale-independent metrics are crucial.

MAPE is a scale-independent metric that is easy to interpret and particularly useful in this context. Since the dataset contains no zero values that would complicate its calculation, MAPE serves as a highly effective measure for comparing model accuracy across different investment categories. Lastly, R^2 quantifies the proportion

Table 7.2 The search space and the selected values of the hyper-parameters for the tree-based regression models considered for predicting energy savings

Model	Hyper-parameter	Min value	Max value	Step	Selected
RF	Number of estimators	100	500	100	100
	Maximum depth	1	7	2	3
	Maximum features	0.2	1	0.2	1
	Bootstrap	Values tested: True, False			True
XGBoost	Number of estimators	100	500	100	300
	Maximum depth	2	10	2	6
	Learning rate	0.1	0.9	0.2	0.3
	Complexity control	0	10	2	0
LightGBM	Number of estimators	20	100	20	80
	Maximum depth	2	10	2	6
	Number of leaves	10	50	10	30
	Learning rate	0.02	0.1	0.02	0.08

of variance in the target variable explained by the predictions, providing a measure of how well the regression models fit the data.

The formulas for calculating these metrics are provided below:

$$MAE = \frac{1}{n}\sum_{1}^{n}|y - \hat{y}|, \tag{7.1}$$

$$RMSE = \sqrt{\frac{1}{n}\sum_{1}^{n}(y - \hat{y})^2}, \tag{7.2}$$

$$MAPE = \frac{100\%}{n}\sum_{1}^{n}\left|\frac{y - \hat{y}}{y}\right|, \tag{7.3}$$

$$R^2 = 1 - \frac{\sum_{1}^{n}(y - \hat{y})^2}{\sum_{1}^{n}(y - \bar{y})^2}, \tag{7.4}$$

where y and $\hat{y}$ is the real and the predicted value of the energy savings for the examined EE investment in the test set, respectively, $\bar{y}$ is the average value of the real observations, and n is the total number of the samples included in the evaluation.

7.4.3 Results and Discussion

The proposed ML methodology was evaluated using a test set of 754 EE investments, representing 20% of the entire dataset. The test set includes samples from all investment categories to assess the models' robustness across different types of EE projects. The experimental results are summarized in Table 7.3.

The findings demonstrate that all three tree-based models effectively handle the complexities of predicting energy savings for a diverse range of renovation actions across different categories and countries. Among the models, XGBoost outperformed the others in terms of mean absolute error (MAE) at 25,613 kWh and mean absolute percentage error (MAPE) at 32.6%. However, LightGBM and RF achieved a slightly higher R^2 score of 78.7%.

While RF exhibited the highest MAE and MAPE among the three models, it performed better in terms of root mean squared error (RMSE) at 50,459 kWh. This suggests that the RF model has fewer large prediction errors compared to the other models.

The results show that boosting-based models (XGBoost and LightGBM) generally perform better on metrics that do not penalize large errors, while the bagging-based RF model excels on metrics like RMSE, which emphasize large mispredictions. In this context, the proposed equal-weight (EW) ensemble model surpasses all three individual models across all error metrics. Specifically, the EW ensemble model achieves an RMSE of 49,806 kWh, an MAE of 25,339 kWh, a MAPE of 32.2%, and an R^2 score of 79.2%. These results confirm that the proposed ensemble model

Table 7.3 Results on test set

Metric	RF	XGBoost	LightGBM	EW ensemble
MAE (kWh)	26,394	25,613	25,778	**25,339**
$RMSE$ (kWh)	50,459	50,947	51,002	**49,806**
$MAPE$ (%)	33.5	32.6	32.9	**32.2**
R^2 (%)	78.7	78.3	78.7	**79.2**

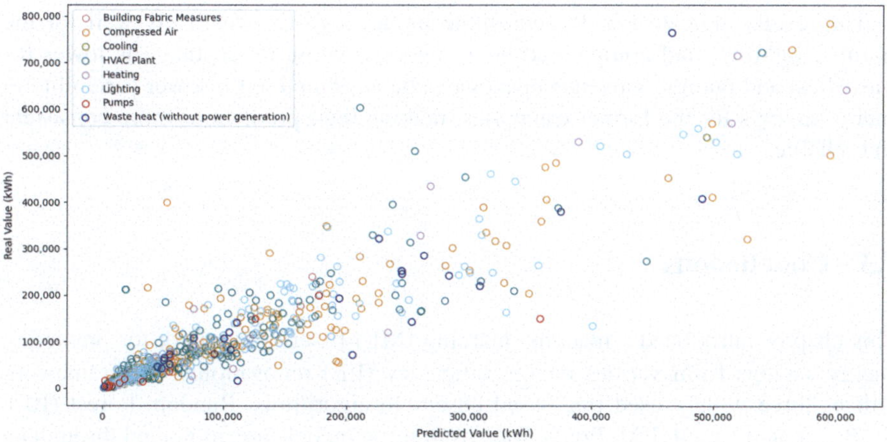

Fig. 7.5 Scatter plot visualization of the real and the predicted values based on the EW ensemble model for the EE renovation projects of the test set

Table 7.4 Forecasting accuracy of the ensemble model on the test set. The results are presented both per investment category and in total (average)

Category	$MAPE$ (%)	R^2 (%)
Heating	44.1	83.6
Cooling	27.7	98.0
Lighting	27.3	85.1
Compressed air	29.8	81.9
Waste Heat	48.7	63.8
Pumps	43.9	65.9
Building fabric measures	35.3	87.6
HVAC plant	35.4	59.9
Average	32.2	79.2

offers superior predictive accuracy, successfully reducing the generalization error of the individual tree-based models on which it is built.

An important aspect of the empirical evaluation is assessing the predictive performance of the ensemble model across different investment categories. A scatter plot comparing the actual and predicted values for the EW ensemble model is shown in Fig. 7.5. As expected, some categories, such as lighting, are generally easier to predict, while others, like heating, are more challenging due to the inherent variability in energy consumption.

To evaluate the model's performance by category, Table 7.4 presents the MAPE and R^2 metrics for the EW ensemble model across different investment categories. These metrics were chosen for comparison because they are scale-independent, unlike RMSE and MAE, which are not suitable for datasets with varying scales.

The results indicate that the ensemble model performs relatively well for the cooling, lighting, and compressed air categories, while its accuracy is lower for waste heat and pumps. This discrepancy can be attributed to the lower variability in energy savings for the former categories, making their predictions more consistent and reliable.

7.5 Conclusions

This chapter introduced a machine learning (ML)-based framework for predicting energy savings from various energy efficiency (EE) renovations. The framework utilizes three widely used tree-based ensemble algorithms: Random Forest (RF), XGBoost, and LightGBM. Predictions from these models are combined through an ensembling layer with equal weights to reduce uncertainty and enhance forecasting accuracy.

The performance of both the individual models and the ensemble was evaluated using a comprehensive database of EE renovation investments. The findings show that the ensemble model consistently outperforms the individual ML models across four accuracy metrics. Moreover, the analysis reveals that certain types of EE measures are more predictable than others, offering valuable insights for stakeholders assessing investment risks. Importantly, none of the EE categories exhibited notably poor predictive performance.

This research provides a foundation for accurately estimating the energy savings potential of EE renovations. However, there is significant room for further improvement. Future work could focus on establishing a structured process for collecting and validating EE data, enabling the creation of a standardized, open-source database that spans a broader range of EE projects, diverse actions, and wider geographical coverage to account for varying climatic conditions. Additionally, integrating detailed project characteristics-such as floor area, construction year, occupancy, operating hours, and building type-could further enhance the accuracy of predictions.

Given the inherent uncertainties in predictions, future research could also explore methods to link forecast errors to the uncertainty associated with EE investments. This would equip decision-makers with better tools to assess risks in EE financing [62]. Building on these advancements, the development of a web-based "predictions-as-a-service" platform could provide stakeholders with user-friendly tools for making informed decisions about EE investments and financing.

References

1. M. Aguirre, G. Ibikunle, Determinants of renewable energy growth: a global sample analysis. Energy Policy **69**, 374–384 (2014)
2. E. Al Daoud, Comparison between xgboost, lightgbm and catboost using a home credit dataset. Int. J. Comput. Inf. Eng. **13**(1), 6–10 (2019)
3. M. Al-Rakhami, A. Gumaei, A. Alsanad, A. Alamri, M.M. Hassan, An ensemble learning approach for accurate energy load prediction in residential buildings. IEEE Access **7**, 48328–48338 (2019)
4. U. Ali, M.H. Shamsi, M. Bohacek, C. Hoare, K. Purcell, E. Mangina, J. O'Donnell, A data-driven approach to optimize urban scale energy retrofit decisions for residential buildings. Appl. Energy **267**, 114861 (2020)
5. F.A. Anifowose, J. Labadin, A. Abdulraheem, Ensemble model of non-linear feature selection-based extreme learning machine for improved natural gas reservoir characterization. J. Nat. Gas Sci. Eng. **26**, 1561–1572 (2015)
6. M. Aria, C. Cuccurullo, A. Gnasso, A comparison among interpretative proposals for random forests. Mach. Learn. Appl. **6**, 100094 (2021)
7. ASHRAE, Measurement of energy, demand, and water savings. ASHRAE Guid. **4**, 1–150 (2014)
8. L. Breiman, Bagging predictors. Mach. Learn. **24**(2), 123–140 (1996)
9. L. Breiman, Random forests. Mach. Learn. **45**(1), 5–32 (2001)
10. T. Chen, C. Guestrin, XGBoost: a scalable tree boosting system, in *Proceedings of the 22nd ACM SIGKDD International Conference on Knowledge Discovery and Data Mining*, KDD '16 (ACM, New York, NY, USA, 2016), pp. 785–794
11. L. Cheng, J. De Vos, P. Zhao, M. Yang, F. Witlox, Examining non-linear built environment effects on elderly's walking: a random forest approach. Transp. Res. Part D: Transp. Environ. **88**, 102552 (2020)
12. J.-S. Chou, D.-K. Bui, Modeling heating and cooling loads by artificial intelligence for energy-efficient building design. Energy Build. **82**, 437–446 (2014)
13. U. DOE, International performance measurement & verification protocol (2002). https://www.nrel.gov/docs/fy02osti/31505.pdf
14. H. Doukas, P. Xidonas, N. Mastromichalakis, How successful are energy efficiency investments? a comparative analysis for classification & performance prediction. Comput. Econ. **59**(2), 579–598 (2022)
15. European Commission, Directive 2010/31/eu of the European parliament and of the council dated may 19, 2010 about the overall energy efficiency of buildings (2018). https://eur-lex.europa.eu/legal-content/EN/TXT/?uri=CELEX%3A02010L0031-20100708
16. European Commission, Directive 2010/31/eu of the European parliament and of the council of 19 may 2010 on the energy performance of buildings (2021). https://eur-lex.europa.eu/legal-content/EN/TXT/?uri=CELEX%3A02010L0031-20210101
17. A.A. Freitas, Comprehensible classification models: a position paper. ACM SIGKDD Explor. Newsl **15**(1), 1–10 (2014)
18. C.V. Gallagher, K. Leahy, P. O'Donovan, K. Bruton, D.T. O'Sullivan, Development and application of a machine learning supported methodology for measurement and verification (M&V) 2.0. Energy Build. **167**, 8–22 (2018)
19. P. Geurts, D. Ernst, L. Wehenkel, Extremely randomized trees. Mach. Learn. **63**(1), 3–42 (2006)
20. P. Geyer, A. Schlüter, S. Cisar, Application of clustering for the development of retrofit strategies for large building stocks. Adv. Eng. Inform. **31**, 32–47 (2017)
21. P. Geyer, S. Singaravel, Component-based machine learning for performance prediction in building design. Appl. Energy **228**, 1439–1453 (2018)
22. B. Grillone, S. Danov, A. Sumper, J. Cipriano, G. Mor, A review of deterministic and data-driven methods to quantify energy efficiency savings and to predict retrofitting scenarios in buildings. Renew. Sustain. Energy Rev. **131**, 110027 (2020)

23. I. Hamilton, H. Kennard, O. Rapf, J. Kockat, S. Zuhaib, T. Abergel, M. Oppermann, M. Otto, S. Loran, N. Steurer et al., Global status report for buildings and construction: towards a zero-emission. Nairobi, Kenya, United Nations Environment Programme, Efficient and Resilient Buildings and Construction Sector (2020)
24. Y. He, N. Liao, J. Bi, L. Guo, Investment decision-making optimization of energy efficiency retrofit measures in multiple buildings under financing budgetary restraint. J. Clean. Prod. **215**, 1078–1094 (2019)
25. Hernández, J.L., Serroni, S., Stamatopoulos, S., Sarmas, E., Revel, G.M., Vélez, F., Multi-criteria building performance assessment, in *2024 15th International Conference on Information, Intelligence, Systems & Applications (IISA)* (IEEE, 2024), pp. 1–7
26. IEA, Tracking buildings (2021). https://www.iea.org/reports/tracking-buildings-2021
27. M. Isa, M.M.G.M.A. Rahman, I. Sipan, T.K. Hwa, Factors affecting green office building investment in Malaysia. Procedia Soc. Behav. Sci. **105**, 138–148 (2013)
28. T. Januschowski, Y. Wang, K. Torkkola, T. Erkkilä, H. Hasson, J. Gasthaus, Forecasting with trees. Int. J. Forecast. (2021)
29. D. Jin, Y. Lu, J. Qin, Z. Cheng, Z. Mao, SwiftIDS: real-time intrusion detection system based on LightGBM and parallel intrusion detection mechanism. Comput. Secur. **97**, 101984 (2020)
30. G. Ke, Q. Meng, T. Finley, T. Wang, W. Chen, W. Ma, Q. Ye, T.-Y. Liu, LightGBM: a highly efficient gradient boosting decision tree. Adv. Neural Inf. Process. Syst. **30** (2017)
31. D. Koutsandreas, E. Spiliotis, F. Petropoulos, V. Assimakopoulos, On the selection of forecasting accuracy measures. J. Oper. Res. Soc. **73**(5), 937–954 (2022)
32. Y. Liu, T. Liu, S. Ye, Y. Liu, Cost-benefit analysis for energy efficiency retrofit of existing buildings: a case study in china. J. Clean. Prod. **177**, 493–506 (2018)
33. S. Makridakis, E. Spiliotis, V. Assimakopoulos, M5 accuracy competition: results, findings, and conclusions. Int. J. Forecast. (2022)
34. M. Manfren, P.A. James, L. Tronchin, Data-driven building energy modelling-an analysis of the potential for generalisation through interpretable machine learning. Renew. Sustain. Energy Rev. **167**, 112686 (2022)
35. M. Manfren, B. Nastasi, Parametric performance analysis and energy model calibration workflow integration-a scalable approach for buildings. Energies **13**(3), 621 (2020)
36. D.E. Marasco, C.E. Kontokosta, Applications of machine learning methods to identifying and predicting building retrofit opportunities. Energy Build. **128**, 431–441 (2016)
37. R. Mitchell, E. Frank, Accelerating the XGBoost algorithm using GPU computing. Peer J. Comput. Sci. **3**, e127 (2017)
38. S. Naji, A. Keivani, S. Shamshirband, U.J. Alengaram, M.Z. Jumaat, Z. Mansor, M. Lee, Estimating building energy consumption using extreme learning machine method. Energy **97**, 506–516 (2016)
39. A. Natekin, A. Knoll, Gradient boosting machines, a tutorial. Front. Neurorobot. **7**, 21 (2013)
40. N.-T. Ngo, Early predicting cooling loads for energy-efficient design in office buildings by machine learning. Energy Build. **182**, 264–273 (2019)
41. NREL, The uniform methods project: methods for determining energy efficiency savings for specific measures (2016). https://www.nrel.gov/docs/fy18osti/70472.pdf
42. A. Papapostolou, F.D. Mexis, E. Sarmas, C. Karakosta, J. Psarras, Web-based application for screening energy efficiency investments: a MCDA approach, in *2020 11th International Conference on Information, Intelligence, Systems and Applications (IISA* (IEEE, 2020), pp. 1–7
43. G. Paterson, D. Mumovic, P. Das, J. Kimpian, Energy use predictions with machine learning during architectural concept design. Sci. Technol. Built Environ. **23**(6), 1036–1048 (2017)
44. F. Pedregosa, G. Varoquaux, A. Gramfort, V. Michel, B. Thirion, O. Grisel, M. Blondel, P. Prettenhofer, R. Weiss, V. Dubourg, J. Vanderplas, A. Passos, D. Cournapeau, M. Brucher, M. Perrot, E. Duchesnay, Scikit-learn: machine learning in Python. J. Mach. Learn. Res. **12**, 2825–2830 (2011)
45. W. Poortinga, S. Jiang, C. Grey, C. Tweed, Impacts of energy-efficiency investments on internal conditions in low-income households. Build. Res. Inf. **46**(6), 653–667 (2018)

46. A.M. Prasad, L.R. Iverson, A. Liaw, Newer classification and regression tree techniques: bagging and random forests for ecological prediction. Ecosystems **9**(2), 181–199 (2006)
47. J. Rosenow, N. Eyre, Reinventing energy efficiency for net zero. Energy Res. Soc. Sci. **90**, 102602 (2022)
48. O. Sagi, L. Rokach, Ensemble learning: a survey. Wiley Interdiscip. Rev.: Data Min. Knowl. Discov. **8**(4), e1249 (2018)
49. E. Sarmas, A. Fragkiadaki, V. Marinakis, Explainable AI-based ensemble clustering for load profiling and demand response. Energies **17**(22), 5559 (2024)
50. E. Sarmas, V. Marinakis, H. Doukas, A data-driven multicriteria decision making tool for assessing investments in energy efficiency. Oper. Res. 1–20 (2022a)
51. E. Sarmas, E. Spiliotis, N. Dimitropoulos, V. Marinakis, H. Doukas, Estimating the energy savings of energy efficiency actions with ensemble machine learning models. Appl. Sci. **13**(4), 2749 (2023)
52. E. Sarmas, E. Spiliotis, V. Marinakis, T. Koutselis, H. Doukas, A meta-learning classification model for supporting decisions on energy efficiency investments. Energy Build. **258**, 111836 (2022b)
53. R.E. Schapire, The boosting approach to machine learning: an overview. Nonlinear Estim. Classif. 149–171 (2003)
54. A. Severinsen, Ø. Myrland, Statistical learning to estimate energy savings from retrofitting in the Norwegian food retail market. Renew. Sustain. Energy Rev. **167**, 112691 (2022)
55. S. Singaravel, J. Suykens, P. Geyer, Deep-learning neural-network architectures and methods: using component-based models in building-design energy prediction. Adv. Eng. Inform. **38**, 81–90 (2018)
56. E. Spiliotis, Decision trees for time-series forecasting. Foresight: Int. J. Appl. Forecast. **64**, 30–44 (2022)
57. E. Stamatopoulos, A. Forouli, D. Stoian, P. Kouloukakis, E. Sarmas, V. Marinakis, An adaptive framework for assessing climate resilience in buildings. Build. Environ. **264**, 111869 (2024)
58. C. Strobl, J. Malley, G. Tutz, An introduction to recursive partitioning: rationale, application, and characteristics of classification and regression trees, bagging, and random forests. Psychol. Methods **14**(4), 323 (2009)
59. L. Toderean, T. Cioara, I. Anghel, E. Sarmas, V. Michalakopoulos, V. Marinakis, Demand response optimization for smart grid integrated buildings: review of technology enablers landscape and innovation challenges. Energy Build. 115067 (2024)
60. C. Wade, *Hands-On Gradient Boosting with XGBoost and Scikit-Learn: Perform Accessible Machine Learning and Extreme Gradient Boosting with Python* (Packt Publishing Ltd., 2020)
61. Z. Wen, J. Shi, B. He, J. Chen, K. Ramamohanarao, Q. Li, Exploiting GPUS for efficient gradient boosting decision tree training. IEEE Trans. Parallel Distrib. Syst. **30**(12), 2706–2717 (2019)
62. P. Xidonas, H. Doukas, E. Sarmas, A python-based multicriteria portfolio selection DSS. RAIRO-Oper. Res. **55**, S3009–S3034 (2021)

Part IV
Advances in Energy Management

Chapter 8
Optimization Model for Scheduling Flexible Loads to Mitigate Energy Peaks

Abstract This chapter presents an energy management method designed to optimize electricity consumption by mitigating peak demand through the strategic scheduling of an irrigation pumping system's operating hours on an isolated island. Initially, Machine Learning (ML) models are employed to accurately predict island-wide electricity consumption and hourly energy production from Renewable Energy Sources (RES). Subsequently, these predictions are integrated into a heuristic optimization algorithm that establishes the optimal pump operating schedule, thereby minimizing consumption peaks while adhering to operational constraints. The efficacy of the proposed method is evaluated using the case study of Tilos, a remote Greek island equipped with an energy management system that oversees and regulates local water pumping stations for household water supply and irrigation. The findings suggest that intelligent pump scheduling in a small-scale island context can decrease daily and weekly electricity consumption variability by over 15% without incurring any associated monetary costs. Furthermore, this chapter concludes that the potential advantages of this approach are contingent upon three critical factors: the volume of load that can be shifted daily, the accuracy of the forecasts, and the extent of electricity generated by photovoltaic (PV) systems.

Keywords Artificial intelligence · Peak shaving · Energy management · Forecasting · Flexibility management · Machine learning · Water-energy nexus

8.1 Introduction

In this chapter, a comprehensive approach to intelligent energy management is introduced, aimed at smoothing electricity consumption in remote island systems. The proposed methodology is founded on two essential components: the development of ML models for precise forecasting of renewable energy production and electricity demand, and the formulation of a scheduling algorithm to optimally reschedule the operational hours of water pumping stations to align with periods of lower anticipated load.

E. Sarmas et al., *Artificial Intelligence for Energy Systems*, Learning and Analytics in Intelligent Systems 46, https://doi.org/10.1007/978-3-031-85209-1_8

Ensuring the proper operation of water supply systems is essential for providing communities with access to water. Water and energy are two of the most critical resources, and their integrated management can yield substantial economic and environmental benefits. The United Nations Sustainable Development Goals, particularly Goals 6 and 11, emphasize the importance of ensuring sustainable water availability and management, with actions aimed at securing safe water access and enhancing resource efficiency. By 2030, the goal is to implement integrated water resource management systems that promote local community participation in improving water management. Since 2014, global energy consumption in the water sector has reached approximately 120 million tons of oil equivalent (Mtoe) annually, with electricity accounting for about 4% of global consumption. Processes such as pumping, filtration, coagulation, and chemical dosing dominate electricity use in this sector. Water demand is projected to increase by 44% by 2050 due to growing needs in manufacturing, thermal energy production, agriculture, and housing. According to the International Energy Agency's Water-Energy Nexus report, economic and population growth, along with climate change, are intensifying water availability challenges, particularly in emerging economies, deepening the interdependence between water and energy in the coming decades.

Water utilities and municipalities typically aim to ensure a secure water supply while minimizing energy consumption, operational costs, and environmental impacts. Various sustainability efforts have been implemented in the water sector, such as integrating renewable energy sources (RES) to decentralize and diversify electricity production. Other measures, such as retrofitting water system equipment, are also being explored, although these often require substantial financial investments, which can hinder their implementation. An alternative approach involves improving pumping efficiency through optimization techniques, particularly digital monitoring systems and advanced optimization modeling. These solutions, which do not require significant infrastructure investments, are becoming increasingly attractive.

The challenge of managing water resources is particularly acute in small, remote areas like islands [57], where limited water resources and seasonal demand spikes, especially during the tourist season, exacerbate the problem. Persistent issues include insufficient regional and national planning and the lack of a coherent strategy for agricultural water use. In response, intelligent energy management algorithms have been developed to smooth loads and achieve peak savings, leveraging dynamic modeling in specialized software systems such as APROS and TRNSYS 17. Additionally, some studies have focused on the direct integration of photovoltaic (PV) systems with water pumping systems to ensure reliable and economical operations or to optimize the design of standalone PV-powered water pumps.

This study centers on intelligent energy management for remote island energy systems, with a focus on mitigating load peaks through optimal scheduling of water pumping station operations. The research objective is to develop a daily operational schedule for water pumping stations that minimizes net load peaks (i.e., total electricity consumption minus RES-based production), while ensuring compliance with constraints related to the minimum operating hours required for water availability.

The effectiveness of the proposed solution is evaluated using specific Key Performance Indicators (KPIs), which measure daily and weekly net electricity consumption variations at the island level. The methodology is applied to Tilos, a small Greek island in the Aegean Sea, characterized by distributed PV electricity production and several municipal water pumping stations.

The proposed peak-saving method offers a data-driven approach to the water system scheduling challenge, characterized by its flexibility in adjusting operating hours and accommodating multiple shiftable load constraints. This approach is built upon the following key elements:

- A set of three ML models designed to forecast electricity consumption by incorporating daily and monthly seasonal load patterns along with underlying trends.
- A set of three ML models developed to predict PV electricity generation, leveraging weather forecasts and other explanatory variables.
- A set of constraints governing the daily operating hours necessary for pumping stations to maintain water availability, while also accounting for their operational flexibility.
- A heuristic optimization algorithm designed to dynamically smooth peaks, ensuring that (i) the island's energy system maximizes the use of electricity generated from renewable energy sources (RES), and (ii) the pumping stations' operating hours avoid coinciding with peak electricity consumption periods on the island.

The innovation of the proposed method lies in its autonomous architecture, which operates independently of professional simulation software tools. This method integrates two sets of three core forecasting models, characterized by their high accuracy, ease of implementation, minimal training data requirements, and robustness against overfitting and uncertainty. Additionally, it features a heuristic optimization algorithm that is adaptable to various inputs and load-shifting rules. Unlike conventional peak-saving solutions that heavily rely on battery energy storage systems, this method achieves peak savings without requiring energy storage, making it suitable for a wide range of isolated microgrids.

Furthermore, the method approaches peak savings from a novel perspective, directly linking water availability to energy savings. It provides empirical insights into how peak-saving benefits are influenced by (i) the daily volume of load that can be shifted, (ii) the accuracy of forecasts used in the optimization process, and (iii) the quantity of electricity generated from renewable energy sources (RES). These considerations are particularly significant in small-scale island settings, where the local water supply system's electricity demand constitutes a substantial portion of the island's total energy consumption. Additionally, such environments often exhibit significant variability in load between off-peak and peak periods as well as across different seasons.

The remainder of this work is structured as follows: Sect. 8.2 reviews related research and highlights existing gaps in the field. Section 8.3 details the proposed peak-saving method, including the three core forecasting models and the optimization algorithm. Section 8.4 presents a real-world case study applying the method to the Greek island of Tilos. Finally, Sect. 8.5 summarizes the findings and outlines key directions for future research.

8.2 Problem Formulation and Related Work

8.2.1 Problem Statement

The volume of generated data continues to grow at an unprecedented rate. In the energy sector, real-time data is increasingly produced by Internet of Things (IoT) technologies, such as smart meters for monitoring energy consumption and renewable energy production [51], sensor-based systems [31], energy efficiency investment records [53], and network-based assets like transformer feeders [23]. Moreover, access to open data, including historical weather and energy datasets, has become more accessible, creating new opportunities for developing advanced models and pattern recognition techniques. Data from other domains, such as water pump operations, are also readily available and can be integrated into analytical algorithms.

These diverse data sources present significant potential for creating multi-scale, multi-stakeholder approaches through innovative analytical methods that provide stakeholders in the energy sector with more accurate and actionable insights, enhancing data-driven decision-making [17, 44]. Accordingly, this study aims to harness machine learning models and optimization techniques to unlock the value of data related to electricity consumption, PV production, weather conditions, and variable loads such as water pumps. The proposed framework offers a comprehensive planning approach to automate water supply system management, addressing the unique challenges of small-scale island environments.

8.2.2 Water Supply System Optimization Techniques

The improvement of water supply systems is increasingly supported by digital tools and software solutions. These tools are designed to optimize the planning and operation of water supply systems by minimizing resource usage, reducing water losses, and maximizing energy savings during water distribution. Although considerable energy savings can also be attained in water treatment processes, this discussion will concentrate specifically on optimization solutions for water distribution.

Traditional hydraulic modeling tools, such as EPANET [9], have been widely used to analyze the behavior of water distribution networks by allocating user demands to computational nodes. EPANET provides capabilities for simulating both hydraulics and water quality in pressurized pipe networks. These simulations enhance understanding of water flow dynamics within distribution systems. Leveraging EPANET, numerous studies have been conducted to design water supply systems capable of meeting daily demands in various regions [2, 27, 49].

Beyond commercial software, efforts to improve the operational efficiency of water systems have led to the development of optimization techniques focused on pump scheduling. These approaches determine the optimal hours for pump operation, often leveraging off-peak electricity tariffs and the storage capacity of water distribution systems. Early advancements in this area utilized genetic algorithms to minimize pump operating costs [36]. Heuristic and metaheuristic methodologies have also been integrated with software like EPANET to optimize water delivery strategies, reducing costs and energy consumption in real-world distribution networks [7]. These methods have demonstrated energy savings of up to 10%, highlighting the potential of data-driven optimization solutions.

Further studies have addressed the integration of pump scheduling with the management of water storage tank levels. By coordinating pumping schedules with tank activation levels, these approaches aim to minimize peak pumping and reduce pumping head. One such application of genetic algorithms, incorporating historical data and integer decision variables, achieved a 20% reduction in energy costs [20]. Additional strategies include managing pump station operations based on water levels in multiple tanks or aligning operations with time-of-day pricing to reduce pumping during peak tariff periods [39].

While heuristic methods have shown promise in achieving cost savings and reducing CO_2 emissions, their integration of stored water level management with pump scheduling optimization significantly increases the complexity of the algorithms, leading to higher computational demands [65]. The inclusion of multiple decision variables further constrains the convergence of these heuristic approaches. In response, hybrid methods, such as those incorporating modified mutation mechanisms, have been developed to mitigate these challenges.

8.2.3 Production Forecasting

Over the years, numerous forecasting methods have been developed to predict renewable energy production and support decisions related to energy management, pricing, and optimization, with a particular focus on wind and solar energy-two of the most common and cost-effective renewable sources [47, 54]. In the domain of solar energy forecasting, regression-based approaches, such as neural networks (NNs) [NNs; 13, 24, 40], decision tree models [14], and support vector machines [8], have proven to be highly effective. These methods excel at incorporating multiple factors influencing PV production and capturing non-linear relationships. Key influencing factors include historical production data, meteorological forecasts, and installation-specific conditions, among others. However, quantitatively comparing these forecasting methods is challenging, as their accuracy depends significantly on the quality and resolution of historical data, the forecasting horizon, and the reliability of meteorological input data [52, 58].

Nespoli et al. [42] conducted a comparative analysis of two widely used machine learning methods for solar energy forecasting: one relying solely on historical data and the other integrating meteorological forecasts alongside historical observations. Their findings revealed that while the first model exhibited consistent performance, it was often less accurate than the second model on sunny days. Conversely, both models showed reduced accuracy on cloudy days, likely due to inaccuracies in the meteorological forecasts used as inputs. These results underscore the critical role of accurate weather data-such as ambient temperature, solar radiation, and cloud cover-in enhancing solar energy forecasts. They also highlight the growing adoption of hybrid methods and ensemble techniques [25], which are particularly effective at addressing uncertainties in models, data, and parameters [48].

8.2.4 Load Forecasting

Load forecasting plays a pivotal role in ensuring the efficient operation and strategic planning of power systems. Depending on the forecasting horizon, which can vary from minutes to years, different datasets and modeling approaches are required to achieve accurate predictions [47]. In operational management, short-term forecasts-covering periods from 1 to 168 h (one week) ahead-are particularly valuable. These forecasts support critical decision-making processes, including bidding, portfolio optimization, tariff design, and energy-saving strategies such as load shifting, peak shaving, energy storage, and load balancing [64]. Over the years, numerous methods have been developed for short-term load forecasting [28, 55, 60, 61]. A focused review of energy forecasting methodologies for buildings was conducted by Sun et al. [62]. Broadly, forecasting methods can be grouped into three categories: time series methods, ML methods, and hybrid methods.

Time series methods, including ARIMA and exponential smoothing, utilize statistical models to identify patterns in historical load data and extrapolate them into the future [12]. Although these methods are computationally efficient and intuitive, they exhibit limited flexibility and face challenges when modeling non-linear relationships.

ML-based forecasting methods address these limitations by incorporating additional factors that influence demand, including weather conditions, calendar effects, holidays, and special events. Among machine learning models, neural networks (NNs) [29] and, more recently, deep learning techniques [43] have gained prominence. Furthermore, regression tree-based models, such as XGBoost, have exhibited strong performance in various load forecasting applications [1, 3, 32, 66]. These models are computationally more efficient and easier to configure than NNs, rendering them an appealing option for numerous applications.

Finally, hybrid forecasting methods integrate time series and machine learning models to address model uncertainties and enhance forecasting accuracy [5]. This integrative approach utilizes the strengths of both methodologies to yield robust and reliable predictions.

8.3 Proposed Methodology

The proposed energy management method is based on two fundamental pillars. The first pillar comprises two sets of three core forecasting models, which are responsible for predicting the electricity generated by renewable energy sources (RES) as well as the total electricity consumed by the system. The second pillar incorporates a heuristic optimization algorithm that allocates the daily pump operating hours optimally in order to minimize predicted peaks. These pillars are dynamically integrated to formulate a comprehensive scheduling plan for the water supply system, which not only mitigates peak loads but also ensures the availability of water.

8.3.1 Forecasting Models

This section introduces the three core ML models employed in this study for forecasting PV electricity generation and consumption, outlining their training and optimization processes. As elaborated later in this section, the final forecasts are generated through an ensemble approach that combines the predictions from these core models. While each forecasting variable is predicted independently, the core models for both PV generation and consumption share the same underlying principles and design. Following recent advancements in the field, decision tree-based models and NNs were selected for their proven accuracy and popularity in energy forecasting applications [47].

Decision Tree Models partition the data space recursively based on specific features (independent or explanatory variables) and fit simple predictive models within each partition [33]. Each leaf node in the tree corresponds to a class that represents the most suitable target value. Predictions for new data points are obtained by navigating from the tree's root to a leaf, following the outcomes of comparisons at each decision node [50]. Although originally developed for classification tasks, decision trees are also well-suited for regression tasks involving continuous dependent variables [37]. In regression applications, the models are typically trained using a loss function, such as mean squared error, which measures the difference between actual and predicted values [34]. This study evaluates two popular implementations of regression trees: the tree boosting system XGBoost and the gradient boosting regression tree algorithm LightGBM. Both are widely regarded, with LightGBM emerging as the standard in recent forecasting competitions on platforms such as Kaggle [4].

Neural Networks (NNs) are composed of multiple interconnected nodes, arranged in layers, that process and transmit information. Each neuron receives numerical inputs, applies specific weights and non-linear activation functions, and outputs the results to subsequent layers. The weights, which represent the trainable parameters of the model, are iteratively updated during the training process to minimize forecasting error. Although various types of NNs, such as recurrent neural networks (RNNs) [16], have been proposed for time series forecasting, multilayer perceptrons (MLPs) continue to demonstrate high effectiveness. They have shown strong performance in applications such as deep learning for load forecasting [43]. Consequently, this study focuses on multilayer feed-forward NNs, meticulously tailored to the specific characteristics of the dataset.

XGBoost [6] is an ensemble learning method based on gradient boosting, designed to deliver accurate and reliable forecasts across various tasks, including regression, classification, and ranking. Renowned for its efficiency, XGBoost has been applied to numerous real-world problems while maintaining minimal resource requirements [35]. Similar to other decision tree-based models, XGBoost encompasses several hyperparameters that significantly impact its performance. The most influential hyperparameters include the learning rate (eta), which regulates the shrinkage of feature weights and controls the conservativeness of the boosting process; the minimum split loss (gamma), which specifies the minimum loss reduction required to execute a node split; and the maximum depth of the trees, which balances model complexity with the risk of overfitting. Grid search was utilized to optimize these hyperparameters, testing learning rates from 0.1 to 0.6 in increments of 0.1, minimum split losses from 2 to 20 in increments of 2, and maximum depths from 1 to 25 in increments of 5.

LightGBM [21], a more recent gradient boosting decision tree algorithm, is recognized for its efficiency and high accuracy across various machine learning tasks. It has been extensively utilized in applications within the energy sector, such as wind energy forecasting [19] and electricity load prediction [45]. Similar to XGBoost, LightGBM is dependent on several key hyperparameters that significantly influence its performance. These hyperparameters include the learning rate, the number of leaves (which controls model complexity), the bagging fraction (which is used to reduce variance), and the feature fraction (which facilitates random feature selection during training). Grid search was employed to optimize these hyperparameters, examining values for the number of leaves ranging from 10 to 50 in increments of 5, and values for the other parameters from 0.1 to 0.9 in increments of 0.1.

The Multilayer Perceptron (MLP) is a supervised learning algorithm within the family of feed-forward neural networks [15], capable of modeling complex nonlinear relationships. For this study, the implementation of MLP Regressor from scikit-learn [46] was utilized, employing the Adam optimizer for stochastic gradient descent. The architecture comprises two hidden layers, each containing 64 nodes. Although deeper architectures were tested, they did not improve forecasting results, likely due to the limited feature set and small dataset size. Grid search was conducted to optimize the learning rate and batch size for the MLP model, with the tested parameter ranges summarized in Table 8.1.

In all instances, optimal hyperparameter values were identified utilizing 20% of the training data as a validation set. To ensure robust outcomes across diverse time periods and to simulate real-world forecasting performance, the validation set comprised observations spanning various months, days, and hours within the dataset. This methodology assisted in mitigating the risk of overfitting. Mean absolute error (MAE) and mean squared error (MSE) were employed as metrics to evaluate and select the optimal hyperparameter combinations for each model. Table 8.1 presents a summary

Table 8.1 The search space and the selected values of the hyper-parameters for the three ML forecasting models used in the present study. The values are reported for the PV production and electricity consumption series separately

Model	Hyper-parameter	Min value	Max value	Step	Selected	
					PV production	Consumption
XGBoost	Learning rate	0.1	0.6	0.1	0.5	0.3
	Minimum split loss	2	20	2	2	4
	Maximum tree depth	1	25	5	5	5
LightGBM	Learning rate	0.1	0.9	0.1	30	30
	Number of leaves	10	50	10	0.1	0.3
	Bagging fraction	0.1	0.9	0.1	0.7	0.7
	Feature fraction	0.1	0.9	0.1	0.9	0.7
MLP	Learning rate	0.001	0.01	0.002	0.001	0.001
	Batch size	Values tested: 32, 64, 28			32	64

of the hyperparameters optimized for each model, along with the corresponding search space and the final selected values.

The final forecasts are generated by integrating the predictions from each of the three core models. The weights employed to aggregate these base predictions are determined individually for each series during the validation phase (hyperparameter optimization). This methodology utilizes ensembling, a well-established technique for enhancing forecast accuracy, minimizing model and parameter uncertainty, and alleviating overfitting through diversification [48, 59]. Insights from recent major forecasting competitions underscore the efficacy of ensembling [38], particularly in the context of integrating regression tree-based models and neural networks, as examined in this study [18, 41]. Although an alternative approach could involve the selection of the best-performing model for each hourly interval, ensembling has proven to offer superior accuracy and reliability for the objectives of this study.

8.3.2 Optimization Models

The objective of the algorithm is to optimally adjust the operating hours of water pumping stations to coincide with hourly intervals of low expected electricity consumption on the island, thereby reducing peak loads. In the design of the optimization algorithm, two operating modes are considered: manually operated and automated operation scenarios.

The first mode, termed "flexible," permits unrestricted scheduling of pumping stations, allowing them to commence and cease operations at any hour throughout the day. The second mode, referred to as "constrained," establishes a continuous operational schedule in which a pumping station must operate uninterrupted until the required daily operating hours have been fulfilled.

For both operating modes, the optimization algorithm depends on the following inputs:

- The number of available pumping stations(N).
- The number of required operating hours for each pumping station during a day ($PHours$).
- The hourly electricity consumption (installed capacity) of each pumping station ($PCons$).
- The forecasted net electrical load of the island (excluding the consumption of the pumping stations) for each hourly interval during a day($TCons$).

The pseudocode for the optimization algorithm developed for the flexible operating mode is presented in Algorithm 8.1. This algorithm is designed to strategically schedule pump operations to minimize peaks in the net load, defined as total electricity consumption minus PV generation, at the island network level. Specifically, the algorithm identifies the pumping station contributing to the maximum net electrical load and reallocates its operation to the interval with the lowest forecasted hourly load, based on predictions of the island's energy consumption and PV generation.

After assigning the selected pumping station to the optimal interval, the algorithm updates two key variables: $TCons$, representing the island's total electricity consumption, and $PHours$, indicating the remaining hours that each pumping station needs to operate. This process is repeated iteratively until all required operating hours for every pumping station are allocated to the most favorable intervals. The overall workflow of the optimization algorithm is depicted in the flowchart provided in the appendix. Importantly, in this flexible operating mode, the algorithm achieves a global minimum in peak load reduction.

In the second operating mode, where pump operation is constrained, the optimization algorithm is adapted to address the continuous operation requirement. While this approach may not always yield the optimal solution, it approximates the optimal outcome in terms of peak load reduction. As in the first mode, the algorithm begins by identifying the pumping station with the highest hourly consumption. However, in this mode, the algorithm selects a multi-hour interval equal to the station's required daily operating hours, reflecting the constraint of uninterrupted operation.

The algorithm evaluates all possible multi-hour intervals to find the interval containing the lowest maximum hourly energy consumption. It then assigns the pumping station to this interval, minimizing the potential for peak load occurrences. This process is repeated iteratively until all pumping stations are scheduled into multi-hour intervals. Throughout, the algorithm dynamically updates the net load for the selected intervals after each pump assignment, ensuring that no new peaks are introduced.

Algorithm 8.1: Pump scheduling algorithm (flexible mode of operation)

Inputs : N: number of pumps
$PCons$:$\{x_1, x_2, \ldots, x_N\}$, consumption per pump
$PHours$:$\{y_1, y_2, \ldots, y_N\}$, operating hours per pump
$TCons$:$\{z_1, z_2, \ldots, z_{24}\}$, consumption per hour interval

Output: $OperPlan[i, j]$: the operating plan; 1 if pump i is mapped to interval j; 0 otherwise

```
while PHours != 0 do
    for PHours[i] > 0 do
        utilized pump ← max(PCons);
    end
    max consumption ← PCons[utilized pump];
    for OperPlan[utilized pump, j] == 0 do
        utilized interval ← min(TCons);
    end
    OperPlan[utilized pump, utilized interval] ← 1;
    TCons[utilized interval] ← TCons[utilized interval] + max consumption;
    PHours[utilized pump] ← PHours[utilized pump] − 1;
end
```

The optimization algorithm was implemented in Python 3.0 and thoroughly evaluated through a case study, detailed in Sect. 8.4.

It is important to note that for more accurate identification and smoothing of peaks, the net electrical load should ideally account for the configuration of the electrical grid and the associated losses. While this information was not available for the case studied here, it should be incorporated into similar applications whenever accessible. However, since the proposed optimization method targets relatively small, remote islands, the impact of losses is expected to be minimal due to the typically short length of the distribution network in these settings.

8.4 Case Study

8.4.1 Dataset

To evaluate the effectiveness of the proposed optimization method for reducing peak loads, the case of Tilos, a small Greek island in the Aegean Sea and part of the Dodecanese island chain, is examined. Tilos serves as a representative example of a remote island that has made notable strides in transitioning to green energy, largely due to the TILOS Horizon 2020 project [63]. As part of this initiative, Greece's first fully licensed hybrid power station, combining renewable energy sources (RES) and battery storage, was developed to meet a significant portion of the island's electricity demand. The project also introduced advancements in smart metering and demand management, leveraging local water pumping stations as a flexible and impactful resource for these efforts.

In addition to the central hybrid power station, the island has seen the emergence of distributed RES in recent years. Historical operational data from these distributed PV systems were utilized as input for the optimization method and forecasting models proposed in this study. An empirical evaluation of comparable systems integrating local PV power with fixed-characteristic water pumping infrastructure is also presented. The water pumping assets included in this study consist of five pumps on Tilos, all monitored and managed through a centralized platform.

The dataset used spans approximately two years and three months (19,008 hourly data points), covering the period from March 1, 2019, to April 30, 2021. It includes the total electricity consumed on Tilos, electricity generated by PV systems[1], and electricity consumption by the five water pumping stations. These time series serve as the primary input variables for the proposed optimization method.

The electricity consumption of the water pumping stations is determined by their operating schedules. Each station operates during specific hours of the day, as illustrated in Fig. 8.1. These schedules are predefined and consistent within each day of a given month but vary between months to align with the island's water supply needs for residents and irrigation.

The installed capacities of the pumping stations vary, with the following specifications: 6, 10, 16, 2, and 11 kW for stations 1, 2, 3, 4, and 5, respectively. This variation reflects factors such as water table levels, available head, and other relevant parameters. These factors collectively influence the volume of water transported to local storage tanks for distribution and end use. To ensure the proposed optimization method is both effective and practical, it must maintain the total daily and monthly operating hours currently established by the station operators, thereby ensuring a consistent and adequate water supply.

Total electricity consumption experiences peaks during the summer months, notably in July and August, likely attributable to heightened tourist activity and increased use of air conditioning. In contrast, consumption remains significantly lower throughout the remainder of the year. However, December and January exhibit marked spikes. Similarly, PV electricity generation reaches its zenith in the summer and during the months of April, May, and September, while it experiences significantly in winter.

Figure 8.2 further demonstrates pronounced daily seasonality in the data, influenced by human behavior and weather conditions. For example, PV electricity generation follows the diurnal cycle of solar radiation, peaking around midday and dropping to zero during early morning and nighttime hours. Conversely, energy consumption increases after 9:00 AM, peaks around 8:00 PM, and declines overnight. The distinct daily patterns of PV electricity generation and island consumption underscore that even high PV energy production cannot naturally offset consumption peaks

[1] To facilitate comparisons and present findings more clearly, the PV capacity factor (solar power divided by installed capacity, expressed as a percentage) is reported instead of absolute solar power (measured in kW).

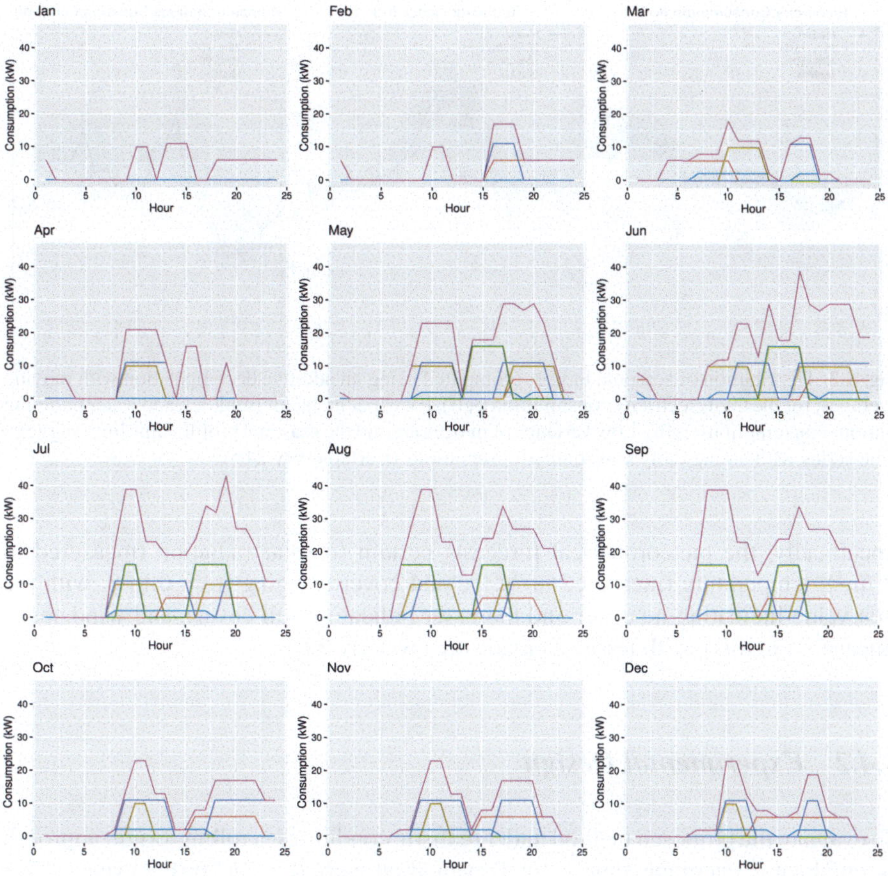

Fig. 8.1 Daily electricity consumption of the five pumping stations considered in this study. The daily schedule of operation changes across different months but remains the same within each month. The total electricity consumed by the stations is displayed in purple

(real-time self-consumption) without targeted load-shifting interventions. Notably, water pumping stations predominantly operate during midday and afternoon, presenting an opportunity for energy managers to reduce peaks by adjusting their operating schedules.

The mismatch between total electricity consumption and PV generation (Pearson correlation coefficient: 0.1) and the moderate correlation between island-wide electricity consumption and water pumping station usage (Pearson correlation coefficient: 0.5) further emphasize the relevance of this study. These insights highlight the potential impact of the proposed optimization approach, particularly in small-scale environments where the sensitivity to such mismatches is pronounced.

In addition to the three series presented in Fig. 8.2, the dataset includes hourly meteorological forecasts that will be utilized for accurately predicting solar power.

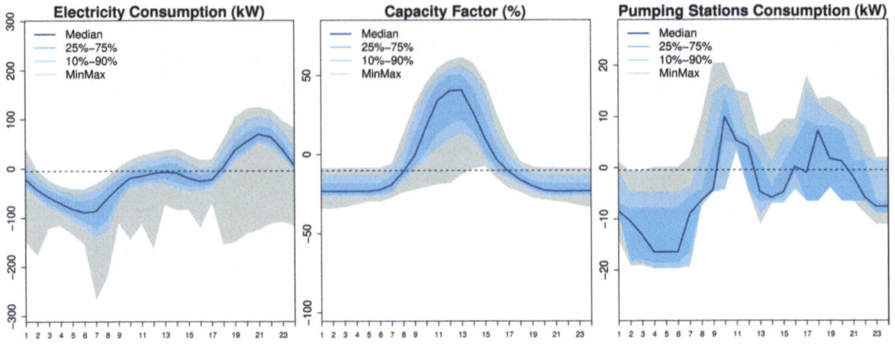

Fig. 8.2 Distribution of seasonal indices (computed using an additive decomposition with moving averages; for the total electricity consumption (left), PV capacity factor (middle), and water pumping stations consumption (right). Low variance of indices around the seasonal profile (blue line) suggests time series with strong daily seasonality (observations overlap every 24 h)

Specifically, the meteorological forecasts pertain to solar radiation (measured in W/m^2), temperature (measured in °C), cloud cover (ranging from 0 to 8, with the first value indicating no cover and the latter indicating full cover), and wind speed (measured in m/s), with forecasts made on a weekly basis.

8.4.2 *Experimental Design*

To evaluate the proposed method under realistic conditions, a rolling-origin approach is employed, encompassing a total evaluation period of 8,760 h (1 year). This approach facilitates a long-term assessment of the method's benefits while considering calendar and weather-related factors that may influence performance throughout the year. Within this framework, the first 10,248 observations are utilized to train the forecasting models, as outlined in Sect. 8.3.1. Subsequently, forecasts for the island's total electricity consumption and solar power generation are produced for the subsequent 168 h (1 week).

These forecasts serve as inputs to the optimization algorithm described in Sect. 8.3.2, which determines the optimal operating schedule for the island's water pumping stations. Following this, the realized electricity consumption and PV generation data for the previously forecasted week are analyzed to evaluate the impact of the proposed schedules. This is done using the following key performance indicators (KPIs):

1. SD_{day}: Average standard deviation of electricity consumption, measured on a daily basis.
2. SD_{week}: Average standard deviation of electricity consumption, measured on a weekly basis.

3. $Max - Min_{day}$: Average difference between maximum and minimum electricity consumption, measured on a daily basis.
4. $Max - Min_{week}$: Average difference between maximum and minimum electricity consumption, measured on a weekly basis.
5. $Max - Mean_{day}$: Average difference between maximum and average electricity consumption, measured on a daily basis.
6. $Max - Mean_{week}$: Average difference between maximum and average electricity consumption, measured on a weekly basis.

The key performance indicators (KPIs) used in this study are inspired by Feng et al. [11], which proposed the standard deviation and peak-valley difference (the difference between maximum and minimum values) as metrics to evaluate a mixed-integer linear programming model for peak-saving in thermoelectric plants. These metrics have also been employed as KPIs or objective functions in various similar peak-saving problems, such as the peak-saving operation of hydropower plants [30] and hydrothermal-nuclear plants in multi-grid power environments [10], demonstrating their relevance and effectiveness for assessing the performance of the proposed optimization algorithm.

For comparison, these KPIs are also calculated for the realized energy consumption under a "business-as-usual" (BaU) scenario, where no optimization is applied, and the water pumps operate according to their regular schedules.

After completing the first evaluation round, the forecasting origin is shifted forward by 168 h, and the first 10,416 observations are used to retrain the forecasting models and reapply the optimization algorithm for the second week of the test set. This process is repeated until no observations remain, yielding 52 evaluation periods that comprehensively represent the optimization method's performance.

The frequency of updating the operating schedule was selected with practical constraints in mind. For example, weather forecasts are currently supplied to the energy management system via an online service on a weekly basis, preventing daily recalculation of solar power forecasts. Additionally, while many energy management systems can automatically adjust pumping station schedules, changes often require approval from energy managers, which may not always be feasible on a regular basis. However, the proposed method can easily be adapted for more frequent updates, which could enhance energy forecasting accuracy and improve optimization outcomes.

Two modes of applying the optimization method were evaluated. In the first, "flexible" mode, pump operation is entirely flexible, allowing pumping stations to start and stop at any time of day. In the second, "constrained" mode, pump operation is restricted, such that once a station starts, it must run continuously until its required operating hours are completed.

The flexible mode simulates fully automated systems capable of remotely controlling water pumps, while the constrained mode represents systems where water pumps are manually controlled by energy managers. Comparing the results from these two modes provides insights into the potential benefits of automated systems over manually controlled ones.

8.4.3 Accuracy of Forecasting Models

As discussed in Sect. 8.3, the optimization algorithm relies on forecasts of future electricity consumption and PV generation. Generally, higher forecasting accuracy leads to more effective energy management. In this study, the forecasts required by the algorithm are for the island's total electricity consumption and solar energy generation. These forecasts are generated using an ensemble of ML models, as detailed in Sect. 8.3.1. However, given the unique characteristics of each series, different input features are used for each case to improve model training and enhance forecasting accuracy.

Electricity consumption demonstrates pronounced seasonal patterns on both daily and annual scales. Autoregression, which uses past electricity consumption values as predictors for future values, is a highly effective approach, particularly when combined with categorical variables such as month, week, day of the week, and hour of the day. These features facilitate pattern recognition and improve forecast accuracy. As a result, the core ML models for forecasting electricity consumption use only these predictors (168-hour lag and calendar variables) as input features.

Solar energy is inherently stochastic and heavily influenced by weather conditions. Consequently, unlike electricity consumption models, past observations are not included as features for forecasting solar energy. Instead, predictions rely on meteorological forecasts provided by the energy management system (EMS). As expected, solar radiation is strongly correlated with PV generation, making it a key predictive feature. Temperature and wind speed also exhibit positive but weaker correlations with solar power. Additionally, higher cloud cover generally corresponds to reduced energy production, although its predictive value is limited, likely due to the uncertainty inherent in cloud cover forecasts. To enable the forecasting models to account for calendar effects and daily seasonality, categorical variables such as month, week, and hour of the day are included as additional features.

The accuracy of the proposed ensembles is assessed using three metrics: the root mean squared error (RMSE) and the mean absolute error (MAE) of the forecasts over the evaluation period, as well as the Pearson correlation coefficient between the forecasted and actual values. These metrics are calculated as follows:

$$RMSE = \sqrt{\frac{1}{8760}\sum_{t=1}^{8760}(y_t - f_t)^2},$$

$$MAE = \frac{1}{8760}\sum_{t=1}^{8760}|y_t - f_t|,$$

Table 8.2 Forecasting performance of the ensemble of ML models utilized in the case study

Measure	Electricity consumption	Solar power
$RMSE$	32.58	0.34
MAE	22.54	0.14
$r_{y,f}$	0.97	0.96

$$r_{y,f} = \frac{\sum_{t=1}^{8760}(y_t - \bar{y})(f_t - \bar{f})}{\sqrt{\sum_{t=1}^{8760}(y_t - \bar{y})^2}\sqrt{\sum_{t=1}^{8760}(f_t - \bar{f})^2}},$$

where y_t is the actual future value of the time series under examination at hour t of the evaluation period, f_t is the forecast provided by the model used to generate the predictions, while $\bar{y}$ and $\bar{f}$ are the mean values of the actual and predicted values, respectively. The choice of the selected performance metrics can be justified as follows:

Squared error metrics, such as RMSE, are optimized for the mean [22], assessing how well the forecasting models estimate the average electricity production or consumption while penalizing larger errors, which often occur during peak hours. In contrast, absolute error metrics, such as MAE, are optimized for the median [56], offering insight into the models' accuracy in predicting the most common production or consumption values. MAE is more robust against large outliers and focuses on non-peak hours. Additionally, the Pearson correlation coefficient evaluates the linear relationship between forecasts and actual values, indicating whether the models can capture the variance of the target series. While no single accuracy metric is perfect, each offering distinct strengths and limitations [26], this combination of metrics provides a comprehensive evaluation of the models' performance across the most critical aspects of the forecasting task.

The results, summarized in Table 8.2, demonstrate that both ensembles achieve sufficient accuracy for the one-week forecasting horizon, effectively capturing the daily patterns of the key variables addressed by the proposed optimization method. Furthermore, as illustrated in Fig. 8.3, the ensembles successfully model the seasonality, trends, and weather-driven variations in the series during both summer and winter periods.

8.4.4 Optimization Results

Once forecasts for each week of the simulation are generated, the proposed optimization algorithm is applied to adjust the pump operating schedule, with the objective

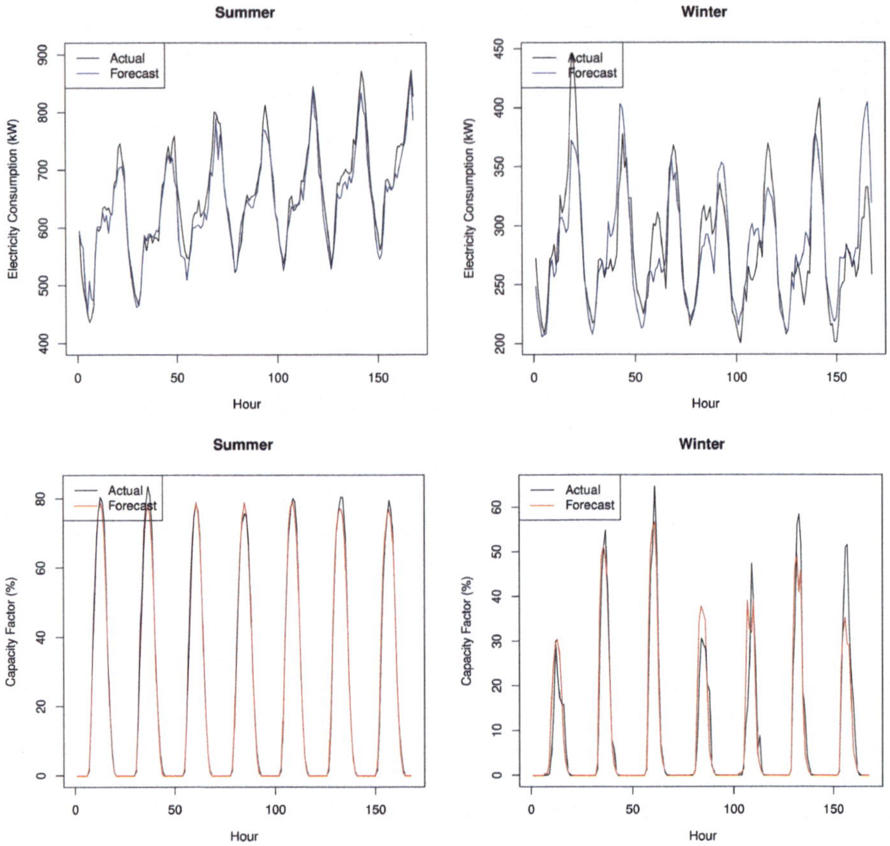

Fig. 8.3 Examples illustrating how the electricity consumption (top) and solar power (bottom) forecasting models perform. The examples refer to two randomly selected summer (2020-07-31–2020-08-06) and winter (2020-12-04–2020-12-10) weeks of the evaluation period

of achieving peak savings. Figure 8.4 illustrates two examples of this optimization process, showcasing electricity consumption during both summer and winter weeks under the business-as-usual (BaU) mode, as well as under flexible and constrained operating modes.

The results indicate that both optimization modes achieve comparable reductions in variance, effectively mitigating peaks and increasing consumption during lower-intensity hours. As anticipated, the impact of peak savings is more pronounced during the summer week, due to higher PV-based electricity generation and greater potential for load shifting compared to the winter week.

Figure 8.5 summarizes the daily adjustments made to the initial pump operating hours for each optimization mode. In both cases, the algorithm recommends halting pump operation between 08:00 and 14:00 and from 18:00 to 23:00. These time

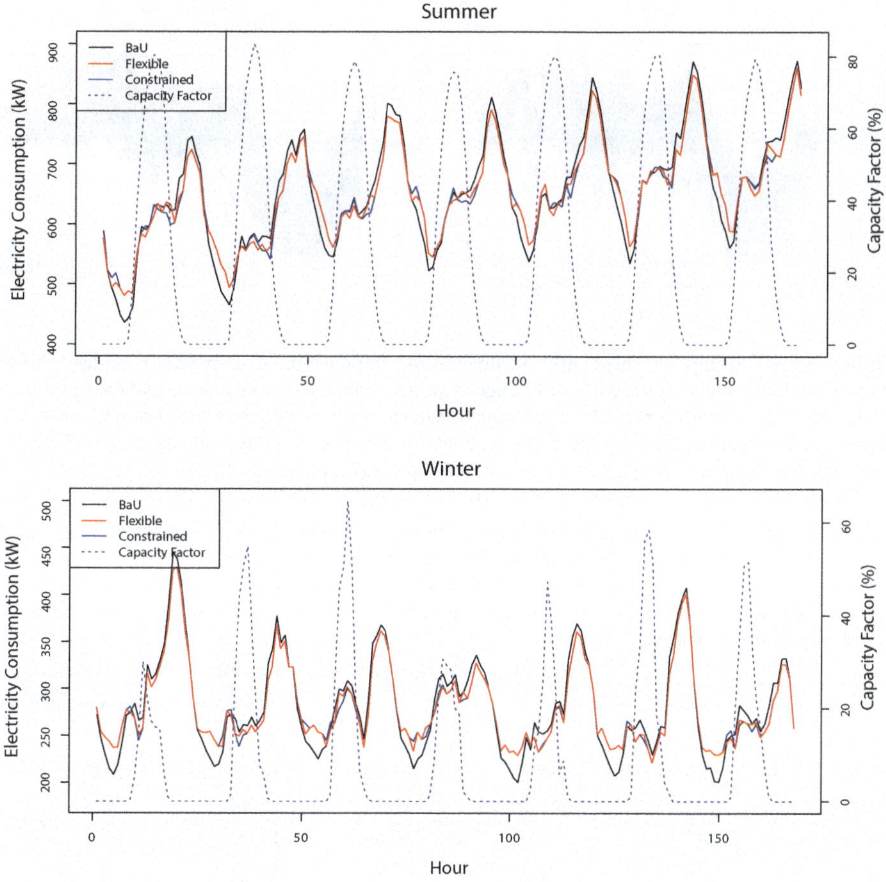

Fig. 8.4 Examples illustrating how the total electricity consumption of the island is adjusted when the flexible and the constrained optimization modes of operation are considered for shaving peaks. The examples refer to two randomly selected summer (2020-07-31–2020-08-06) and winter (2020-12-04–2020-12-10) weeks of the evaluation period. Solar power is also presented to facilitate comparisons

intervals correspond to periods of high expected electricity consumption combined with insufficient PV generation to adequately meet the demand. Instead, the algorithm suggests scheduling pump operation during the early morning hours (01:00–07:00), when electricity consumption is at its lowest, or in the afternoon (15:00–17:00), when solar energy generation remains relatively significant.

The performance of the proposed optimization method over the entire simulation period is summarized in Table 8.3, using the KPIs defined in Sect. 8.4.2 for evaluation. Under the flexible operating mode, the daily and weekly deviations in electricity consumption are reduced by approximately 20 and 17%, respectively. In the constrained operating mode, the results are slightly lower, with reductions of about 19

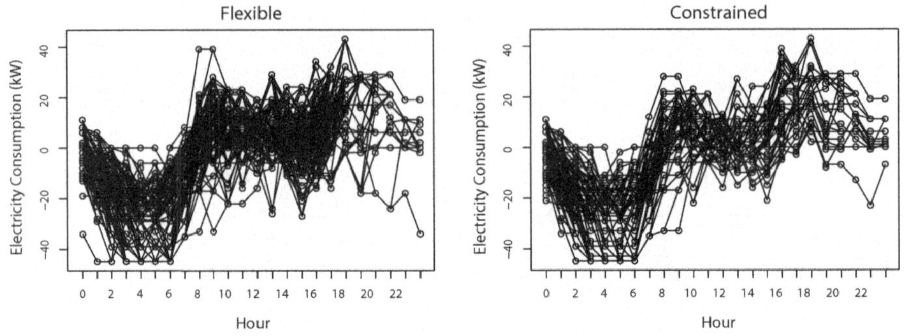

Fig. 8.5 Load shifting performed under the flexible and the constrained optimization modes of operation. Each line represents the daily differences reported between the BaU mode and the optimized modes (flexible or constrained). Positive values indicate hours where the algorithm has suggested against the operation of the water pumping stations, while negative values indicate hours where the algorithm has suggested in favor of the operation of the water pumping stations

Table 8.3 Performance of the proposed method in terms of peak shaving. Various KPIs are used to evaluate the daily and weekly deviation of the electricity consumption of Tilos when no loads are shifted (BaU), flexible load shifting is allowed, and constrained load shifting is performed. The percentage improvements of the two load shifting optimization modes over the BaU mode of operation are also provided.

KPI	BaU	Flexible	Constrained	Flexible	Constrained
	Absolute values			Percentage improvements (%)	
SD_{day}	51.58	41.23	41.81	20.07	18.94
SD_{week}	54.93	45.73	46.19	16.75	15.91
$Max - Min_{day}$	175.01	143.79	147.74	17.84	15.58
$Max - Min_{week}$	242.12	212.90	216.66	12.07	10.52
$Max - Mean_{day}$	98.99	87.23	87.27	11.88	11.84
$Max - Mean_{week}$	142.02	130.40	130.01	8.18	8.46

and 16%, respectively. Similarly, load-shifting actions in the flexible mode result in a reduction of the average peak-to-valley difference in electricity consumption by over 17% on a daily basis and 12% on a weekly basis.

Considering that the installed capacity of the water pumping stations (45 kW) accounts for only 13% of average electricity consumption and 5% of maximum consumption during the simulation period, these results highlight the significant potential of the proposed method. Its application can substantially improve grid stability, particularly when energy-intensive loads can be managed with flexibility.

Considering that the proposed peak-saving method is partly influenced by the amount of electricity generated by renewable energy sources (RES), which in this case is relatively small compared to the total electricity consumption of the system

Table 8.4 Performance of the proposed method in terms of peak shaving when PV electricity production is increased by a factor of 10 and 40. To facilitate comparisons, the improvements of the two load shifting modes of operation over the BaU mode are provided as percentages

KPI	Flexible	Constrained
	Percentage improvements (%)	
Original scale	20.07	18.94
Scale up ($\times 10$)	16.68	15.87
Scale up ($\times 40$)	11.96	11.34

and the water pumping stations, we extended the analysis by artificially scaling the installed capacity of the existing PV system. Specifically, the historical PV generation data was scaled by factors of 10 and 40.[2]

Using the scaled PV generation data, the entire experiment was repeated, and the KPIs previously introduced were recalculated to evaluate the impact of increased electricity generation on the achieved peak savings. The results are summarized in Table 8.4.

The proposed method consistently demonstrates a positive impact on the daily deviation of electricity consumption, with improvements ranging from approximately 12% (with a scaling factor of 40) to 20% (at the original scale, i.e., a scaling factor of 1). As anticipated, the constrained operating mode is slightly less effective than the flexible mode, with performance reductions of about 1–2%. Interestingly, the improvements diminish in a dampened, non-linear manner as the scaling factor increases. This is attributed to the notable mismatch between the peak consumption hours and PV generation times, as discussed earlier and illustrated in Fig. 8.4.

As PV production increases, consumption reductions tend to occur during hours when demand is already relatively low and eligible for load shifting. Consequently, while the quantity of generated electricity and its temporal distribution partially influence the method's performance, greater RES production does not necessarily result in enhanced peak-saving performance. Given these findings, remote islands like Tilos should carefully evaluate the scale of their RES production in relation to their energy demand profile and the amount of load that can be shifted. Determining the most profitable level of RES expansion in such settings is a complex task, influenced by factors such as expected return on investment. For Tilos, our analysis indicates that the optimal scaling factor for PV in achieving peak savings corresponds to the average annual electricity consumption of the water pumps. Moreover, the current installed capacity of the PV system is adequate to facilitate the optimal distribution of the water pump load.

[2] The factor of 10 was chosen to align the generated electricity with the scale of the shifted load, while the factor of 40 represents a scenario where generated electricity exceeds the shifted load.

8.5 Conclusions

This study introduced a comprehensive approach to intelligent energy management, designed to smooth electricity consumption in remote island power systems. The method combines two key components: an ensemble of ML models for accurate forecasting of RES production and electricity consumption, and a scheduling algorithm to optimally shift the operating hours of water pumping stations to periods of low expected load. The findings demonstrate significant peak load reductions while maintaining water availability and adhering to operational constraints of the water supply system.

The developed forecasting models exhibited computational efficiency, ease of implementation, and necessitated a minimal set of input variables. These variables included autoregressive features for load forecasting and weather data for RES production forecasting. The models effectively captured daily and monthly patterns in the forecasted variables and achieved satisfactory accuracy. Using these forecasts, the proposed optimization algorithm scheduled pump operating hours based on weekly forecasts and the operational profiles of the pumping stations. The algorithm supported both flexible and constrained operating modes.

The method was evaluated through a year-long simulation, achieving reductions of more than 15% in daily and weekly consumption deviation across both operating modes. The optimal daily pump schedule prioritized early morning hours, when electricity demand is typically low, and afternoon hours, when PV generation is significant. The results further indicated that the potential benefits of the method are contingent upon the amount of shiftable load and the accuracy of forecasts, particularly regarding PV electricity production. Furthermore, the alignment of electricity generation with consumption patterns plays a critical role, as higher electricity production does not necessarily translate to improved peak-saving performance.

Future improvements to the method may encompass the development of a user-friendly interface, enabling energy managers to simulate various operating scenarios and utilize the developed algorithms more effectively. Expanding the optimization algorithm to integrate systems with energy storage or additional RES types, such as wind, biomass, and geothermal energy, represents another promising avenue. Finally, exploring alternative ensembling strategies and incorporating variations of the core ML models could further enhance forecasting accuracy.

References

1. R.A. Abbasi, N. Javaid, M.N.J. Ghuman, Z.A. Khan, S. Ur Rehman, Amanullah, Short term load forecasting using XGBoost, in *Web, Artificial Intelligence and Network Applications*, ed. by L. Barolli, M. Takizawa, F. Xhafa, T. Enokido (Springer International Publishing, Cham, 2019), pp. 1120–1131
2. A. Adeniran, M. Oyelowo, An EPANET analysis of water distribution network of the University of Lagos Nigeria. J. Eng. Res. **18**(2), 69–83 (2013)

3. E. Aguilar Madrid, N. Antonio, Short-term electricity load forecasting with machine learning. Information **12**(2) (2021)
4. C.S. Bojer, J.P. Meldgaard, Kaggle forecasting competitions: an overlooked learning opportunity. Int. J. Forecast. **37**(2), 587–603 (2021)
5. O.O. Bozkurt, G. Biricik, Z.C. Taysi, Artificial neural network and Sarima based models for power load forecasting in Turkish electricity market. PLoS ONE **12**(4), 1–24 (2017)
6. T. Chen, C. Guestrin, XGBoost: a scalable tree boosting system, in *Proceedings of the 22nd ACM SIGKDD International Conference on Knowledge Discovery and Data Mining* (2016), pp. 785–794
7. L. Costa, H. Ramos, M. De Castro, Hybrid genetic algorithm in the optimization of energy costs in water supply networks. Water Sci. Technol.: Water Supply **10**(3), 315–326 (2010)
8. A.T. Eseye, J. Zhang, D. Zheng, Short-term photovoltaic solar power forecasting using a hybrid Wavelet-PSO-SVM model based on SCADA and meteorological information. Renew. Energy **118**, 357–367 (2018)
9. G. Farina, E. Creaco, M. Franchini, Using EPANET for modelling water distribution systems with users along the pipes. Civ. Eng. Environ. Syst. **31**(1), 36–50 (2014)
10. Z.-K. Feng, W.-J. Niu, C.-T. Cheng, J.-Z. Zhou, Peak shaving operation of hydro-thermal-nuclear plants serving multiple power grids by linear programming. Energy **135**, 210–219 (2017)
11. Z.-K. Feng, W.-J. Niu, W.-C. Wang, J.-Z. Zhou, C.-T. Cheng, A mixed integer linear programming model for unit commitment of thermal plants with peak shaving operation aspect in regional power grid lack of flexible hydropower energy. Energy **175**, 618–629 (2019)
12. L. Ferbar Tratar, E. Strmčnik, The comparison of holt-winters method and multiple regression method: a case study. Energy **109**, 266–276 (2016)
13. N. Ghadami, M. Gheibi, Z. Kian, M.G. Faramarz, R. Naghedi, M. Eftekhari, A.M. Fathollahi-Fard, M.A. Dulebenets, G. Tian, Implementation of solar energy in smart cities using an integration of artificial neural network, photovoltaic system and classical Delphi methods. Sustain. Urban Areas **74**, 103149 (2021)
14. M.A. Hassan, A. Khalil, S. Kaseb, M. Kassem, Exploring the potential of tree-based ensemble methods in solar radiation modeling. Appl. Energy **203**, 897–916 (2017)
15. S. Haykin, *Neural Networks: A Comprehensive Foundation* (Prentice Hall PTR, 1994)
16. H. Hewamalage, C. Bergmeir, K. Bandara, Recurrent neural networks for time series forecasting: current status and future directions. Int. J. Forecast. **37**(1), 388–427 (2021)
17. L. Ilias, E. Sarmas, V. Marinakis, D. Askounis, H. Doukas, Unsupervised domain adaptation methods for photovoltaic power forecasting. Appl. Soft Comput. **149**, 110979 (2023)
18. Y. In, J.-Y. Jung, Simple averaging of direct and recursive forecasts via partial pooling using machine learning. Int. J. Forecast. (2021)
19. Y. Ju, G. Sun, Q. Chen, M. Zhang, H. Zhu, M.U. Rehman, A model combining convolutional neural network and LightGBM algorithm for ultra-short-term wind power forecasting. IEEE Access **7**, 28309–28318 (2019)
20. M. Kazantzis, A. Simpson, D. Kwong, S. Tan, A new methodology for optimizing the daily operations of a pumping plant, in *Proceedings of 2002 Conference on Water Resources Planning* (ASCE, Roanoke, USA, 2002)
21. G. Ke, Q. Meng, T. Finley, T. Wang, W. Chen, W. Ma, Q. Ye, T.-Y. Liu, LightGBM: a highly efficient gradient boosting decision tree. Adv. Neural. Inf. Process. Syst. **30**, 3146–3154 (2017)
22. S. Kolassa, Evaluating predictive count data distributions in retail sales forecasting. Int. J. Forecast. **32**(3), 788–803 (2016)
23. X. Kong, C. Liu, Y. Shen, W. Hu, T. Ma, Power supply reliability evaluation based on big data analysis for distribution networks considering uncertain factors. Sustain. Urban Areas **63**, 102483 (2020)
24. D. Korkmaz, SolarNet: a hybrid reliable model based on convolutional neural network and variational mode decomposition for hourly photovoltaic power forecasting. Appl. Energy **300**, 117410 (2021)

25. N. Kourentzes, D.K. Barrow, S.F. Crone, Neural network ensemble operators for time series forecasting. Expert Syst. Appl. **41**(9), 4235–4244 (2014)
26. D. Koutsandreas, E. Spiliotis, F. Petropoulos, V. Assimakopoulos, On the selection of forecasting accuracy measures. J. Oper. Res. Soc. (2021)
27. A. Kumar, K. Kumar, B. Bharanidharan, N. Matial, E. Dey, M. Singh, V. Thakur, S. Sharma, N. Malhotra, Design of water distribution system using EPANET. Int. J. Adv. Res. **3**(9), 789–812 (2015)
28. C. Kuster, Y. Rezgui, M. Mourshed, Electrical load forecasting models: a critical systematic review. Sustain. Urban Areas **35**, 257–270 (2017)
29. C. Li, Designing a short-term load forecasting model in the urban smart grid system. Appl. Energy **266**, 114850 (2020)
30. S. Liao, H. Liu, Z. Liu, B. Liu, G. Li, S. Li, Medium-term peak shaving operation of cascade hydropower plants considering water delay time. Renew. Energy **179**, 406–417 (2021)
31. X. Liu, Z. Xiao, R. Zhu, J. Wang, L. Liu, M. Ma, Edge sensing data-imaging conversion scheme of load forecasting in smart grid. Sustain. Urban Areas **62**, 102363 (2020)
32. Y. Liu, H. Luo, B. Zhao, X. Zhao, Z. Han, Short-term power load forecasting based on clustering and XGBoost method, in *2018 IEEE 9th International Conference on Software Engineering and Service Science (ICSESS)* (2018), pp. 536–539
33. W.-Y. Loh, Classification and regression trees. Wiley Interdiscip. Rev.: Data Min. Knowl. Discov. **1**(1), 14–23 (2011)
34. W.-Y. Loh, Fifty years of classification and regression trees. Int. Stat. Rev. **82**(3), 329–348 (2014)
35. F. Ma, X. Yan, Research on the energy consumption estimation method of pure electric vehicle based on XGBoost, in *2019 3rd International Conference on Electronic Information Technology and Computer Engineering (EITCE)* (IEEE, 2019), pp. 1021–1026
36. G. Mackle, G. Savic, G.A. Walters, Application of genetic algorithms to pump scheduling for water supply, in *First International Conference on Genetic Algorithms in Engineering Systems: Innovations and Applications* (IET, 1995), pp. 400–405
37. S. Makridakis, E. Spiliotis, V. Assimakopoulos, Statistical and machine learning forecasting methods: concerns and ways forward. PLoS ONE **13**(3), 1–26 (2018)
38. S. Makridakis, E. Spiliotis, V. Assimakopoulos, The M4 competition: 100,000 time series and 61 forecasting methods. Int. J. Forecast. **36**(1), 54–74 (2020)
39. A. Marchi, A.R. Simpson, M.F. Lambert, Pump operation optimization using rule-based controls. Procedia Eng. **186**, 210–217 (2017)
40. V. Michalakopoulos, E. Sarantinopoulos, E. Sarmas, V. Marinakis, Empowering federated learning techniques for privacy-preserving PV forecasting. Energy Rep. **12**, 2244–2256 (2024)
41. I. Nasios, K. Vogklis, Blending gradient boosted trees and neural networks for point and probabilistic forecasting of hierarchical time series. Int. J. Forecast. (2022)
42. A. Nespoli, E. Ogliari, S. Leva, A. Massi Pavan, A. Mellit, V. Lughi, A. Dolara, Day-ahead photovoltaic forecasting: a comparison of the most effective techniques. Energies **12**(9) (2019)
43. B.N. Oreshkin, G. Dudek, P. Pełka, E. Turkina, N-beats neural network for mid-term electricity load forecasting. Appl. Energy **293**, 116918 (2021)
44. D.P. Panagoulias, E. Sarmas, V. Marinakis, M. Virvou, G.A. Tsihrintzis, H. Doukas, Intelligent decision support for energy management: a methodology for tailored explainability of artificial intelligence analytics. Electronics **12**(21), 4430 (2023)
45. S. Park, S. Jung, S. Jung, S. Rho, E. Hwang, Sliding window-based LightGBM model for electric load forecasting using anomaly repair. J. Supercomput. 1–22 (2021)
46. F. Pedregosa, G. Varoquaux, A. Gramfort, V. Michel, B. Thirion, O. Grisel, M. Blondel, P. Prettenhofer, R. Weiss, V. Dubourg, et al., Scikit-learn: machine learning in Python. J. Mach. Learn. Res. **12**, 2825–2830 (2011)
47. F. Petropoulos, D. Apiletti, V. Assimakopoulos, M.Z. Babai, D.K. Barrow, S.B. Taieb, C. Bergmeir, R.J. Bessa, J. Bijak, J.E. Boylan, J. Browell, C. Carnevale, J.L., Castle, P. Cirillo, M.P. Clements, C. Cordeiro, F.L.C. Oliveira, S.D. Baets, A. Dokumentov, J. Ellison, P. Fiszeder, P.H. Franses, D.T. Frazier, M. Gilliland, M.S. Gönül, P. Goodwin, L. Grossi, Y. Grushka-Cockayne,

M. Guidolin, M. Guidolin, U. Gunter, X. Guo, R. Guseo, N. Harvey, D.F. Hendry, R. Hollyman, T. Januschowski, J. Jeon, V.R.R. Jose, Y. Kang, A.B. Koehler, S. Kolassa, N. Kourentzes, S. Leva, F. Li, K. Litsiou, S. Makridakis, G.M. Martin, A.B. Martinez, S. Meeran, T. Modis, K. Nikolopoulos, D. Önkal, A. Paccagnini, A. Panagiotelis, I. Panapakidis, J.M. Pavía, M. Pedio, D.J. Pedregal, P. Pinson, P. Ramos, D.E. Rapach, J.J. Reade, B. Rostami-Tabar, M. Rubaszek, G. Sermpinis, H.L. Shang, E. Spiliotis, A.A. Syntetos, P.D. Talagala, T.S. Talagala, L. Tashman, D. Thomakos, T. Thorarinsdottir, E. Todini, J.R.T. Arenas, X. Wang, R.L. Winkler, A. Yusupova, F. Ziel, Forecasting: theory and practice (2021)
48. F. Petropoulos, R.J. Hyndman, C. Bergmeir, Exploring the sources of uncertainty: why does bagging for time series forecasting work? Eur. J. Oper. Res. **268**(2), 545–554 (2018)
49. G.V. Ramana, V.S. Chekka, Validation and examination of existing water distribution network for continuous supply of water using EPANET. Water Resour. Manage **32**(6), 1993–2011 (2018)
50. L. Rokach, O. Maimon, Decision trees, in *Data Mining and Knowledge Discovery Handbook* (Springer, Berlin, 2005), pp. 165–192
51. M. Saadi, M.T. Noor, A. Imran, W.T. Toor, S. Mumtaz, L. Wuttisittikulkij, IoT enabled quality of experience measurement for next generation networks in smart cities. Sustain. Urban Areas **60**, 102266 (2020)
52. E. Sarmas, N. Dimitropoulos, V. Marinakis, Z. Mylona, H. Doukas, Transfer learning strategies for solar power forecasting under data scarcity. Sci. Rep. **12**(1), 1–13 (2022a)
53. E. Sarmas, E. Spiliotis, V. Marinakis, T. Koutselis, H. Doukas, A meta-learning classification model for supporting decisions on energy efficiency investments. Energy Build. 111836 (2022b)
54. E. Sarmas, E. Spiliotis, E. Stamatopoulos, V. Marinakis, H. Doukas, Short-term photovoltaic power forecasting using meta-learning and numerical weather prediction independent long short-term memory models. Renew. Energy **216**, 118997 (2023)
55. E. Sarmas, S. Strompolas, V. Marinakis, F. Santori, M.A. Bucarelli, H. Doukas, An incremental learning framework for photovoltaic production and load forecasting in energy microgrids. Electronics **11**(23), 3962 (2022)
56. N.C. Schwertman, A.J. Gilks, J. Cameron, A simple noncalculus proof that the median minimizes the sum of the absolute deviations. Am. Stat. **44**(1), 38–39 (1990)
57. P. Skaloumpakas, E. Sarmas, M. Rachmanidis, V. Marinakis, Reshaping the energy landscape of Crete through renewable energy valleys. Sci. Rep. **14**(1), 8038 (2024)
58. S. Sobri, S. Koohi-Kamali, N.A. Rahim, Solar photovoltaic generation forecasting methods: a review. Energy Convers. Manage. **156**, 459–497 (2018)
59. E. Spiliotis, F. Petropoulos, N. Kourentzes, V. Assimakopoulos, Cross-temporal aggregation: improving the forecast accuracy of hierarchical electricity consumption. Appl. Energy **261**, 114339 (2020)
60. D. Stoian, E. Spiliotis, E. Stamatopoulos, E. Sarmas, P. Rekomaa, V. Marinakis, A neural network-based approach for hierarchical electricity consumption forecasting in buildings, in *2024 IEEE International Workshop on Metrology for Living Environment (MetroLivEnv)* (IEEE, 2024), pp. 274–279
61. L. Suganthi, A.A. Samuel, Energy models for demand forecasting—a review. Renew. Sustain. Energy Rev. **16**(2), 1223–1240 (2012)
62. Y. Sun, F. Haghighat, B.C. Fung, A review of the-state-of-the-art in data-driven approaches for building energy prediction. Energy Build. **221**, 110022 (2020)
63. TILOS, Tilos Horizon 2020 Program (GA No (646529)). Technology innovation for the local scale, Optimum integration of battery energy storage (2015). https://www.tiloshorizon.eu/
64. L. Toderean, T. Cioara, I. Anghel, E. Sarmas, V. Michalakopoulos, V. Marinakis, Demand response optimization for smart grid integrated buildings: review of technology enablers landscape and innovation challenges. Energy Build. 115067 (2024)
65. J.E. Van Zyl, D.A. Savic, G.A. Walters, Operational optimization of water distribution systems using a hybrid genetic algorithm. J. Water Resour. Plan. Manag. **130**(2), 160–170 (2004)
66. Y. Wang, S. Sun, X. Chen, X. Zeng, Y. Kong, J. Chen, Y. Guo, T. Wang, Short-term load forecasting of industrial customers based on SVMD and XGBoost. Int. J. Electr. Power Energy Syst. **129**, 106830 (2021)

Chapter 9
Optimization Model for Electric Vehicle Integration and Energy Storage to Achieve Energy Autonomy

Abstract This chapter proposes an integrated methodology to enhance energy autonomy at the microgrid level and mitigate the challenges associated with reverse power flow. Reverse power flow, which occurs when electricity flows in the opposite direction of standard grid operations, typically arises when distributed energy production exceeds local demand. This phenomenon can lead to risks such as inefficient operations, equipment damage, grid instability, and energy losses. To address reverse power flow and promote energy autonomy, an algorithm based on predictive models is introduced. The proposed framework employs deep learning models to forecast electricity generation from photovoltaic (PV) systems and consumption within the microgrid. These forecasts are subsequently integrated into an optimization algorithm that schedules flexible loads, including electric vehicles (EVs), to align with anticipated energy patterns. A payback study is conducted to evaluate the effects of optimized load scheduling and energy storage on the microgrid's autonomy. The analysis also examines the impact of varying battery capacities and EV fleet sizes. Results indicate that forecast-based load shifting can significantly reduce reverse power flow, particularly when managing larger flexible loads. Energy storage is found to complement load shifting, further enhancing microgrid performance. Nonetheless, the effectiveness of these strategies is contingent upon the accuracy of forecasts and the capacity of storage systems.

Keywords Artificial intelligence · Energy autonomy · Microgrid · Forecasting · Reverse power flow · Electric vehicles · Energy storage · Optimization · Deep learning

9.1 Introduction

In this chapter, we present a methodological framework designed to minimize Reverse Power Flow (RPF) at the microgrid level. Although the primary focus is on mitigating RPF, the framework also offers additional benefits, including a reduction in energy consumption from the main grid, a decrease in peak demand, and an enhancement of energy autonomy. The approach consists of two main components:

E. Sarmas et al., *Artificial Intelligence for Energy Systems*, Learning and Analytics in Intelligent Systems 46, https://doi.org/10.1007/978-3-031-85209-1_9

(i) a predictive modeling module that forecasts energy production and consumption for the subsequent day, and (ii) an energy management module incorporating two heuristic optimization algorithms. The first algorithm within the energy management module schedules flexible loads based on the predictions, while the second ahe charge and discharge processes of Energy Storage Systems (ESS).

Towards ensuring access to affordable, reliable, and sustainable energy, the energy sector faces the challenge of enhancing global energy efficiency and autonomy through the integration of Renewable Energy Sources (RES) [67]. This constitutes a critical challenge that aligns with Goal 7 of the United Nations' Sustainable Development Goals (i.e., "Affordable and Clean Energy") [62] and relies heavily on the effective and scalable operation of electric microgrids. Microgrids, which are small-scale local energy networks, can function independently or in conjunction with the main grid [34]. They integrate multiple RES, such as solar panels and wind turbines, often supported by ESS like batteries, to store surplus energy for use when generation does not meet demand [60].

The design and operation of microgrids aim to provide a reliable and resilient energy source, especially in regions with limited or no main grid access [23, 56]. Grid-connected microgrids also play a crucial role in powering buildings or communities while maintaining the flexibility to draw energy from the main grid when needed [6]. Energy autonomy, defined as the capacity to produce and use local energy rather than depending on external sources, is a key advantage of microgrids [17]. To achieve autonomy, microgrids must generate and store sufficient electricity to meet local demand, even when isolated from the main grid [37]. RES integration ensures a sustainable energy supply, while ESS enhances flexibility, supporting autonomy and resilience.

However, the widespread implementation of microgrids introduces significant operational challenges, particularly the issue of RPF. RPF occurs when surplus electricity flows back to the main grid, potentially damaging equipment, causing power quality issues, and reducing network efficiency [35]. It undermines microgrid energy autonomy and increases reliance on the main grid. Proper balancing of energy production and demand is essential to address RPF, achievable through flexible generation, optimal scheduling of shiftable loads (e.g., EVs and air conditioning systems), and effective utilization of ESS [21, 29, 59].

This study investigates the role of predictive models and optimization algorithms in enhancing microgrid efficiency by addressing RPF and improving energy autonomy. A representative microgrid comprising an office building, a photovoltaic (PV) system, an ESS, and a fleet of EVs is analyzed. Simulations evaluate RPF and Demand from Grid (DfG) across scenarios optimizing EV charging schedules, considering ESS support and the impact of varying battery capacities and EV fleet sizes. The analysis also assesses how prediction errors affect optimization outcomes.

The proposed framework offers a prediction-based solution for microgrid load scheduling, characterized by flexibility in incorporating various RES and shiftable loads. Its key components include:

- *Two Predictive Models (PMs)* which forecast electricity generation by the PV system and microgrid consumption on an hourly basis for the next day.
- *A Heuristic Optimization Algorithm* which schedules EV charging sessions to minimize RPF based on PV generation and consumption forecasts.
- A *Rule-Based ESS Control* which automates the charging and discharging of batteries as required.

The remainder of the chapter is organized as follows: Sect. 9.2 reviews pertinent literature and identifies existing research gaps. Section 9.3 presents the proposed framework, including predictive models, optimization algorithms, and ESS control methods. Section 9.4 applies the framework to a real-world case study in an Italian microgrid. Finally, Sect. 9.5 summarizes findings and suggests directions for future research.

9.2 Problem Formulation and Related Work

The proposed framework relies on ESS management methods and load-shifting techniques, which are driven by energy consumption and RES generation forecasts. Therefore, the following review focuses on the aforementioned research areas.

9.2.1 Forecasting RES Production

RES are becoming increasingly significant in the global energy mix [27]. Consequently, accurate forecasting of RES generation, such as PV production, is vital for the operation and management of microgrids [11]. PV generation is inherently stochastic and influenced by various factors, including weather conditions such as solar irradiance, temperature, and cloud cover, as well as the physical properties of the PV system, such as its size and orientation [4, 10]. As a result, accurately predicting the electricity generated by PV systems over a specific time period becomes practically challenging.

Several scientific studies have attempted to provide accurate PV generation forecasts [9, 30, 44, 50]. Although various categorizations have been proposed for the methods used, the most widely accepted classification divides PV forecasting methods into physical models, statistical models, and hybrid models. Physical models, such as the clear-sky model, rely on the physical properties of the PV system and weather conditions [13]. In contrast, data-driven models use historical PV production and weather data to predict future PV generation [16]. Finally, hybrid models combine physical models and data-driven models to enhance the accuracy of individual approaches and mitigate uncertainty [47].

Recent advancements in PV production forecasting indicate that the development of predictive models can further improve forecasting accuracy [40, 65]. This improvement can be attributed to the increased availability of data, including both RES generation data and satellite imagery, as well as advancements in weather forecasting methods and neural network (NN) training strategies [42]. For instance, [1] introduced a novel NN architecture for day-ahead PV energy forecasting based on a variation of the gated recurrent unit model. Similarly, [32] proposed a framework based on physical constraints, utilizing LSTMs and incorporating domain knowledge of PV systems to forecast hourly PV energy generation.

9.2.2 Forecasting Building Consumption

Buildings are a significant part of microgrids as they consume a large portion of the generated electricity. For accurate forecasting of building electricity consumption and optimizing their operation, load forecasting methods can be utilized [58, 72]. Although data-driven models commonly used for building load forecasting and PV generation forecasting share some common principles, they also exhibit significant differences in terms of the input features considered. Unlike PV generation forecasting models, which heavily depend on weather predictions, building load forecasting models rely more on historical energy consumption data, which effectively represent the underlying usage patterns and behaviors of the building [33].

Several factors contribute to the electricity demand of a building. One key factor is the building type, as it is clear that, for example, an office building unoccupied on weekends will exhibit a different electricity usage pattern compared to a residential building. Furthermore, the building's size and occupancy levels significantly influence its energy consumption [55]. A building housing a large number of occupants or extensive electrical equipment will likely have higher energy requirements. Forecasting building energy loads is a well-researched area that predominantly utilizes statistical and predictive modeling approaches. In [68], various data-driven models were compared, including machine learning (ML), PM, and heuristic methods, for thermal load forecasting, concluding that the LSTM model performs best for short-term, hourly forecasting, while tree-based models, such as XGBoost, are more suitable for day-ahead predictions. Similarly, Fan et al. [12] explored the capabilities of PM in forecasting next-day cooling load profiles of buildings, with results indicating that PM models achieve better cooling load predictions compared to conventional statistical models.

9.2.3 Flexible Load Management

According to Graditi et al. [15], shiftable loads are electrical loads that can be adjusted or temporarily switched off to align with variations in electricity supply or demand, without causing equipment damage or compromising operational performance. Examples include water pumps, electric currents, electric water heaters, and air conditioning systems. As discussed in Sect. 9.1, managing shiftable loads is crucial for the effective operation of a microgrid, as they enhance system reliability and efficiency by enabling load adjustments based on available electricity generation [14]. In the literature, three different ways are proposed to manage the shiftable loads of a microgrid and balance electricity supply and demand:

- **Load shedding**: This refers to deliberately turning off or reducing certain electrical loads. Load shedding is required in times of energy demand approaching the available supply which creates a potential risk of a dangerous imbalance [7]. In such situations and to stabilize the grid and prevent larger and longer-lasting outages, certain electrical loads are curtailed.
- **Load Shifting**: This refers to shifting the use of electrical loads to periods when more abundant electricity supply is available [38]. For example, an EV could be charged during hours when PV systems produce excess electricity, rather than during peak demand periods or at night when RES do not produce electricity [63].
- **Demand Response Programs**: These programs involve motivating customers to reduce electricity usage during periods of high demand or emergencies via financial rewards and/or other incentives or financial rewards [36, 41, 51]. These programs can also lead to a reduction of the overall demand on the microgrid and prevent RPF.

Electric vehicles (EVs) are one of the most prominent types of shiftable loads within a microgrid, as their charging (or discharging) can be easily aligned with the availability of electricity from RES. They offer several opportunities to serve as a flexibility asset in microgrid systems. One of the most straightforward approaches is to schedule EV charging during periods of high electricity availability, such as when solar panels produce excess energy, rather than during times of peak demand [64].

An illustrative example can be found in Mohsenian-Rad et al. [39], where a methodology was proposed for the stochastic optimization problem of EV charging scheduling under uncertain departure times, taking into account time-of-use pricing tariffs. In the context of microgrids, Van Der Kam and van Sark [63] proposed three control algorithms to manage the charging profile of multiple EVs, either in real time or using linear optimization. The simulated microgrid in their study consisted of a PV installation, an office, internet servers, three households, and two EVs.

Another possibility is to take advantage of Vehicle-to-Grid (V2G) technology, which allows EVs to be charged and discharged, functioning as ESS that return surplus electricity stored in the EV back to the grid when needed [31, 69]. Finally, considering behavioral aspects, EVs are the perfect candidate asset for implementing demand response programs. Such programs incentivize EV owners to charge their vehicles during low-demand periods or discharge their vehicles back to the grid during high-demand periods or emergencies [5, 28].

9.2.4 *Energy Storage Sizing*

Energy Storage Systems (ESS) play a crucial role in achieving high levels of energy autonomy by storing surplus energy during periods of excess generation and supplying it back to the microgrid when needed. This ensures energy independence during periods of low RES production. ESS offer numerous benefits to microgrids, including balancing energy supply and demand, enhancing power quality, and supporting ancillary services such as frequency and voltage regulation [18]. These advantages collectively contribute to greater energy autonomy. Specifically, ESS improve microgrid reliability (even during nighttime or cloudy days), enhance resilience to power outages, and lower energy costs by utilizing stored energy during peak price periods [20].

Battery Energy Storage Systems (BESS) are among the most commonly used types of energy storage systems, capable of storing electrical energy in chemical form to provide a portable and reliable power source. A key application of BESS is in microgrids, where they store excess energy generated by RES for later use [71]. The capacity of a BESS is typically determined by the energy demands of the microgrid it serves, along with various factors such as daily or hourly load profiles and environmental conditions (e.g., temperature, humidity, altitude), which can affect both performance and lifespan [19].

Several works have addressed the problem of optimizing the sizing of BESS in the context of a microgrid, also comsidering techno-economic factors [22]. For instance, [48] presented a mechanism for optimal power management in grid-connected PV systems, aiming to improve the penetration rate of PV generation into the grid while exploiting BESS. Along similar lines, [3] utilized a BESS for primary frequency control of a microgrid, concluding that the optimal size primarily depends on the permissible overload duration.

BESS have been widely utilized to address the issue of RPF. For instance, Unahalekhaka and Sripakarach [61] demonstrated that BESS can effectively reduce RPF and smooth out the distribution load curve. However, Zhou et al. [73] highlighted through experimental evidence that many BESS systems were fully charged too early to effectively mitigate microgrid peaks. This often occurred on days with high PV generation and low daily loads, where the BESS were not fully discharged during the night. Therefore, a detailed investigation into the optimal sizing of BESS in microgrids is essential, as it can enable more stable, flexible, cost-efficient, and sustainable microgrid operations.

9.3 Proposed Methodology

The proposed methodological framework consists of (i) two forecasting modules that provide predictions of PV energy production and building consumption (presented in Sect. 9.3.1), (ii) a heuristic scheduling algorithm that optimizes the charging periods of an electric vehicle fleet (discussed in Subsect. 9.3.2.1), and (iii) a rule-based method that manages an ESS accumulator (explored in Subsect. 9.3.2.2). The proposed framework, attempts to address the problem of managing the distributed energy resources of a microgrid. It places particular emphasis on mitigating issues related to RPF while simultaneously increasing the energy autonomy of the microgrid. In this context, a general overview of the proposed framework is presented in Fig. 9.1.

9.3.1 Forecasting Models

The CNN-LSTM model is employed to forecast both PV production and electricity consumption. This model was chosen based on recent studies demonstrating its superior performance compared to other ML and Deep Learning (DL) models for similar forecasting tasks [2, 43, 54, 66]. The architecture of the proposed CNN-LSTM model is illustrated in Fig. 9.2.

The model operates in two distinct stages. In the first stage, the CNN layers extract spatial features from the input variables through convolutional and pooling operations, partially reducing noise in the process. This output is then passed to the LSTM layer. In the second stage, the LSTM layer captures irregular temporal patterns using the spatial features provided by the CNN. As a result, the CNN-LSTM model effectively predicts the target variable (PV production or electricity consumption) through a fully connected architecture. The predictions from the two forecasting models were assessed using three widely used metrics in energy forecasting.

The features used by the CNN-LSTM models were selected based on the recent findings of [53], who analyzed various ML and DL models for forecasting PV production and electricity consumption under similar conditions. This selection was further informed by best practices highlighted in the literature, as discussed in Subsects. 9.2.1 and 9.2.2, and by additional tests conducted by the authors.

Table 9.1 provides an overview of the variables used as input for the PV production and electricity consumption models. These features include past observations (lags) to capture seasonal patterns and trends, Numerical Weather Predictions (NWPs), and calendar data (hour, day, and month of the forecasted period).

Both neural networks (NNs) were trained using the Keras library (version 2.2.4) with the TensorFlow-GPU backend (version 1.15.0), leveraging the CUDA Toolkit (version 10.0.130) and CuDNN (version 7.6.5). The training was conducted on a DELL Latitude 3510 computer equipped with an Intel Core i7-10510U processor and 16 GB of RAM.

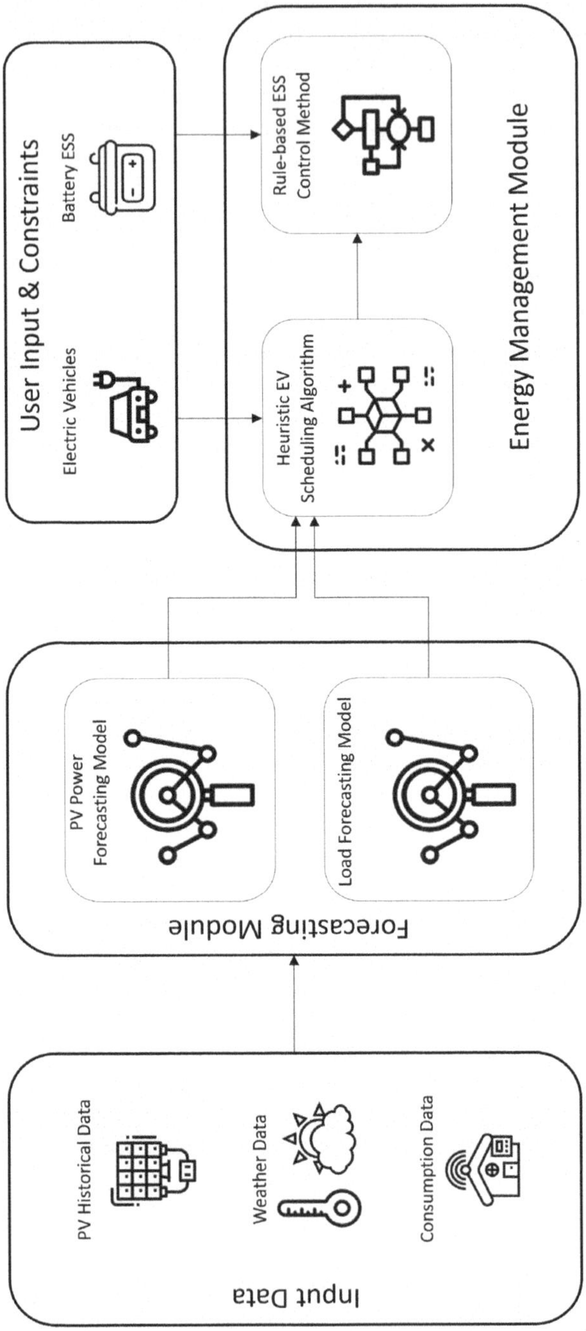

Fig. 9.1 Overview of the building blocks of the proposed framework, illustrating the required input, as well as the forecasting and energy management modules developed in this study

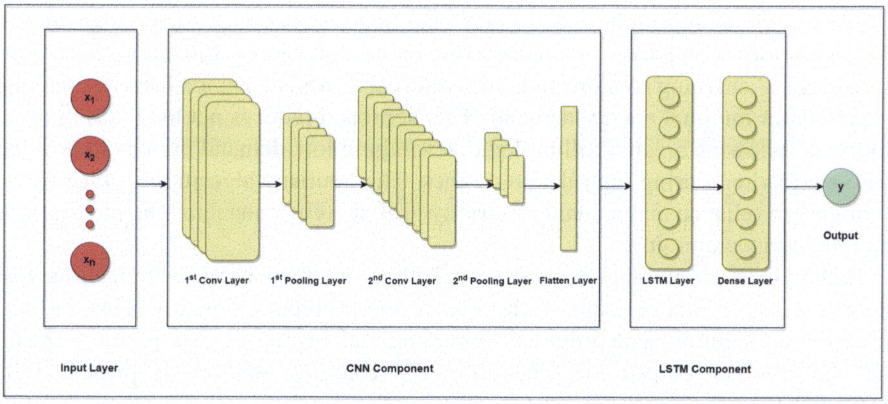

Fig. 9.2 The architecture of the utilized CNN-LSTM forecasting model

Table 9.1 Features used as input variables in the PV production and the electricity consumption models

Model	#	Feature	Description
PV production	1	Air temperature	Float NWP value of air temperature
	2	Relative humidity	Float NWP value of relative humidity
	3	Global radiation	Float NWP value of global radiation
	4	Hour of the day	Integer value between 0 and 23
	5	Month of the year	Integer value between 0 and 11
Electricity consumption	1	24 h lag	Consumption; same hour one days ago
	2	48 h lag	Consumption; same hour two days ago
	3	168 h lag	Consumption; same hour one week ago
	4	Hour of the day	Integer value between 0 and 23
	5	Day of the week	Integer value between 0 and 6
	6	Month of the year	Integer value between 0 and 11

9.3.2 Optimization Models

9.3.2.1 Scheduling EV Charging Sessions

This subsection introduces the heuristic algorithm developed to schedule EV charging periods. Heuristic and metaheuristic algorithms are part of a broader class of optimization techniques designed to provide approximate solutions to complex optimization problems [8]. While both aim to identify optimal or near-optimal solutions, they differ in their approach. Heuristic algorithms are rule-based methods that tackle relatively simple and well-defined problems by following a set of predefined rules to guide the search process. However, the solutions they generate are not always guaranteed to be optimal [24].

The proposed algorithm schedules EV charging sessions daily to optimize microgrid operations with a focus on maximizing renewable energy utilization. Each day, it assigns EV charging sessions to hourly intervals with low net load, corresponding to periods of minimal energy demand. This approach reduces net load during peak hours by following a valley-filling logic, leveraging low-demand periods to prevent grid overloading during high-demand times. The heuristic algorithm's objective is to minimize reliance on the main electricity grid, thereby enhancing the microgrid's overall energy autonomy.

Rather than employing exhaustive search or mathematical optimization, the algorithm uses a practical approach to solve the problem efficiently. This ensures an optimal[1] solution with minimal computational overhead. The pseudocode of the algorithm developed for scheduling EV charging sessions is presented in Algorithm 9.1.

Algorithm 9.1: Heuristic algorithm for optimally scheduling the charging sessions of EVs

---

Inputs : N: number of EVs to be charged
TotalCons:$\{x_1, x_2, \ldots, x_N\}$, electricity required per EV to be fully charged
$HourlyCons$:$\{x_1, x_2, \ldots, x_N\}$, hourly electricity required per EV while being charged
$NetLoad$:$\{z_1, z_2, \ldots, z_{24}\}$, hourly predicted net load (total consumption minus RES production);
Output: $EVSchedule[i, j]$: the day-ahead plan of the charging sessions (boolean variable); 1 if EV i is scheduled to charge at interval j; 0 otherwise

for *all available EVs* **do**
  /* Find EV with the greatest hourly consumption */
  utilisedEV ← max(*HourlyCons*);
  /* Calculate total charging time (in hours) */
  chargingHours ← $TotalCons$[utilizedEV] ÷ $HourlyCons$[utilizedEV];
  /* Find interval with minimum net load */
  for $i \in$ chargingHours *hourly intervals within the day* **do**
    intervalNetLoad[i] ← sum($NetLoad[i : i +$ chargingHours]);
    selectedInterval ← min(*intervalNetLoad*);
  end
  /* Update schedule by scheduling the EV in the selected interval */
  $EVSchedule$[utilizedEV, selectedInterval] ← 1;
  /* Update the hourly intervals and the available EVs */
  for $i \in$ selectedInterval **do**
    $NetLoad[i] \leftarrow NetLoad[i] + HourlyCons$[utilizedEV];
  end
  Erase utilisedEV from available EVs;
end

---

[1] The solution is considered optimal in terms of peak shaving, as it prioritizes EVs with the highest hourly energy requirements.

As outlined in Algorithm 9.1, the algorithm begins by iterating through all available EVs. It selects the EV with the highest hourly energy consumption and assigns it to the variable *utilisedEV*. The total charging time required for the selected EV is then calculated by dividing its total energy demand for full charging by its hourly energy consumption.

Subsequently, the algorithm identifies the interval with the lowest net load within the required charging duration by summing the net load for all possible hourly intervals. For instance, if the EV requires three hours to charge fully, all three-hour intervals (e.g., 00:00-03:00, 01:00-04:00, ..., 21:00-00:00) are evaluated. The EV is then scheduled for the interval with the lowest net load, and the schedule is updated in the *EV Schedule* matrix. Finally, the net load of the selected intervals is updated by adding the EV's hourly energy consumption, and the selected EV is removed from the list of unassigned EVs. This process is repeated until all EVs are allocated to their optimal charging intervals.

The output is a two-dimensional matrix referred to as *EV Schedule*[i,j]. Setting *EV Schedule*$[i, j] = 1$, indicates that EV i is scheduled for charging during interval j. Otherwise, *EV Schedule*$[i, j]$is set to 0. The optimization algorithm is fed with four inputs, namely:

- N: The number of EVs to be charged on a given day.
- *TotalCons* $\{x_1, x_2, \ldots, x_N\}$: A list containing the amount of electricity required per EV to fully charge.
- $HourlyCons$ $\{x_1, x_2, \ldots, x_N\}$: A list containing the hourly electricity required per EV during the charging process.
- $NetLoad$ $\{z_1, z_2, \ldots, z_{24}\}$: A list containing the hourly predicted net electrical load of the microgrid, i.e., the forecasted total consumption minus the forecasted RES generation.

All the required input information for the algorithm can easily be retrieved from the specifications of each EV. Indeed, the energy demand for fully charging an EV can be determined from its battery capacity and state of charge. The battery capacity, typically expressed in kilowatt-hours (kWh), represents the maximum energy the EV's battery can store. The state of charge, given as a decimal value, indicates the proportion of the battery capacity that has already been used. Therefore, the energy demand for full charging is calculated as the product of the battery capacity and the state of charge.

Meanwhile, the hourly electrical energy required for charging each EV (commonly referred to as the charging rate) is measured in kilowatts (kW). This rate reflects the amount of electrical energy the EV can accept during charging and can vary based on factors such as charging technology, battery condition, and temperature. The charging rate for a specific EV is typically available in the vehicle's documentation or on the manufacturer's website [57].

9.3.2.2 ESS Charging/Discharging Management

This subsection outlines the rule-based method employed to manage the ESS within the microgrid framework. To achieve this, a unified system at the microgrid level must be established [45, 46], encompassing generation sources, controllable and uncontrollable loads, and the ESS, as depicted in Fig. 9.3.

In this study, the PV installation serves as the primary source of electricity generation within the microgrid. The uncontrollable loads are represented by the office building, which experiences peak energy demand during daytime hours, while the controllable loads consist of EV charging sessions. These sessions are scheduled using the heuristic algorithm described in Subsect. 9.3.2.1, designed to achieve peak shaving of the net load. During periods of high PV production, the generated power is utilized to supply both flexible and inflexible loads [49]. This relationship is expressed in the following power balance equation:

$$P_{PV}(t) + P_{ESS}(t) = P_{EV}(t) + P_{Load}(t) + P_{Grid}(t), \tag{9.1}$$

where P_{PV} P_{PV} is the power generated by the PV system, P_{Load} is the electricity consumption of the office building, P_{EV} is the power consumption of the EV charging sessions, and P_{ESS} is the battery power. P_{PV}, P_{Load}, and P_{EV} can only take positive values, but P_{ESS} and P_{Grid} can take both positive and negative values. A positive P_{ESS} value indicates that the battery is supplying energy to the microgrid (battery discharging), while a negative P_{ESS} value indicates that the battery is storing excess energy generated by RES (battery charging).

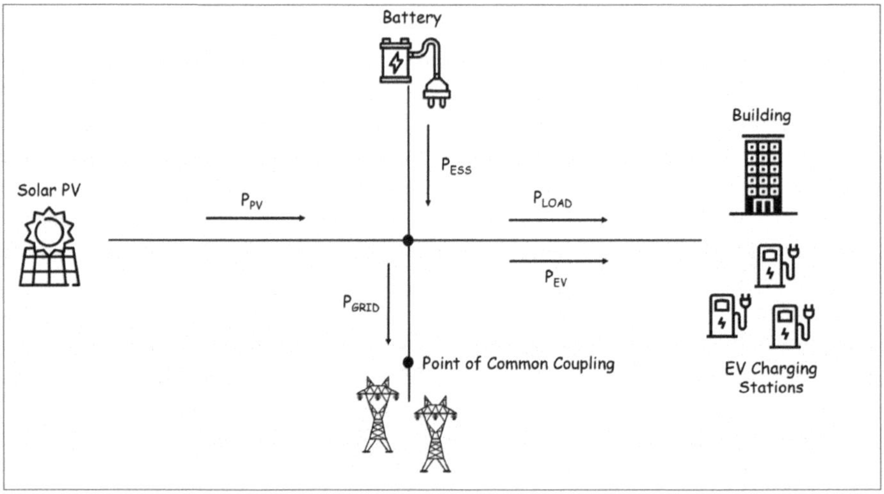

Fig. 9.3 Overview of the microgrid system topology considered in this study, illustrating the direction of positive power flows

The value P_{Grid} delineates the energy exchange dynamics between the microgrid and the main electricity grid, presenting two potential scenarios:

- When the PV system's production exceeds the total energy demand, resulting in RPF to the main grid.
- When the PV system cannot meet the total energy demand, leading to reliance on the main grid and reduced energy autonomy.

More specifically:

- In the first scenario, if production surpasses consumption, the surplus energy is first used to charge the battery. Once the battery is fully charged, any remaining excess energy is fed into the grid.
- In the second scenario, when energy demand exceeds generation, the microgrid initially draws energy from the battery. If the battery cannot supply the full demand, the deficit is covered by the main grid.

These scenarios are visually represented in Fig. 9.4.

With these considerations in mind, the energy management equation of the microgrid can be expressed as:

$$P_{PV}(t) + P_{ESS}(t) + P_{DfG}(t) = P_{EV}(t) + P_{Load}(t) + P_{RPF}(t), \tag{9.2}$$

where P_{DfG} P_{DfG} is the power demand from the grid, and P_{RPF} is the reverse power flow. These two variables take only positive values and replace P_{Grid} in the previous equation. In the case of surplus PV production, P_{DfG} equals zero, while in the case where demand exceeds production, P_{RPF} equals zero. Additionally, the state of charge (SoC) of the ESS must remain within its minimum and maximum limits to ensure proper battery operation. Consequently, the following constraint must be satisfied:

$$P_{ESS_{min}}(t) \leq P_{ESS}(t) \leq P_{ESS_{max}}(t). \tag{9.3}$$

Building on the previously described microgrid model and the optimized scheduling of EV charging periods, Algorithm 9.2 outlines the pseudocode for the rule-based method to control the ESS for the following day. This algorithm uses the generation and demand data of network assets as inputs and determines the battery's state of charge (SoC) at each hourly interval. Additionally, it incorporates the calculation of RPF and DfG during the simulation process.

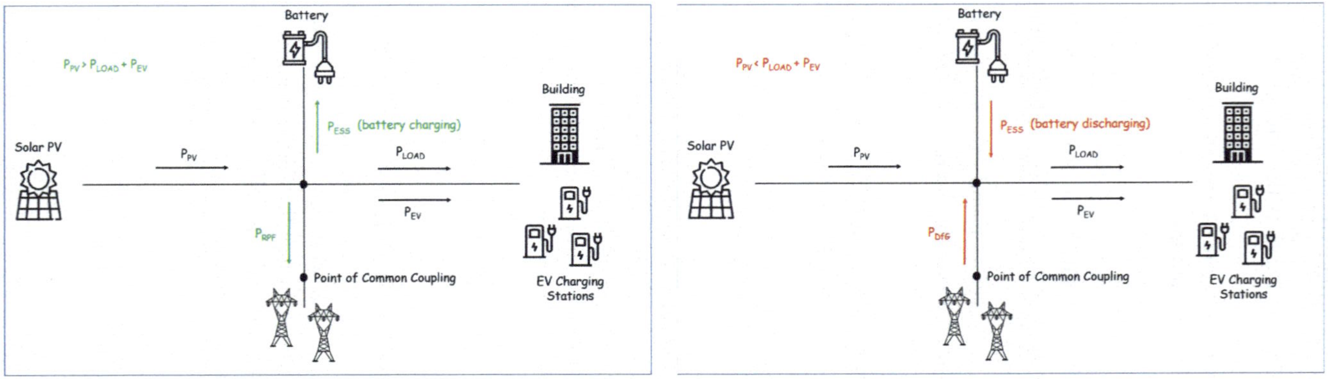

(a) PV-based power generation is greater than the sum of building and EV energy demand.

(b) PV-based power generation is lesser than the sum of building and EV energy demand.

Fig. 9.4 Graphical representation of the power flows during the two possible scenarios: **a** power generation is greater than the demand (left figure), **b** demand is greater than the power generation (right figure)

Algorithm 9.2: Rule-based algorithm for controlling the ESS.

Inputs : *Production*:$\{x_1, x_2, \ldots, x_{24}\}$, hourly production of solar PV;
Demand:$\{y_1, y_2, \ldots, y_{24}\}$, hourly demand of building and EV charging sessions;
Output: *ESS_end*:$\{z_1, z_2, \ldots, z_{24}\}$ the state of charge of the battery at the end of each hour;
Obviously, the following equity applies for all intervals: $ESS\_end[z_i] = ESS\_start[z_{i+1}]$

```
for i in range(24 hours) do
    if Production[i] ≥ Demand[i] then
        /* Case 1: Excess of solar PV production */
        difference ← Production[i] − Demand[i];
        ESS_potential ← ESS_max − ESS_start[i];
        if difference ≥ ESS_potential then
            /* Case 1.1: Excess energy is more than ESS capacity */
            ESS_end[i] ← ESS_max;
            DfG ← 0; RPF ← difference - ESS potential;
        else
            /* Case 1.2: Excess energy is less than ESS capacity */
            ESS_end[i] ← ESS_start[i] + difference;
            DfG ← 0; RPF ← 0;
        end
    else
        /* Case 2: Demand is more than solar PV production */
        difference ← Demand[i] − Production[i] ;
        if difference ≤ ESS_start[i] then
            /* Case 2.1: Net demand can be satisfied by the ESS */
            ESS_end[i] ← ESS_start[i] − difference;
            DfG ← 0; RPF ← 0;
        else
            /* Case 2.2: Net demand cannot be satisfied by the ESS */
            ESS_end[i] ← 0;
            DfG ← difference - ESS_start[i]; RPF ← 0;
        end
    end
end
```

9.4 Case Study

To illustrate the practical application of the proposed framework and assess its benefits across four key performance indicators (KPIs), we analyze the case of ASM Terni S.p.A., an organization based in Terni, central Italy. The KPIs include: (i) annual RPF, (ii) annual DfG, (iii) energy autonomy, measured as the number of days,[2] and (iv) peak shaving, defined as the average daily range of DfG [52].

[2] In Figs. 9.7 and 9.8, energy autonomy is expressed in days rather than hours to facilitate comparison. These results represent aggregated individual hours grouped into 24-hour units rather than actual consecutive days with zero DfG.

9.4.1 Dataset

The microgrid analyzed in this study covers an area of approximately 3 ha in the suburban region of Terni. It includes a three-story office building, three multipurpose warehouses supporting electricity, natural gas, and water supply operations, and two parking lots equipped with PV system canopies. The microgrid is linked to the external grid via two secondary substations, each fitted with oil-filled transformers rated at 250 and 400 kVA, respectively.

Electricity consumption in the building and warehouses, amounting to The microgrid consumes approximately 650 MWh annually, primarily due to lighting and the operation of electrical devices such as computers and servers. Its PV generation system consists of 1100 monocrystalline silicon panels, each rated at 150 W and 12 V, with dimensions of 98.5 × 98.5 cm. Located in the parking lot, the PV system connects to the microgrid through a 200 kVA inverter. Both loads and PV generators are equipped with real-time monitoring sensors powered by power quality analyzers, enabling the measurement of voltage, current, and active/reactive power values for various devices. The data is stored in a database with a granularity of one second.

The parking area includes three EV charging stations, each offering two charging points capable of delivering up to 22 kW AC and 32 A, supporting both single-phase and three-phase configurations. These stations feature a user interface for initiating and managing charging sessions, real-time smart meters, and remote power adjustment functionality. Currently, the EV fleet using these facilities comprises five company-owned vehicles, with plans for significant fleet expansion in the near future to support sustainable urban transportation.

The average EV battery capacity is around 40 kWh. Assuming drivers recharge when the battery drops to about 20%, each charging session typically consumes approximately 33 kWh and lasts an average of three hours. These estimates align with actual usage data, which summarizes charging session durations and energy consumption. Despite current demand levels, the charging infrastructure exceeds requirements. Consequently, it is common for users to plug in their vehicles upon arriving at work (around 08:00) and unplug them at the end of the workday (around 17:00).

9.4.2 Forecasting Model Performance

The performance of the forecasting models for predicting electricity consumption and solar energy production was assessed using two metrics, namely the Root Mean Squared Error (RMSE) and the Pearson correlation coefficient, as detailed below:

$$RMSE = \sqrt{\frac{1}{N}\sum\nolimits_{t=1}^{N}(y_t - f_t)^2},$$

$$r_{y,f} = \frac{\sum\nolimits_{t=1}^{N}(y_t - \bar{y})(f_t - \bar{f})}{\sqrt{\sum\nolimits_{t=1}^{N}(y_t - \bar{y})^2}\sqrt{\sum\nolimits_{t=1}^{N}(f_t - \bar{f})^2}},$$

where N is the number of observations considered for the evaluation (8760 h), y_t is the actual future value of the examined timeseries at hour t of the evaluation period, f_t is the corresponding forecast, and $\bar{y}$ and $\bar{f}$ are the average actual and predicted values, respectively.

Because squared errors are optimized for the mean [25] and disproportionately penalize larger errors-often occurring during peak hours [26]-the selected metrics are effective for identifying models that deliver improved forecasts both on average and during the most critical times of the day for energy managers and producers. However, to enable comparison with standard benchmarks in the field [70], the Forecast Skill Indicator (FSI) was also computed alongside the two accuracy metrics mentioned earlier. This metric illustrates the relative performance of the developed forecasting models compared to persistence models, which forecast each value as equal to the last known observation for the corresponding seasonal period, as shown below:

$$\text{FSI} = 1 - \frac{RMSE_{proposed}}{RMSE_{persistence}}, \tag{9.4}$$

where the persistence model predictions p are equivalent to those of the seasonal Naive model, as follows:

$$p_t = y_{t-s}. \tag{9.5}$$

Given that in short-term forecasting settings, solar energy can be considered to exhibit a 24 h seasonal cycle (intra-daily seasonality), while electricity consumption exhibits a 168 h seasonal cycle (intra-weekly seasonality), the parameter s was set to 24 and 168 h, respectively. Table 9.2 summarizes the performance of the forecasting models for electricity consumption and solar energy used in the case study. As shown, both models manage to significantly outperform the persistence models,

Table 9.2 Forecasting performance of the proposed models utilized in the case study

Measure	Electricity consumption	Solar power
RMSE	8.73	9.59
FSI	0.34	0.51
$r_{y,f}$	0.89	0.97

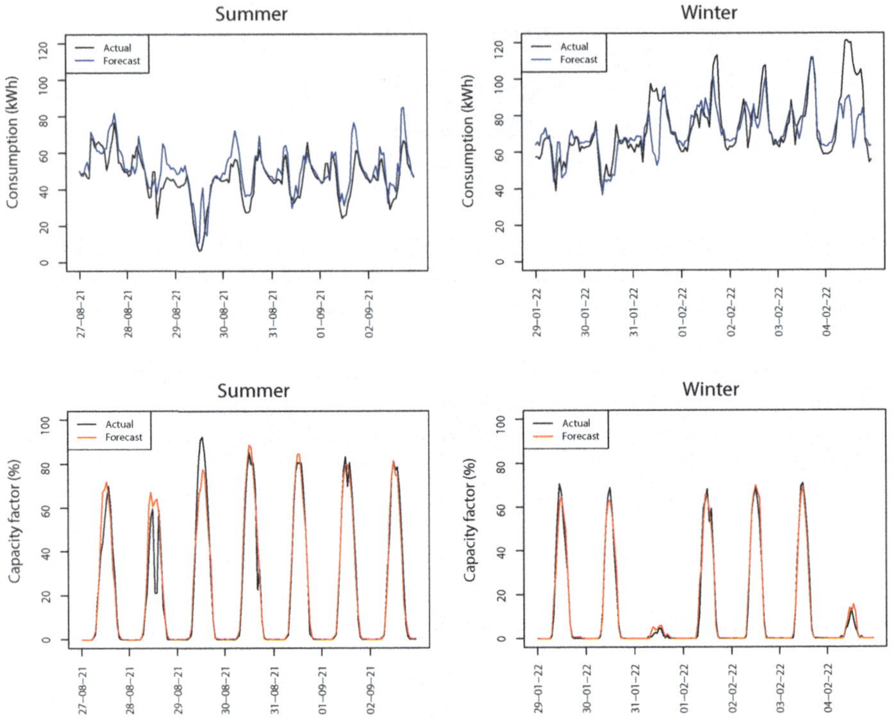

Fig. 9.5 Examples illustrating how the electricity consumption (top) and solar power (bottom) forecasting models perform. The examples refer to two randomly selected summer (2021-08-27 to 2021-09-02) and winter (2022-01-29 to 2022-02-04) weeks of the evaluation period

improving accuracy by 34 and 51%, respectively. Moreover, based on the $r_{y,f}$ metric, we observe that the forecasts are highly correlated with the actual future values, thereby facilitating the optimizations conducted in our study. This is further confirmed by Fig. 9.5 which provides indicative examples of forecasts for a randomly selected summer and winter week from the evaluation period.

9.4.3 Energy Management Optimization

To evaluate the potential impact of varying EV fleet sizes on the performance metrics in the future, we simulate the daily influx of $N \in \{5, 10, \ldots, 35, 40\}$ EVs over a one-year period (365 d). To achieve this, we begin by randomly selecting N numbers from a uniform distribution $U(0.1, 0.9) * 100\%$, representing the initial battery charge level of each EV. Next, assuming that all drivers whose vehicle's charge is below 20% are willing to recharge, we calculate the EV influx for the first day of the simulation period.

For the subsequent days, we estimate the EV influx based on the assumption that each vehicle travels an average of 20 km per day, which reduces its previous charge level. Similar to the first day, any time an EV's battery charge drops below 20%, it is assumed to be fully recharged on the following day. To ensure consistent and fair comparisons across different simulations, the EV charge state initialization is done only once.

To facilitate comparisons and assess the potential impact of the proposed framework under different usage conditions, we examine the following three scenarios:

- **Business as usual (BaU)**: In this scenario, all drivers intending to charge their electric vehicles plug them into the chargers upon arriving at work (08:00). Furthermore, no battery is available to store the electricity generated by the PV system.
- **Optimal Scheduling of EV Charging Sessions**: In this scenario, drivers intending to charge their EVs connect their vehicles to the chargers at the times recommended by the optimization algorithm outlined in Subsect. 9.3.2.1. However, as in the BaU scenario, no battery is available to store the electricity produced by the PV system.
- **Optimal Scheduling of EV Charging Sessions and Automation of Energy Storage**: Similar to the previous scenario, drivers connect their EVs to the chargers at the times recommended by the proposed algorithm. Additionally, once the EV charging schedule is determined, a battery is employed to store surplus electricity generated by the PV system when appropriate and to supply energy as needed, in accordance with the algorithm described in Subsect. 9.3.2.2. To assess the impact of batteries with different capacities on the evaluated KPIs, simulations are conducted for batteries with capacities $C \in \{5, 10, 15, \ldots, 90, 95, 100\}$ kWh.

The data used to implement each scenario and its variations are hourly and span approximately 2 years and 11 months (25,536 h), from September 6, 2019, to August 3, 2022. The period from August 4, 2021, to August 3, 2022, is used for evaluation purposes, while the remaining data are used for training the models that will forecast the building's electricity consumption and PV energy generation for the next day (24 h). Thus, the initial dataset includes the total electricity consumed by the building and the electricity generated by the PV installation, complemented by simulated data indicating the electricity consumed by electric vehicles. These variables essentially serve as inputs to the proposed optimization algorithms.

To illustrate the operation of the proposed framework and evaluate its maximum potential benefits, we start by applying the approach using the building's actual electricity consumption and PV generation data as inputs, effectively assuming perfect forecasting. Figure 9.6 demonstrates an example of how the optimization framework affects EV charging schedules, net load, battery state of charge, Demand for Grid (DfG), and Renewable Power Factor (RPF) on a randomly selected date within the evaluation period.

As depicted, PV generation peaks around midday, prompting the scheduling algorithm to shift the EV charging load by four hours. This adjustment significantly

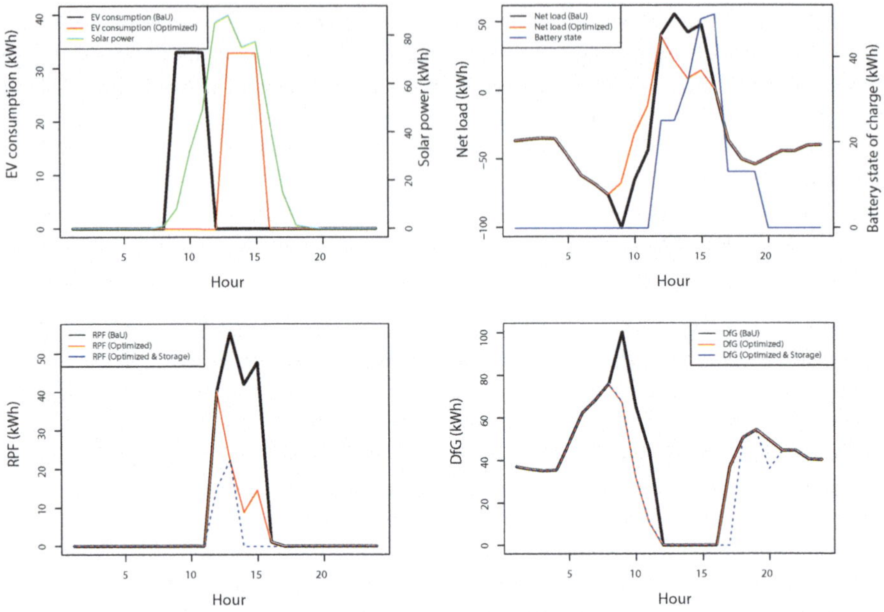

Fig. 9.6 Example (randomly selected date) illustrating how the proposed optimization framework affects the timing of the EV charging sessions (top left), net load and battery state of charge (top right), DfG (bottom left), and RPF (bottom right). A 50 kWh battery is used and 5 EVs are considered

reduces the RPF, substantially flattens the net load peak, and nearly eliminates the DfG. Furthermore, when energy storage is available, a portion of the surplus PV energy generated during the afternoon is stored and used during the evening, further reducing both the RPF and DfG.

Figure 9.7 illustrates the effects of the scheduling algorithm in the BaU scenario, where no battery is available to store energy generated by the PV system. As expected, grid energy consumption increases with the size of the EV fleet. However, optimization still achieves energy savings ranging from 1.5% for 5 EVs to 7.2% for 40 EVs. Interestingly, while the percentage savings increase as more EVs utilize the facility, the improvements are not linear and gradually plateau beyond a certain point. This suggests the potential to identify an optimal ("golden") number of EV users based on the facility's capacity for inflexible loads and solar energy generation.

For the Renewable Power Factor (RPF), optimal scheduling proves particularly effective, reducing RPF values from 7.2% for 5 EVs to 44.3% for 40 EVs. Larger EV fleets see greater improvements, as more charging sessions reduce surplus solar energy. However, RPF cannot be entirely eliminated because EV charging is limited to office hours, leaving some PV energy generated in early morning and late afternoon unused. Extending EV charging to non-office hours, potentially involving external users, could further reduce RPF.

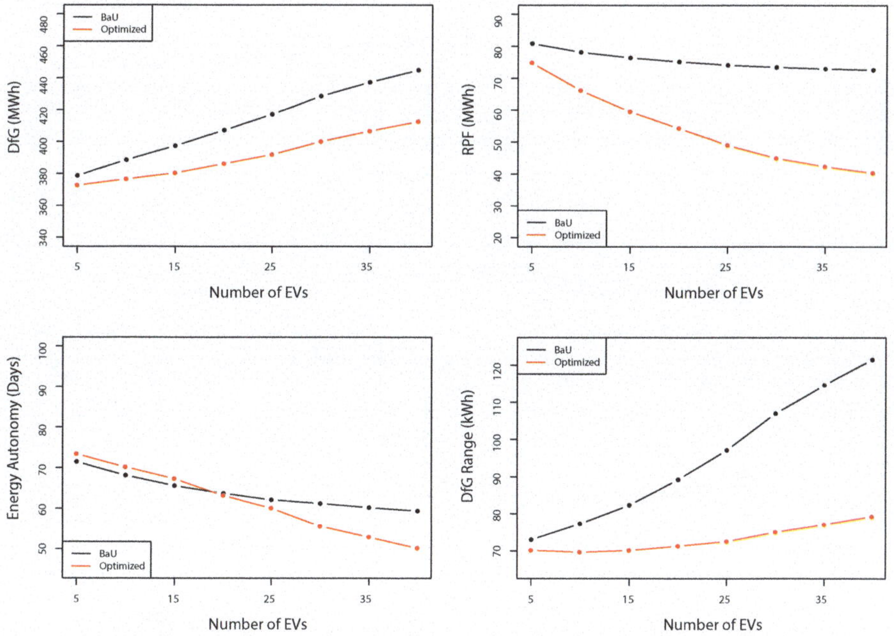

Fig. 9.7 Impact (full year) of optimally scheduling the EV charging sessions when no battery is available for storing the energy produced by the PV system. Performance is measured both for the BaU and optimized scenario assuming different numbers of EVs

As anticipated, the scheduling algorithm also substantially improves peak shaving, reducing the average daily DfG range from 3.7% for 5 EVs to 34% for 40 EVs. Conversely, the microgrid's energy self-sufficiency remains nearly constant for smaller EV fleets but declines slightly for larger ones. This occurs because when the flexible load exceeds a certain threshold, PV generation becomes insufficient to sustain energy self-sufficiency, particularly as the scheduling algorithm prioritizes utilizing PV production during peak demand periods.

Figure 9.8 presents comparisons similar to those in Fig. 9.7, this time emphasizing the role of the energy storage system in the optimized scheduling of EV charging. The results indicate that larger-capacity batteries contribute to reductions in both Demand for Grid (DfG) and Renewable Power Factor (RPF). The maximum percentage improvements in DfG (comparing scenarios without a battery to those with a 100 kWh battery) remain relatively consistent for a given number of EVs, ranging from 5% for 40 EVs to 7% for 5 EVs. A similar pattern is observed in peak shaving, with the DfG range reduced by up to 2

In contrast, the impact of energy storage on RPF is more pronounced and varies significantly with the number of EVs, with improvements ranging from 35% for 5 EVs to 52% for 40 EVs. Regarding energy self-sufficiency, the results are also promising, showing that the proportion of zero-DfG instances can increase from 33% for 5 EVs to 45% for 40 EVs.

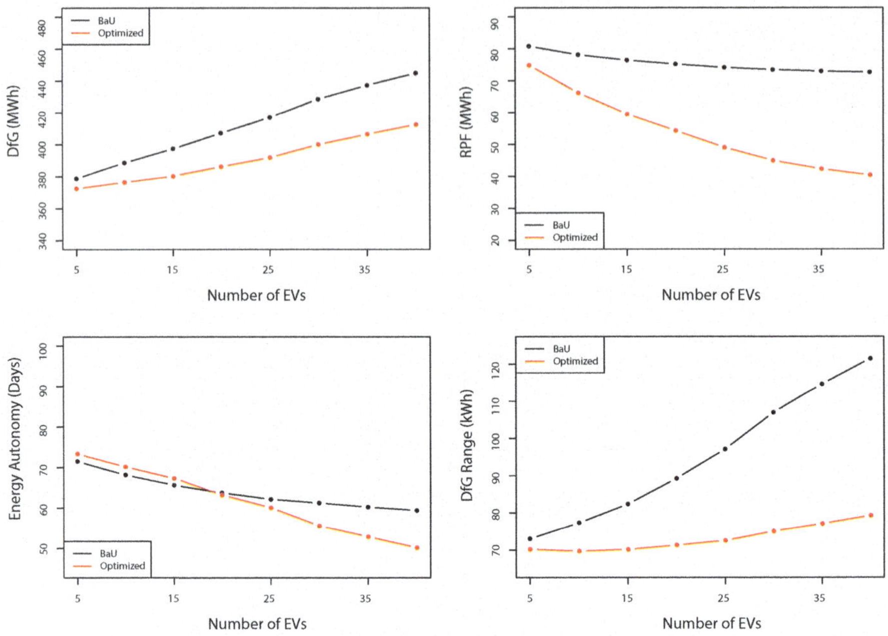

Fig. 9.8 Impact (full year) of the ESS on the optimal scheduling of the EV charging sessions. Performance is measured assuming various numbers of EVs and battery capacities

These findings highlight that energy storage effectively complements the scheduling algorithm, further enhancing its performance by reducing RPF and improving energy self-sufficiency.

In the final step of our analysis, recognizing that real-world models for forecasting electricity consumption and PV generation inherently include some prediction errors, we re-ran the simulations for the third scenario using forecasted values instead of actual data. For each variation of the scenario (a combination of battery capacity and number of EVs), we evaluated the relative performance of the two approaches based on the defined KPIs.

Figure 9.9 illustrates the impact of prediction error on the optimized scheduling of EV charging sessions. The findings reveal that imperfect forecasts have a minimal adverse effect on DfG, with performance decreasing by <2% compared to simulations assuming "perfect forecasting." This negative impact is notably smaller for scenarios with fewer EV users (<0.7% for 10 EVs) and larger-capacity batteries. Similarly, energy self-sufficiency remains nearly unchanged.

However, prediction error has a more pronounced effect on RPF, particularly when many EVs are using the facility. In these cases, RPF can increase by up to 20%. For smaller EV fleets, this impact is much less significant (<5% for 10 EVs). A similar trend is observed for the peak shaving KPI, which can degrade by up to 11% for larger EV fleets but remains largely unaffected (<2%) for smaller numbers of EVs.

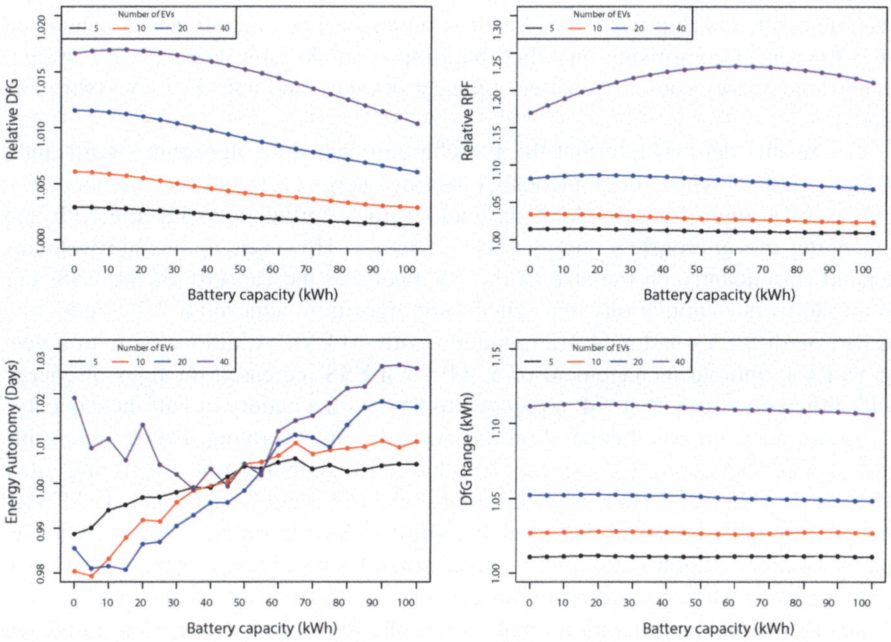

Fig. 9.9 Impact (full year) of forecast error on the optimal scheduling of EV charging sessions. Relative performance is computed assuming various numbers of EVs and battery capacities

While the forecasting models used in this study are relatively accurate, improving prediction accuracy could enable even greater energy savings and lower energy costs. This is especially critical for facilities accommodating a large number of EVs daily.

9.5 Conclusions

Smart energy management systems play a crucial role in enhancing the energy self-sufficiency of microgrids, reducing energy costs, and facilitating smoother operation. Leveraging advancements in AI for energy forecasting, this study investigated the impact of forecast-driven optimization algorithms on the efficient management of microgrids incorporating shiftable loads and energy storage systems (ESS).

We introduced a comprehensive framework composed of two primary modules: a forecasting module for predicting electricity generation and consumption for the forthcoming day, and an energy management module featuring two heuristic optimization algorithms. The first algorithm schedules flexible loads based on forecast data, while the second manages the charging and discharging of the ESS. The framework's primary objective is to minimize the Renewable Power Factor (RPF) of the microgrid, with additional benefits including reduced grid energy consumption, lower peak demand, and improved energy self-sufficiency. The evaluation was conducted on a microgrid comprising an office building (non-shiftable loads), a PV system (renewable energy source), a battery (energy storage), and a fleet of EVs (shiftable loads).

The results demonstrate that the EV charging schedule algorithm significantly reduces the RPF while simultaneously lowering energy costs and peak demand. The ESS control algorithm complements load shifting, further reducing the RPF and enhancing the microgrid's energy self-sufficiency. However, these improvements depend significantly on the size of the EV fleet and the capacity of the ESS. For example, in our simulations, the scheduling algorithm achieved a 7.2% reduction in RPF with 5 EVs and a 44.3% reduction with 40 EVs. Additionally, with a fleet of 10 EVs, optimal management of a 100 kWh ESS increased the days of energy self-sufficiency from 70 to 94, compared to 85 d with a battery of half the capacity.

Future research could expand on this work by incorporating diverse renewable energy sources beyond PV systems for electricity generation or integrating other types of flexible loads, such as water pumps, electric water heaters, and HVAC systems. These variations would expand the optimization algorithm's search space and introduce more complex system dynamics, potentially yielding greater performance improvements while necessitating more generalized and adaptable solutions.

Another promising direction involves refining the objective functions employed by optimization algorithms to incorporate additional considerations affecting energy systems. These could include factors such as the preferences and comfort of building occupants and EV drivers, as well as dynamic energy pricing, to create more holistic and user-centric energy management strategies.

References

1. M. Abdel-Basset, H. Hawash, R.K. Chakrabortty, M. Ryan, PV-net: an innovative deep learning approach for efficient forecasting of short-term photovoltaic energy production. J. Clean. Prod. **303**, 127037 (2021)
2. A. Agga, A. Abbou, M. Labbadi, Y. El Houm, I.H.O. Ali, CNN-LSTM: an efficient hybrid deep learning architecture for predicting short-term photovoltaic power production. Electr. Power Syst. Res. **208**, 107908 (2022)
3. M.R. Aghamohammadi, H. Abdolahinia, A new approach for optimal sizing of battery energy storage system for primary frequency control of islanded microgrid. Int. J. Electr. Power Energy Syst. **54**, 325–333 (2014)
4. R. Ahmed, V. Sreeram, Y. Mishra, M. Arif, A review and evaluation of the state-of-the-art in PV solar power forecasting: techniques and optimization. Renew. Sustain. Energy Rev. **124**, 109792 (2020)
5. P. Aliasghari, B. Mohammadi-Ivatloo, M. Alipour, M. Abapour, K. Zare, Optimal scheduling of plug-in electric vehicles and renewable micro-grid in energy and reserve markets considering demand response program. J. Clean. Prod. **186**, 293–303 (2018)
6. E.O. Arwa, K.A. Folly, Reinforcement learning techniques for optimal power control in grid-connected microgrids: a comprehensive review. IEEE Access **8**, 208992–209007 (2020)

7. N.N.A. Bakar, M.Y. Hassan, M.F. Sulaima, M. Na'im Mohd. Nasir, A. Khamis, Microgrid and load shedding scheme during islanded mode: a review. Renew. Sustain. Energy Rev. **71**, 161–169 (2017)
8. Z. Beheshti, S.M.H. Shamsuddin, A review of population-based meta-heuristic algorithms. Int. J. Adv. Soft Comput. Appl **5**(1), 1–35 (2013)
9. U.K. Das, K.S. Tey, M. Seyedmahmoudian, S. Mekhilef, M.Y.I. Idris, W. Van Deventer, B. Horan, A. Stojcevski, Forecasting of photovoltaic power generation and model optimization: a review. Renew. Sustain. Energy Rev. **81**, 912–928 (2018)
10. I.C. Dimitriou, A. Arsenopoulos, G.P. Trachanas, E. Sarmas, V. Marinakis, Assessing the adoption level of solar PV installations, on district scale of urban environment. Renew. Energy **237**, 121676 (2024)
11. S. Dutta, Y. Li, A. Venkataraman, L.M. Costa, T. Jiang, R. Plana, P. Tordjman, F.H. Choo, C.F. Foo, H.B. Puttgen, Load and renewable energy forecasting for a microgrid using persistence technique. Energy Procedia **143**, 617–622 (2017)
12. C. Fan, F. Xiao, Y. Zhao, A short-term building cooling load prediction method using deep learning algorithms. Appl. Energy **195**, 222–233 (2017)
13. K. Gairaa, S. Benkaciali, M. Guermoui, Clear-sky models evaluation of two sites over Algeria for PV forecasting purpose. Eur. Phys. J. Plus **134**(10), 1–17 (2019)
14. H. Golpîra, S. Bahramara, Internet-of-things-based optimal smart city energy management considering shiftable loads and energy storage. J. Clean. Prod. **264**, 121620 (2020)
15. G. Graditi, M.L. Di Silvestre, R. Gallea, E.R. Sanseverino, Heuristic-based shiftable loads optimal management in smart micro-grids. IEEE Trans. Indus. Inform. **11**(1), 271–280 (2014)
16. P. Gupta, R. Singh, PV power forecasting based on data-driven models: a review. Int. J. Sustain. Eng. **14**(6), 1733–1755 (2021)
17. A.M. Haidar, A. Fakhar, A. Helwig, Sustainable energy planning for cost minimization of autonomous hybrid microgrid using combined multi-objective optimization algorithm. Sustain. Cities Soc. **62**, 102391 (2020)
18. S. Hajiaghasi, A. Salemnia, M. Hamzeh, Hybrid energy storage system for microgrids applications: a review. J. Energy Storage **21**, 543–570 (2019)
19. M. Hannan, S. Wali, P. Ker, M. Abd Rahman, M. Mansor, V. Ramachandaramurthy, K. Muttaqi, T. Mahlia, Z. Dong, Battery energy-storage system: a review of technologies, optimization objectives, constraints, approaches, and outstanding issues. J. Energy Storage **42**, 103023 (2021)
20. A.S. Hassan, L. Cipcigan, N. Jenkins, Optimal battery storage operation for PV systems with tariff incentives. Appl. Energy **203**, 422–441 (2017)
21. Y. Huo, F. Bouffard, G. Joós, Decision tree-based optimization for flexibility management for sustainable energy microgrids. Appl. Energy **290**, 116772 (2021)
22. J. Kaldellis, D. Zafirakis, K. Kavadias, Techno-economic comparison of energy storage systems for island autonomous electrical networks. Renew. Sustain. Energy Rev. **13**(2), 378–392 (2009)
23. M.K. Kiptoo, M.E. Lotfy, O.B. Adewuyi, A. Conteh, A.M. Howlader, T. Senjyu, Integrated approach for optimal techno-economic planning for high renewable energy-based isolated microgrid considering cost of energy storage and demand response strategies. Energy Convers. Manag. **215**, 112917 (2020)
24. N. Kokash, An introduction to heuristic algorithms. *Department of Informatics and Telecommunications* (2005), pp. 1–8
25. S. Kolassa, Evaluating predictive count data distributions in retail sales forecasting. Int. J. Forecast. **32**(3), 788–803 (2016)
26. D. Koutsandreas, E. Spiliotis, F. Petropoulos, V. Assimakopoulos, On the selection of forecasting accuracy measures. J. Oper. Res. Soc. **73**(5), 937–954 (2022)
27. O. Kraan, E. Chappin, G.J. Kramer, I. Nikolic, The influence of the energy transition on the significance of key energy metrics. Renew. Sustain. Energy Rev. **111**, 215–223 (2019)
28. K.N. Kumar, K.J. Tseng, Impact of demand response management on chargeability of electric vehicles. Energy **111**, 190–196 (2016)

29. M.T. Lawder, V. Viswanathan, V.R. Subramanian, Balancing autonomy and utilization of solar power and battery storage for demand based microgrids. J. Power Sour. **279**, 645–655 (2015)
30. P. Li, K. Zhou, X. Lu, S. Yang, A hybrid deep learning model for short-term PV power forecasting. Appl. Energy **259**, 114216 (2020)
31. H. Lund, W. Kempton, Integration of renewable energy into the transport and electricity sectors through V2G. Energy Policy **36**(9), 3578–3587 (2008)
32. X. Luo, D. Zhang, X. Zhu, Deep learning based forecasting of photovoltaic power generation by incorporating domain knowledge. Energy **225**, 120240 (2021)
33. W. Ma, S. Fang, G. Liu, R. Zhou, Modeling of district load forecasting for distributed energy system. Appl. Energy **204**, 181–205 (2017)
34. A.A. Mahmoud, A.A. Hafez, A.M. Yousef, M.A. Gaafar, M. Orabi, A.F. Ali,. Fault-tolerant modular multilevel converter for a seamless transition between stand-alone and grid-connected microgrid. IET Power Electron. (2022)
35. I.B. Majeed, N.I. Nwulu, Impact of reverse power flow on distributed transformers in a solar-photovoltaic-integrated low-voltage network. Energies **15**(23), 9238 (2022)
36. V. Michalakopoulos, E. Sarmas, I. Papias, P. Skaloumpakas, V. Marinakis, H. Doukas, A machine learning-based framework for clustering residential electricity load profiles to enhance demand response programs. Appl. Energy **361**, 122943 (2024)
37. Y. Mohammed, M. Mustafa, N. Bashir, I. Ibrahem, Existing and recommended renewable and sustainable energy development in Nigeria based on autonomous energy and microgrid technologies. Renew. Sustain. Energy Rev. **75**, 820–838 (2017)
38. B. Mohandes, S. Acharya, M.S. El Moursi, A.S. Al-Sumaiti, H. Doukas, S. Sgouridis, Optimal design of an islanded microgrid with load shifting mechanism between electrical and thermal energy storage systems. IEEE Trans. Power Syst. **35**(4), 2642–2657 (2020)
39. H. Mohsenian-Rad et al., Optimal charging of electric vehicles with uncertain departure times: a closed-form solution. IEEE Trans. Smart Grid **6**(2), 940–942 (2014)
40. D.P. Panagoulias, E. Sarmas, V. Marinakis, M. Virvou, G.A. Tsihrintzis, H. Doukas, Intelligent decision support for energy management: a methodology for tailored explainability of artificial intelligence analytics. Electronics **12**(21), 4430 (2023)
41. P. Pinson, H. Madsen et al., Benefits and challenges of electrical demand response: a critical review. Renew. Sustain. Energy Rev. **39**, 686–699 (2014)
42. J. Qin, H. Jiang, N. Lu, L. Yao, C. Zhou, Enhancing solar PV output forecast by integrating ground and satellite observations with deep learning. Renew. Sustain. Energy Rev. **167**, 112680 (2022)
43. J. Qu, Z. Qian, Y. Pei, Day-ahead hourly photovoltaic power forecasting using attention-based CNN-LSTM neural network embedded with multiple relevant and target variables prediction pattern. Energy **232**, 120996 (2021)
44. P. Ramsami, V. Oree, A hybrid method for forecasting the energy output of photovoltaic systems. Energy Convers. Manag. **95**, 406–413 (2015)
45. E.L. Ratnam, S.R. Weller, Receding horizon optimization-based approaches to managing supply voltages and power flows in a distribution grid with battery storage co-located with solar pv. Appl. Energy **210**, 1017–1026 (2018)
46. E.L. Ratnam, S.R. Weller, C.M. Kellett, An optimization-based approach to scheduling residential battery storage with solar PV: assessing customer benefit. Renew. Energy **75**, 123–134 (2015)
47. M.Q. Raza, M. Nadarajah, C. Ekanayake, On recent advances in PV output power forecast. Solar Energy **136**, 125–144 (2016)
48. Y. Riffonneau, S. Bacha, F. Barruel, S. Ploix, Optimal power flow management for grid connected PV systems with batteries. IEEE Trans. Sustain. Energy **2**(3), 309–320 (2011)
49. P. Saini, L. Gidwani, An investigation for battery energy storage system installation with renewable energy resources in distribution system by considering residential, commercial and industrial load models. J. Energy Storage **45**, 103493 (2022)
50. E. Sarmas, N. Dimitropoulos, V. Marinakis, Z. Mylona, H. Doukas, Transfer learning strategies for solar power forecasting under data scarcity. Sci. Rep. **12**(1), 1–13 (2022)

51. E. Sarmas, A. Fragkiadaki, V. Marinakis, Explainable AI-based ensemble clustering for load profiling and demand response. Energies **17**(22), 5559 (2024)
52. E. Sarmas, E. Spiliotis, V. Marinakis, M.A. Bucarelli, F. Santori, H. Doukas, Revving up energy autonomy: a forecast-driven framework for reducing reverse power flow in microgrids. Sustain. Energy Grids Netw. **38**, 101376 (2024)
53. E. Sarmas, E. Spiliotis, V. Marinakis, G. Tzanes, J.K. Kaldellis, H. Doukas, Ml-based energy management of water pumping systems for the application of peak shaving in small-scale islands. Sustai. Cities Soc. **82**, 103873 (2022)
54. X. Shao, C.-S. Kim, P. Sontakke, Accurate deep model for electricity consumption forecasting using multi-channel and multi-scale feature fusion CNN-LSTM. Energies **13**(8), 1881 (2020)
55. P. Skaloumpakas, E. Sarmas, Z. Mylona, A. Cavadenti, F. Santori, V. Marinakis, Predicting thermal comfort in buildings with machine learning and occupant feedback, in *2023 IEEE International Workshop on Metrology for Living Environment (MetroLivEnv)* (IEEE, 2023), pp. 34–39
56. P. Skaloumpakas, E. Sarmas, M. Rachmanidis, V. Marinakis, Reshaping the energy landscape of Crete through renewable energy valleys. Sci. Rep. **14**(1), 8038 (2024)
57. P. Skaloumpakas, E. Spiliotis, E. Sarmas, A. Lekidis, G. Stravodimos, D. Sarigiannis, I. Makarouni, V. Marinakis, J. Psarras, A multi-criteria approach for optimizing the placement of electric vehicle charging stations in highways. Energies **15**(24), 9445 (2022)
58. D. Stoian, E. Spiliotis, E. Stamatopoulos, E. Sarmas, P. Rekomaa, V. Marinakis, a neural network-based approach for hierarchical electricity consumption forecasting in buildings. In *2024 IEEE International Workshop on Metrology for Living Environment (MetroLivEnv)* (IEEE, 2024), pp. 274–279
59. D. Thrän, M. Dotzauer, V. Lenz, J. Liebetrau, A. Ortwein, Flexible bioenergy supply for balancing fluctuating renewables in the heat and power sector-a review of technologies and concepts. Energy Sustain. Soc. **5**(1), 1–15 (2015)
60. F. Tooryan, H. HassanzadehFard, E.R. Collins, S. Jin, B. Ramezani, Smart integration of renewable energy resources, electrical, and thermal energy storage in microgrid applications. Energy **212**, 118716 (2020)
61. P. Unahalekhaka, P. Sripakarach, Reduction of reverse power flow using the appropriate size and installation position of a Bess for a PV power plant. IEEE Access **8**, 102897–102906 (2020)
62. United Nations, Sustainable Development Goals (2022), https://www.un.org/sustainabledevelopment/energy/
63. M. Van Der Kam, W. van Sark, Smart charging of electric vehicles with photovoltaic power and vehicle-to-grid technology in a microgrid; a case study. Appl. Energy **152**, 20–30 (2015)
64. F.G. Venegas, M. Petit, Y. Perez, Active integration of electric vehicles into distribution grids: barriers and frameworks for flexibility services. Renew. Sustain. Energy Rev. **145**, 111060 (2021)
65. H. Wang, Z. Lei, X. Zhang, B. Zhou, J. Peng, A review of deep learning for renewable energy forecasting. Energy Convers. Manag. **198**, 111799 (2019)
66. K. Wang, X. Qi, H. Liu, A comparison of day-ahead photovoltaic power forecasting models based on deep learning neural network. Appl. Energy **251**, 113315 (2019)
67. R. Wang, S.-C. Hsu, S. Zheng, J.-H. Chen, X.I. Li, Renewable energy microgrids: economic evaluation and decision making for government policies to contribute to affordable and clean energy. Appl. Energy **274**, 115287 (2020)
68. Z. Wang, T. Hong, M.A. Piette, Building thermal load prediction through shallow machine learning and deep learning. Appl. Energy **263**, 114683 (2020)
69. H. Wei, Y. Zhang, Y. Wang, W. Hua, R. Jing, Y. Zhou, Planning integrated energy systems coupling V2G as a flexible storage. Energy **239**, 122215 (2022)
70. D. Yang, S. Alessandrini, J. Antonanzas, F. Antonanzas-Torres, V. Badescu, H.G. Beyer, R. Blaga, J. Boland, J.M. Bright, C.F. Coimbra, M. David, Â. Frimane, C.A. Gueymard, T. Hong, M.J. Kay, S. Killinger, J. Kleissl, P. Lauret, E. Lorenz, D. van der Meer, M. Paulescu, R. Perez, O. Perpiñán-Lamigueiro, I.M. Peters, G. Reikard, D. Renné, Y.-M. Saint-Drenan, Y. Shuai, R. Urraca, H. Verbois, F. Vignola, C. Voyant, J. Zhang, Verification of deterministic solar forecasts. Solar Energy **210**, 20–37 (2020)

71. Y. Yang, S. Bremner, C. Menictas, M. Kay, Battery energy storage system size determination in renewable energy systems: a review. Renew. Sustain. Energy Rev. **91**, 109–125 (2018)
72. B. Yildiz, J.I. Bilbao, A.B. Sproul, A review and analysis of regression and machine learning models on commercial building electricity load forecasting. Renew. Sustain. Energy Rev. **73**, 1104–1122 (2017)
73. H.S. Zhou, R. Passey, A. Bruce, A.B. Sproul, Impact of residential battery energy storage systems on the peak reverse power flows from distributed photovoltaic systems. J. Energy Storage **52**, 104817 (2022)

Part V
Conclusions

Chapter 10
Future Directions of Intelligent Energy Management and the Role of Generative AI

Abstract This chapter investigates the future of intelligent energy management, highlighting the transformative potential of generative artificial intelligence (GAI) in addressing critical challenges. Building upon the integrated library of learning models and optimization techniques presented in this monograph, it identifies key directions for advancing research and development. The discussion encompasses the expansion of AI applications across various energy domains and the integration of GAI for predictive modeling, anomaly detection, and energy system optimization. Additionally, it examines interdisciplinary approaches and the implications of emerging trends, such as decentralized energy systems, quantum computing, and advanced materials science. The chapter concludes with an exploration of the ethical and societal dimensions of intelligent energy management, aiming to inspire innovative solutions that align with sustainable and equitable energy practices.

Keywords Artificial Intelligence · Generative AI · Energy management · Renewable energy · Optimization · Machine learning · Data privacy · Decentralized systems · Sustainability

10.1 Introduction

The monograph has presented a comprehensive framework for utilizing artificial intelligence (AI) to address challenges in energy management and efficiency. It has focused on distributed energy generation, microgrid flexibility, and building energy efficiency. This chapter synthesizes key findings and charts future research trajectories, emphasizing the pivotal role of Generative AI (GAI) in shaping the future of energy systems.

The need for intelligent energy management has never been more urgent. Global energy systems encounter substantial challenges, encompassing the integration of renewable energy sources, the management of peak demand, and the mitigation of climate change impacts. AI offers a powerful set of tools to address these challenges, yet the field is still in its early stages. Future advancements must not only expand

E. Sarmas et al., *Artificial Intelligence for Energy Systems*, Learning and Analytics in Intelligent Systems 46, https://doi.org/10.1007/978-3-031-85209-1_10

the technical capabilities of AI systems but also ensure their alignment with societal values and ethical principles.

As energy demands rise, driven by population growth and industrial expansion, traditional energy systems struggle to keep pace with evolving requirements. AI provides a dynamic and adaptive approach to this challenge, enabling smarter grid management, predictive maintenance, and optimization of energy distribution. For instance, machine learning models can forecast energy consumption patterns, allowing operators to make proactive adjustments that enhance efficiency and reduce waste [17, 19].

Microgrid flexibility is another domain where AI's impact is transformative. Microgrids, characterized by their decentralized nature, require sophisticated algorithms to manage energy flows effectively. AI can optimize load balancing, integrate renewable energy seamlessly, and ensure resilience during grid disruptions [16]. By leveraging AI, microgrids can serve as models of sustainability and reliability, paving the way for their broader adoption.

Building energy efficiency, a cornerstone of this monograph, exemplifies the direct impact AI can have on individual and systemic levels. From smart thermostats to large-scale HVAC optimizations, AI enables real-time decision-making that reduces energy consumption without compromising comfort or productivity [14, 18]. Furthermore, AI-driven retrofitting solutions can guide energy efficiency upgrades in existing infrastructure, addressing the global challenge of aging building stock [20].

Looking forward, the integration of GAI introduces a paradigm shift. Unlike traditional AI models, GAI possesses the ability to synthesize entirely new scenarios, datasets, and designs, thereby facilitating unparalleled innovation. For example, GAI can create synthetic training data for energy forecasting models, simulate diverse weather and consumption scenarios, or generate novel configurations for energy-efficient systems. These capabilities augment the potential for breakthroughs in energy management, offering scalable and adaptive solutions to tackle the complexities of contemporary energy systems [13].

However, the advancement of intelligent energy management systems must take into account broader ethical and societal contexts. Issues such as data privacy, algorithmic transparency, and equitable access to AI technologies are critical. Addressing these concerns through robust policy frameworks and interdisciplinary collaboration will ensure that the benefits of AI are distributed equitably and contribute to global sustainability objectives.

This chapter establishes the foundation for an in-depth exploration of these themes, emphasizing the synergistic relationship between technological innovation and societal progress. By advancing intelligent energy management systems that incorporate GAI, researchers and practitioners can forge a path toward a resilient, efficient, and equitable energy future.

10.2 Emerging Research Avenues in Intelligent Energy Management

The key conclusions that can be drawn at this level of the monograph's contribution are presented below:

- The methodological framework presented in the monograph consists of a variety of AI models and optimization algorithms and aspires to provide a scalable solution for addressing specific problems in the energy sector, for which there are currently limited solutions.
- The ML and DL models developed can serve as a powerful tool in supporting decision-making for complex and non-trivial processes related to energy efficiency, energy management, and policymaking at various levels.
- The proposed approach aims to provide predictive models and algorithms that enable the pre-assessment of energy savings resulting from building renovation actions, as well as the optimal placement of flexible loads, such as water pumps and EVs, to smooth load peaks in decentralized microgrids.
- The use of the abundant data available in the energy sector can enhance the accuracy and efficiency of the models and algorithms developed, leading to more reliable and effective decisions. This is supported by the results presented in the experimental applications conducted as part of the monograph, which confirm the achievement of very good predictive performance.
- The models and algorithms resulting from this monograph can be beneficial to a wide range of stakeholders, including policymakers, building owners, and energy service providers, as they offer valuable insights for developing effective strategies with useful implications for supporting decision-making across numerous challenges [3].

This monograph proposes a comprehensive framework for addressing ML and optimization problems across various categories. The framework consists of four components that tackle specific challenges in each problem category, such as data collection and preprocessing, feature engineering, model selection, training and evaluation, and model deployment and monitoring.

By providing an integrated methodology for selecting models, algorithms, and input features, the proposed framework has the potential to enhance the efficiency and accuracy of solving energy-related problems. Furthermore, the unified approach of the framework can simplify the problem-solving process and enable more straightforward comparisons across different problem categories. Consequently, the proposed framework has the potential to contribute to the field of ML and optimization related to energy by offering a systematic approach that can be adapted to various problem domains.

The third level of contribution of this monograph focuses on the application of the proposed methodologies to real-world data to verify their scientific validity [1]. The development of learning models and optimization algorithms is closely linked to the validation and verification of the results generated by these models in applications involving real-world data. As mentioned in earlier sections, the availability of such data has created a need for innovative applications. Consequently, all models developed as part of the monograph were trained and evaluated on real-world data to certify their accuracy and their contribution to the decision-making support process.

In each of the experimental application chapters presented above, the applications include extensive analysis of the data used during the experimental verification process. Visualizations of the data, information on their statistical distribution, preprocessing details, and interesting findings from their analysis were presented. Moreover, while the models were initially developed based on a specific dataset, the final design of the developed applications was structured to facilitate their application with alternative datasets.

In conclusion, the methodologies presented were successfully applied to real-world data, and the extensive data analysis provided valuable insights into their characteristics and behavior. As a result, this level of contribution further reinforces the scientific validity and practical applicability of the proposed methodologies, enabling their future use in similar problems.

The energy sector faces numerous challenges regarding energy resource management, energy efficiency, and the integration of renewable energy sources (RES). Over the past decades, ML and DL models have been developed for forecasting energy demand and RES generation, but traditional offline learning approaches may result in reduced model accuracy after their initial development. In contrast, online learning offers a dynamic learning approach that is well-suited for the constantly evolving environment of the energy sector. In a recent study, an online learning framework was proposed for time series forecasting problems related to energy. The findings indicated that online learning models surpass offline learning models in terms of mean absolute error, rendering them advantageous for the forecasting of energy demand and RES generation.

Another challenge in energy forecasting is the lack of sufficient data, especially for newly installed solar plants or facilities without smart meters. In such instances, transfer learning serves as a valuable method for transferring knowledge from a source domain to various target domains to address related problems. The transfer learning approach was used in the monograph for forecasting the production of photovoltaic (PV) installations using the LSTM model. The results indicated that models trained through transfer learning significantly outperformed conventional models, achieving enhanced accuracy, particularly in scenarios with limited training data. This methodology has opened new avenues for forecasting energy production in recently installed PV systems, establishing transfer learning as a reliable tool for prosumers in balancing energy and managing demand response.

Finally, meta-learning can be effectively applied to a variety of problems, such as funding energy efficiency initiatives and forecasting PV production. It was demonstrated that various meta-learning models could be developed, combining classification or regression methods to improve overall performance. The results from the experimental applications presented in previous chapters indicate that the meta-learner outperformed all base classifiers in the tested problems, thereby paving the way for the use of meta-learning in numerous other problems and applications.

10.3 Prospects and Future Research Directions

Upon completing the monograph, several thoughts and proposals for future research activities in this field have emerged, which are outlined below.

10.3.1 Expansion of the AI Library and Enrichment with Models for Different Problems

As the energy sector moves toward digitalization, there is a growing need for effective and cost-efficient methods to address energy-related challenges. In this context, the use of AI has proven to be a powerful tool for tackling some of these problems. Consequently, a promising future extension of this monograph involves the expansion of the library of models and algorithms to include tools that can tackle additional challenges related to the three pillars of energy issues [10].

One such area is the forecasting of energy production from various RES beyond photovoltaics, such as wind, geothermal, hydroelectric, and others [24, 26]. For example, ML models can analyze historical data on wind speed, temperature, and other factors affecting energy production to predict future trends in generation from various RES. These models can then be used to optimize the utilization of diverse RES [21].

Another area that could benefit from AI is consumer segmentation for designing better demand response systems [8]. Demand response programs are designed to incentivize consumers to shift their energy usage to off-peak hours, reducing the strain on the energy grid during peak demand periods [7]. AI can be utilized to develop models that categorize consumers based on their energy usage patterns and preferences, facilitating the creation of personalized demand response programs for each segment [12, 15].

Finally, a third area that could benefit from AI is energy price forecasting. AI can be utilized to develop models capable of predicting energy prices based on factors such as energy demand and supply, as well as geopolitical events [2, 5]. For example, ML models can analyze historical data on energy prices and identify patterns that reveal the impact of various factors on energy pricing.

10.3.2 Future Applications in the Field of Transfer Learning

As discussed in previous chapters, transfer learning is a promising research area that has demonstrated significant potential in various fields, including photovoltaic (PV) forecasting. This approach facilitates accurate predictions of PV production, even in scenarios with limited training data, thereby serving as a valuable tool for addressing a range of energy-related challenges such as balancing supply and demand, enhancing energy efficiency, and planning the operation of distribution networks within the context of smart cities and energy communities [11].

A promising direction for future research in transfer learning involves exploring its application across diverse domains, such as wind energy forecasting and building consumption prediction. Wind energy forecasting is crucial within the context of RES, especially in situations characterized by limited data availability. Similarly, building consumption prediction is another domain where transfer learning can be utilized to produce reliable forecasts for newly constructed buildings by leveraging data from analogous buildings or sectors.

Another future research direction in this field is investigating new transfer learning techniques, such as Domain Adaptation [25]. Domain Adaptation is a subdiscipline of transfer learning that seeks to adapt a model trained on a specific problem for application in a similar problem that may exhibit different data distributions or characteristics. Domain Adaptation could have potential applications in energy forecasting, where data distributions may fluctuate over time due to factors such as varying weather conditions, equipment status, and environmental factors. By adapting models trained on historical data to new domains, Domain Adaptation can enable accurate predictions of energy consumption and RES production, even in dynamic environments [4].

10.3.3 Data Privacy Protection Through Distributed Learning Algorithms

The results obtained in this monmograph indicate that AI models related to energy can be developed using various ML techniques with very high accuracy. However, ensuring the privacy and security of energy data is of paramount importance, particularly when dealing with sensitive information. The technique of Federated Learning is a promising approach to addressing these challenges, allowing models to be trained closer to data sources (Edge Computing), without transferring sensitive data to a central location [6, 22].

Future research in this area could investigate the potential of Federated Learning for developing energy-related models, such as energy demand forecasting, renewable energy production forecasting, and energy consumption prediction in buildings. Federated Learning has the capability to enhance model accuracy by integrating data from multiple sources while simultaneously improving data privacy and protection [9].

Finally, the development of techniques and Federated Learning algorithms that address the increasing challenges associated with energy data could play a pivotal role in fostering trust among various stakeholders, thereby encouraging a greater willingness to share data. For instance, the development of Federated Learning algorithms with varying levels of privacy could provide a more comprehensive approach to this issue, accommodating the needs of diverse users.

10.4 Generative Artificial Intelligence in Energy Systems

GAI has the potential to revolutionize energy management by generating synthetic data, simulating complex scenarios, and optimizing system configurations [23]. Its applications span various domains:

GAI can significantly enhance energy demand forecasting by creating synthetic datasets that complement limited or incomplete real-world data. These datasets can be tailored to reflect a wide range of conditions, including rare or extreme events, ensuring robust and comprehensive model training. Additionally, GAI facilitates the development of scenario-based forecasting tools, enabling grid operators to visualize and prepare for diverse demand patterns influenced by variables such as climate fluctuations, economic trends, or policy changes. Probabilistic simulations, enriched by generative models, empower energy planners to anticipate peak loads, implement preventive measures, and maintain grid stability even under highly uncertain conditions.

In the realm of renewable energy optimization, GAI offers unparalleled capabilities. By simulating intricate weather patterns, seasonal variability, and renewable energy output scenarios, GAI tools enhance the scheduling of energy production and the allocation of resources. These simulations can guide the integration of multiple renewable sources, such as solar, wind, and hydroelectric power, ensuring a balanced and efficient energy mix. Additionally, GAI can aid in the design of energy storage systems by modeling the performance of various storage technologies under different operational scenarios. Smart city planning can also benefit from generative simulations that evaluate infrastructure changes, energy flow dynamics, and the long-term impacts of renewable energy projects.

GAI enhances anomaly detection and predictive maintenance by creating comprehensive datasets that include synthetic examples of rare or unexpected faults. These enriched datasets enable machine learning models to identify anomalies with greater accuracy, particularly in scenarios where actual fault data is scarce. Furthermore, GAI can simulate the progression of equipment degradation, offering insights into failure mechanisms and facilitating proactive maintenance scheduling. For critical infrastructure, such as power plants and distribution networks, these capabilities lead to reduced downtime, improved reliability, and substantial cost savings.

GAI can transform consumer engagement by creating sophisticated models of energy usage behavior. These models can simulate the responses of various consumer segments to pricing incentives, energy-saving programs, or emerging technologies, thereby providing valuable insights for the design of effective demand response initiatives. Virtual energy communities, empowered by GAI, facilitate the simulation of collective behaviors such as energy sharing and collaborative renewable energy investments, thereby fostering a more participatory and sustainable energy ecosystem. Personalized energy recommendations, derived from generative behavioral simulations, can empower consumers to make informed decisions, reduce consumption, and actively contribute to the achievement of energy efficiency goals.

Despite its immense potential, integrating GAI into energy systems presents several challenges. Data privacy and security remain paramount, particularly when synthetic data is derived from sensitive information. Robust encryption methods and privacy-preserving generative algorithms are essential to safeguard against data breaches and misuse. Furthermore, ensuring the reliability and validity of generative models requires rigorous testing and benchmarking against real-world scenarios. Bias in generated data or simulations must be identified and mitigated to avoid skewed results that could compromise decision-making. Finally, the computational intensity of GAI highlights the need for energy-efficient algorithms and the exploration of quantum computing to achieve scalability.

10.5 Conclusions

The integration of GAI into intelligent energy management represents a significant milestone in the pursuit of sustainable and efficient energy systems. By providing solutions that improve system efficiency, resilience, and equity, GAI has the potential to transform the design, management, and optimization of energy systems. Through the synthesis of synthetic data, advanced simulations, and innovative system configurations, GAI provides unparalleled tools for addressing the complex challenges confronted by contemporary energy infrastructure.

However, the journey toward fully realizing the potential of GAI necessitates concerted efforts across multiple dimensions. Interdisciplinary collaboration among energy experts, computer scientists, policymakers, and ethicists is essential to ensure that the development and deployment of GAI technologies align with societal and environmental objectives. Critical issues such as data privacy, algorithmic fairness, and inclusivity must be addressed to guarantee equitable access to the benefits of these innovations. Future research should prioritize addressing technical barriers, which include the computational intensity of generative models and the need for robust validation frameworks. By investigating energy-efficient computational methods and utilizing emerging technologies such as quantum computing, the scalability and reliability of GAI applications can be considerably improved.

The adoption of GAI in energy systems not only addresses immediate operational challenges but also contributes to broader global objectives, such as carbon neutrality and climate resilience. By fostering innovation that is both technologically advanced and ethically sound, the integration of GAI paves the way for a future in which energy systems are intelligent, sustainable, and universally beneficial. Through sustained commitment to research, development, and collaboration, the potential of GAI can be fully realized, shaping a transformative era for energy management and beyond.

References

1. I.C. Dimitriou, G.P. Trachanas, A. Arsenopoulos, E. Sarmas, Future heat risk index: the case of Attica regions, Greece, in *2024 15th International Conference on Information, Intelligence, Systems & Applications (IISA)* (IEEE, 2024), pp. 1–7
2. O. Grothe, F. Kächele, F. Krüger, From point forecasts to multivariate probabilistic forecasts: the Schaake shuffle for day-ahead electricity price forecasting. Energy Econ. **120**, 106602 (2023)
3. J.L. Hernández, S. Serroni, S. Stamatopoulos, E. Sarmas, G.M. Revel, F. Vélez, Multi-criteria building performance assessment, in *2024 15th International Conference on Information, Intelligence, Systems & Applications (IISA)* (IEEE, 2024), pp. 1–7
4. L. Ilias, E. Sarmas, V. Marinakis, D. Askounis, H. Doukas, Unsupervised domain adaptation methods for photovoltaic power forecasting. Appl. Soft Comput. **149**, 110979 (2023)
5. R. Li, X. Song, A multi-scale model with feature recognition for the use of energy futures price forecasting. Expert Syst. App. **211**, 118622 (2023)
6. V. Michalakopoulos, E. Sarantinopoulos, E. Sarmas, V. Marinakis, Empowering federated learning techniques for privacy-preserving PV forecasting. Energy Rep. **12**, 2244–2256 (2024)
7. V. Michalakopoulos, E. Sarmas, I. Papias, P. Skaloumpakas, V. Marinakis, H. Doukas, A machine learning-based framework for clustering residential electricity load profiles to enhance demand response programs. Appl. Energy **361**, 122943 (2024)
8. G.E. Okereke, M.C. Bali, C.N. Okwueze, E.C. Ukekwe, S.C. Echezona, C.I. Ugwu, K-means clustering of electricity consumers using time-domain features from smart meter data. J. Electr. Syst. Inform. Technol. **10**(1), 1–18 (2023)
9. S. Otoum, I. Al Ridhawi, H. Mouftah, A federated learning and blockchain-enabled sustainable energy-trade at the edge: a framework for industry 4.0. IEEE Internet Things J. (2022)
10. D.P. Panagoulias, E. Sarmas, V. Marinakis, M. Virvou, G.A. Tsihrintzis, H. Doukas, Intelligent decision support for energy management: a methodology for tailored explainability of artificial intelligence analytics. Electronics **12**(21), 4430 (2023)
11. C. Panjapornpon, S. Bardeeniz, M.A. Hussain, P. Chomchai, Explainable deep transfer learning for energy efficiency prediction based on uncertainty detection and identification. Energy AI **12**, 100224 (2023)
12. J. Priolkar, E. Sreeraj, Analysis of price based demand response program using load clustering approach. IETE J. Res., 1–14 (2023)
13. N. Rane, Contribution of chatgpt and other generative artificial intelligence (AI) in renewable and sustainable energy (2023), Available at SSRN 4597674
14. P. Samaras, E. Stamatopoulos, A. Arsenopoulos, E. Sarmas, E. Marinakis, Readiness to adopt the smart readiness indicator scheme across europe: a multi-criteria decision analysis approach, in *2024 IEEE International Workshop on Metrology for Living Environment (MetroLivEnv)* (IEEE, 2024), pp. 268–273
15. E. Sarmas, A. Fragkiadaki, V. Marinakis, Explainable AI-based ensemble clustering for load profiling and demand response. Energies **17**(22), 5559 (2024)
16. E. Sarmas, E. Spiliotis, V. Marinakis, M.A. Bucarelli, F. Santori, H. Doukas, Revving up energy autonomy: a forecast-driven framework for reducing reverse power flow in microgrids. Sustain. Energy Grids Netw. **38**, 101376 (2024)

17. E. Sarmas, E. Spiliotis, V. Marinakis, G. Tzanes, J.K. Kaldellis, H. Doukas, ML-based energy management of water pumping systems for the application of peak shaving in small-scale islands. Sustain. Cities Soc. **82**, 103873 (2022)
18. P. Skaloumpakas, E. Sarmas, Z. Mylona, A. Cavadenti, F. Santori, V. Marinakis, Predicting thermal comfort in buildings with machine learning and occupant feedback, in *2023 IEEE International Workshop on Metrology for Living Environment (MetroLivEnv)* (IEEE, 2023), pp. 34–39
19. P. Skaloumpakas, E. Sarmas, M. Rachmanidis, V. Marinakis, Reshaping the energy landscape of Crete through renewable energy valleys. Sci. Rep. **14**(1), 8038 (2024)
20. E. Stamatopoulos, A. Forouli, D. Stoian, P. Kouloukakis, E. Sarmas, V. Marinakis, An adaptive framework for assessing climate resilience in buildings. Buildi. Environ. **264**, 111869 (2024)
21. C. Sweeney, R.J. Bessa, J. Browell, P. Pinson, The future of forecasting for renewable energy. Wiley Interdisc. Rev.: Energy Environ. **9**(2), e365 (2020)
22. L. Tang, H. Xie, X. Wang, Z. Bie, Privacy-preserving knowledge sharing for few-shot building energy prediction: a federated learning approach. Appl. Energy **337**, 120860 (2023)
23. G.A. Tsihrintzis, E. Sarmas, V. Marinakis, D. Panagoulias, E.-A. Tsihrintzi, M. Virvou, Evaluating non-expert stakeholder interaction with artificial intelligence on energy urban domain using VIRTSI: the case of ChatGPT, in *2024 15th International Conference on Information, Intelligence, Systems & Applications (IISA)* (IEEE, 2024), pp. 1–10
24. H. Wang, Z. Lei, X. Zhang, B. Zhou, J. Peng, A review of deep learning for renewable energy forecasting. Energy Convers. Manag. **198**, 111799 (2019)
25. Z. Yu, J. Li, Z. Du, L. Zhu, H.T. Shen, A comprehensive survey on source-free domain adaptation (2023), arXiv preprint arXiv:2302.11803
26. J. Zheng, J. Du, B. Wang, J.J. Klemeš, Q. Liao, Y. Liang, A hybrid framework for forecasting power generation of multiple renewable energy sources. Renew. Sustain. Energy Rev. **172**, 113046 (2023)

The manufacturer's authorised representative in the EU is Springer Nature Customer Service Centre GmbH, Europaplatz 3, 69115 Heidelberg, Germany. If you have any concerns regarding our products, please contact ProductSafety@springernature.com

Printed and bound by CPI Group (UK) Ltd, Croydon, CR0 4YY
30/09/2026
02225568-0006